0 1 1 2 3 5 8 13 21 34 55 89 144 233 377
610 987 1597 2584 4181 6765 10946 17711 28657
46368 75025 121393 196418 317811 514229
832040 1346269 2178309 3524578 5702887
9227465 14930352 24157817 39088169
63245986 102334155 165580141 267914296
433494437 701408733 SECOND EDITION

BEGINNING NUMBER THEORY

NEVILLE ROBBINS

San Francisco State University

JONES AND BARTLETT PUBLISHERS

Sudbury, Massachusetts

BOSTON   TORONTO   LONDON   SINGAPORE

World Headquarters

Jones and Bartlett Publishers	Jones and Bartlett Publishers Canada	Jones and Bartlett Publishers International
40 Tall Pine Drive	6339 Ormindale Way	Barb House, Barb Mews
Sudbury, MA 01776	Mississauga, Ontario L5V 1J2	London W6 7PA
978-443-5000	CANADA	UK
info@jbpub.com		
www.jbpub.com		

Jones and Bartlett's books and products are available through most bookstores and online booksellers. To contact Jones and Bartlett Publishers directly, call 800-832-0034, fax 978-443-8000, or visit our website, www.jbpub.com.

Substantial discounts on bulk quantities of Jones and Bartlett's publications are available to corporations, professional associations, and other qualified organizations. For details and specific discount information, contact the special sales department at Jones and Bartlett via the above contact information or send an email to specialsales@jbpub.com.

Interior photo credits: Euler: © Classic Vision/age fotostock; Fermat: © Mary Evans Picture Library/Alamy Images; Fibonacci: © Bettmann/Corbis; Gauss: © Bettmann/Corbis; Lagrange: © Classic Vision/age fotostock

Production Credits
Acquisitions Editor: Timothy Anderson
Editorial Assistant: Kate Koch
Production Director: Amy Rose
Production Editor: Tracey Chapman
Marketing Manager: Andrea DeFronzo
Manufacturing Buyer: Therese Connell
Composition: Northeast Compositors, Inc.
Cover Design: Kristin E. Ohlin
Printing and Binding: Malloy, Inc.
Cover Printing: John Pow Company

Library of Congress Cataloging-in-Publication Data
Robbins, Neville.
   Beginning number theory / Neville Robbins.— 2nd ed.
      p. cm.
   Includes index.
   ISBN-13: 978-0-7637-3768-9 (casebound)
   ISBN-10: 0-7637-3768-2 (casebound)
   1. Number theory. I. Title.
   QA241.R58 2006
   512.7—dc22

                              2005032047

6048

Printed in the United States of America
10 09 08 07 06   10 9 8 7 6 5 4 3 2 1

Dedication

This book is dedicated to the memory of Fermat, who was the first to call for the establishment of number theory as a special domain within mathematics.

Contents

Preface

Beginning Number Theory is intended for use in an upper-division undergraduate course in elementary number theory. The usual audience for such a course consists of math majors. (In California, knowledge of elementary number theory is required for prospective secondary math teachers.) One semester of calculus is an adequate prerequisite. In reality, calculus is used quite sparingly in elementary number theory. It suffices to be able to compute the derivative of a polynomial in one variable with integer coefficients.

The natural numbers, namely 1, 2, 3, etc. are the first objects that we encounter in our mathematical education. The *theory of numbers* is that branch of mathematics that is concerned with the study of numbers for their own sake. For this reason, number theory, which has a 4000-year history, has traditionally been considered as pure mathematics. Number theory has fascinated some of the best minds of every era. In recent years, as computers have become more powerful and more available, number theory has developed some significant applications, especially to cryptology, which is discussed in Chapter 12.

About This Book

This text attempts to strike a balance between the traditional and algorithmic approaches to elementary number theory. The traditional subject matter, namely: *divisibility, primes, congruences, number theoretic functions, primitive roots, quadratic reciprocity, sums of squares*, is covered logically and thoroughly. The book's first eight chapters could serve as the basis for a one-semester course. For a one-quarter course, one could use the first seven chapters, omitting Sections 2.4, 4.3, 4.6, 4.9, 6.4, 7.5, and 7.6. The algorithmic viewpoint, which has strongly influenced number theory in recent years, appears within the traditional subject matter. For example, in Chapter 7, not only is the existence of solutions to quadratic congruences (mod p) discussed, but in addition an algorithm is presented whereby such solutions can be obtained, when they exist.

In recognition of recent developments in number theory, the last two chapters are devoted to computational number theory and cryptology. The instructor who wishes to teach these exciting new topics could base a course on Chapters 1 through 4, plus Chapters 11 and 12. (It would also be necessary to present the material from Chapter 5 that deals with Euler's function.)

Features

1. *Classical number theory* is presented in a thoroughly logical, well-organized manner. Particular attention has been devoted to making proofs of major theorems more lucid. This is accomplished by (a) presenting necessary preliminary results beforehand; (b) avoiding needlessly elaborate notation; (c) (in certain instances) providing proofs that are simpler, more explicit, or more algorithmic than the conventional proofs.

2. *Computer exercises*, as well as regular exercises, appear at the end of nearly every section. The computer exercises include both (a) implementation of an algorithm that has been previously described, and (b) the "discovery" type, where the student must perform certain computations and then infer an appropriate conclusion.

3. *Computational number theory* is treated in Chapter 11. This chapter begins with an overview of this new branch of number theory and includes definitions of key concepts such as *computational feasibility* and *polynomial time*. After presenting tests for compositeness and primality of large integers, this chapter offers a thorough treatment of factoring, including the *continued fraction* method, the *quadratic sieve* method, and the *Pollard $p - 1$* method, all with examples worked out in detail.

4. *Cryptology* is the subject of Chapter 12. After giving some details of the history of cryptology and its influence on world affairs, this chapter discusses various methods of cryptography, from Caesar ciphers to public-key cryptosystems.

5. *Conjectures, theorems,* and *proofs* are discussed in Section 1.2. This section is intended to help the student make the transition from lower-

division mathematics, where proofs are usually not greatly stressed, to upper-division mathematics, where the understanding of proofs takes on a greater significance.

6. *The Dirichlet product* is used in Chapter 5 in order to derive the multiplicative property of four standard number theoretic functions: the number-of-divisors function ($\tau(n)$), the sum-of-divisors function ($\sigma(n)$), the Moebius function ($\mu(n)$), and Euler's function ($\phi(n)$). The advantages of this approach include (a) a better-motivated definition of the Moebius function; (b) an easy proof of the Moebius inversion formula.

7. *Full biographies* of five important number theorists: Fibonacci, Fermat, Euler, Gauss, and Lagrange. In addition, there are thumbnail sketches of other mathematicians, including Eratosthenes, Euclid, Mersenne, and Pascal. As a result, the student may better appreciate number theory from a historical perspective.

8. *Equivalence relations* and the related concept of equivalence classes are the subject of Section 4.3. This helps prepare the student for the study of a particular equivalence relation: congruence (mod m).

9. *Starred exercises* offer challenges to those readers whose enthusiasm for the subject surpasses the ordinary.

Acknowledgments

I would like to thank all of the people whose efforts contributed to the preparation and development of this book. I am grateful to Wayne Tarrant of Western Kentucky University, Howard Skogman of SUNY Brockport, Kevin Knudson of Mississippi State University, Hui June Zhu of McMaster University, and Emma Previato of Boston University for reading the manuscript and making a number of helpful suggestions and comments.

Neville Robbins

Chapter 1

Preliminaries

"About binomial theorem I'm teeming with a lot 'o news–
With many cheerful facts about the square of the hypotenuse."

—From the song of the Modern Major General
in Gilbert and Sullivan's *Pirates of Penzance*

1.1 Introduction

Most of us begin our mathematical education by learning how to count. We become familiar with the *natural numbers*, (also known as *positive integers*), namely 1, 2, 3, 4, 5, etc. We learn to add, subtract, multiply, and divide. Later, we are exposed to more general number systems: (1) the *integers*, which include zero as well as both positive and negative whole numbers; (2) the *rational numbers*, that is, fractions; (3) the *real* numbers, which include limits of infinite sequences of rational numbers; (4) the *complex* numbers.

Number theory is the branch of mathematics that is concerned with the properties of numbers for their own sake. (A synonym for number theory is *higher arithmetic*.) The term *number theory* (actually *la theorie des nombres*) was coined by the French mathematician Pierre Fermat during the seventeenth century. As an example of a problem in number theory, consider the *Pythagorean equation*:

$$x^2 + y^2 = z^2 \tag{1.1}$$

The Theorem of Pythagoras states that if a right triangle has sides of lengths x, y, and z, where z is the hypotenuse, then the Pythagorean equation holds. *Pythagoras* (569–470 B.C.) was not only a mathematician, but also a philosopher. In fact, he invented the term *philosopher*, which means lover of wisdom.

Consider the problem of finding all solutions, if any, to the Pythagorean equation such that $0 < x < y < z$ and x, y, z are all integers. (It is easily seen that the smallest such solution is $x = 3$, $y = 4$, $z = 5$.)

The problem of finding integer solutions to the Pythagorean equation dates back to ancient times; a Babylonian tablet from the period 1900–1600 B.C. contains a list of such solutions. The existence of this tablet (which is known as Plimpton 322, and belongs to Columbia University) is evidence that number theory is nearly 4,000 years old. The complete solution to the Pythagorean equation is given at the end of Chapter 4.

Let us modify the Pythagorean equation by replacing the exponent 2 with n to obtain

$$x^n + y^n = z^n \tag{1.2}$$

In 1637, Fermat claimed that equation (1.2) has no solution in positive integers x, y, z for any $n \geq 3$. This statement is known as *Fermat's Last Theorem*. Fermat himself was able to prove that equation (1.2) has no solution in positive integers when $n = 4$.

Starting with Euler in the eighteenth century, many prominent mathematicians attempted to prove Fermat's Last Theorem, but achieved only limited results. The quest for a proof of Fermat's Last Theorem had become a "holy grail" of mathematics. In 1993, a proof was announced by Andrew Wiles of Princeton University. After a few months, a flaw in Wiles' proof was discovered. With the help of a colleague, however, Wiles was able to correct the flaw. The revised proof was published in 1995. (see [18]) Wiles' proof is based on *arithmetic algebraic geometry*, an advanced mathematical specialty that is considerably beyond the scope of this text.

The first systematic treatment of number theory was given by Euclid (about 300 B.C.) in Books 7, 8, and 9 of his *Elements*, an influential book that was later translated from Greek into Arabic and Latin. Number theory was held in high esteem by the ancient Greeks. In Book VII of his master work *The Republic*, the philosopher Plato advises: "We must endeavor to persuade those who are to be the principal men of the state to go and learn arithmetic."

Part of number theory's charm lies in its accessibility to the layperson. Many problems in number theory can be stated in terms that are comprehensible to anyone familiar with grade-school arithmetic and elementary algebra. One such problem, which we mentioned earlier, was to find all solutions to the Pythagorean equation in positive integers. As a second example, we briefly discuss *perfect numbers*.

According to Euclid, a positive integer n is called *perfect* if it equals the sum of its own proper divisors. (The only divisor of n that is not proper is n itself.) At present, 42 perfect numbers are known, all of them even. The smallest perfect numbers are

$$6 = 1 + 2 + 3$$

$$28 = 1 + 2 + 4 + 7 + 14$$

(Perfect numbers are discussed in detail in Chapter 5.) Can an odd number be perfect? Although 23 centuries have elapsed since Euclid's time, no one knows.

There are several kinds of number theory: (1) elementary or classical number theory, (2) algebraic number theory, (3) analytic number theory, (4) computational number theory. The origins of *elementary number theory* are found in antiquity. *Algebraic number theory* uses the concepts of abstract algebra, such as groups, rings, fields, mappings, Galois theory, etc. For example, algebraic number theory is used to show that the only integer solution of the equation $x^3 = y^2 + 1$ is $x = 1, y = 0$. *Analytic number theory* uses the tools of real and complex analysis, namely, inequalities, derivatives, integrals, infinite series, residues, etc. The Prime Number Theorem, one of the most important theorems regarding primes, was first proved (in 1896) using analytic number theory. *Computational number theory*, which is a blend of number theory and computer science, has developed rapidly since 1970. Research in this newest branch of number theory has significant applications to data security and has therefore been generously funded by several federal agencies.

Although traditionally considered as belonging to "pure" mathematics, number theory has been applied to acoustics, boundary value problems, Brownian motion, cryptology, digital signal processing, and error-correcting codes.

Every year, numerous conferences are held for the purpose of presenting new research results in number theory. The author has twice been the organizer of such a conference, namely, the West Coast Number Theory Conference, which is held annually in mid-December, usually in California.

This textbook is about *elementary number theory*. One semester of calculus is an adequate prerequisite to most of the subject matter. (The reader should be familiar with functional notation and should be able to find the derivative of a polynomial.) Some exposure to set theory would also be helpful.

1.2 Conjectures, Theorems, and Proofs

In elementary number theory, we are mostly interested in questions concerning $\mathbf{N}$, the set of all natural numbers, but we must occasionally deal with integers, rational numbers, and real numbers. We assume the usual ordering on $\mathbf{N}$, namely, $1 < 2 < 3 < \cdots$.

If we add together the first n natural numbers, what do we get?

For $n = 1$, we have $1 = 1.$
For $n = 2$, we have $1 + 2 = 3.$
For $n = 3$, we have $1 + 2 + 3 = 6.$
For $n = 4$, we have $1 + 2 + 3 + 4 = 10.$

At first, no pattern is apparent. Suppose, however, that we rewrite these four equations as

$$1 = \frac{1}{2}(1 * 2)$$

$$1 + 2 = \frac{1}{2}(2 * 3)$$

$$1 + 2 + 3 = \frac{1}{2}(3 * 4)$$

$$1 + 2 + 3 + 4 = \frac{1}{2}(4 * 5)$$

Generalizing, it appears that

$$1 + 2 + 3 + \cdots + n = \frac{1}{2}n(n + 1) \tag{1.3}$$

Equation (1.3) seems to be valid for any natural number, n. For the moment, call equation (1.3) a *conjecture*, that is, a guess. Our conjecture may be stated as follows: "If n is any natural number, then the sum of all natural numbers from 1 to n is $\frac{1}{2}n(n + 1)$."

If we manage to *prove* our conjecture, then we call it a *theorem*. (A proof of equation (1.3) is presented in the following section.) A *theorem* is a mathematical statement that can be proven by means of prior assumptions, prior knowledge, and logical reasoning. A theorem may be stated in the form: "If A, then B." A is called the *hypothesis* of the theorem, while B is called the *conclusion*. The *proof* of a theorem consists of a logical sequence of assertions that lead from the hypothesis to the conclusion.

Methods of proof include: (1) *direct substitution*; (2) *case analysis*; (3) *indirect proof*, that is, proof by contradiction, and (4) *inductive proof*. We now present examples of the first three kinds of proofs. Inductive proof is discussed at length in the following section.

Let P, Q be integers such that both P and $D = P^2 - 4Q$, are positive. Let the roots of the equation

$$x^2 = Px - Q$$

be

$$a = \frac{1}{2}(P + \sqrt{D})$$

$$b = \frac{1}{2}(P - \sqrt{D}).$$

Let sequences $\{u_n\}$, $\{v_n\}$ be defined for $n \geq 0$ as follows:

$$u_n = \frac{a^n - b^n}{a - b}$$

$$v_n = a^n + b^n.$$

The following theorems are proved by direct substitution:

Theorem 1.1 $u_{2n} = u_n v_n$

Proof:

$$u_n v_n = (\frac{a^n - b^n}{a - b})(a^n + b^n)$$

$$= \frac{a^{2n} - b^{2n}}{a - b} = u_{2n} \quad \blacksquare$$

Theorem 1.2 $v_n^2 - Du_n^2 = 4Q^n$

Proof: First note that $a - b = \sqrt{D}$, so $(a - b)^2 = D$.
  Also:

$$ab = \frac{1}{4}(P + \sqrt{D})(P - \sqrt{D})$$

$$= \frac{1}{4}(P^2 - D)$$

$$= \frac{1}{4}(4Q)$$

$$= Q$$

Now:

$$v_n^2 - Du_n^2 = (a^n + b^n)^2 - (a - b)^2 (\frac{a^n - b^n}{a - b})^2$$

$$= (a^n + b^n)^2 - (a^n - b^n)^2$$

$$= a^{2n} + 2a^n b^n + b^{2n} - (a^{2n} - 2a^n b^n + b^{2n})$$

$$= 4a^n b^n$$

$$= 4(ab)^n$$

$$= 4Q^n \quad \blacksquare$$

Two integers are said to have the same *parity* if they are both even or both odd. Next, we prove by case analysis that if the integers a and b have the same parity, then $a + b$ is even.

Theorem 1.3   If the integers a and b have the same parity, then $a + b$ is even.

Proof:

Case 1:   If a and b are both even, then $a = 2m$ and $b = 2n$ for some integers, m, n. Therefore $a + b = 2m + 2n = 2(m + n)$, so $a + b$ is even.

Case 2:   If a and b are both odd, then $a = 2m+1$ and $b = 2n+1$ for some integers, m, n. Therefore $a+b = (2m+1)+(2n+1) = 2(m + n + 1)$, so again $a + b$ is even. $\blacksquare$

Now we prove that every odd integer is one more, or one less, than a multiple of 4.

Theorem 1.4   If the integer a is odd, then $a = 4m + 1$ for some integer m, or $a = 4n - 1$ for some integer, n.

Proof: Since a is odd by hypothesis, we have $a = 2k + 1$ for some k.

Case 1:   If k is even, then $k = 2m$, so $a = 2(2m) + 1 = 4m + 1$.

Case 2:   If k is odd, then $k = 2m + 1$, so $a = 2(2m + 1) + 1 = 4m + 3 = 4(m + 1) - 1 = 4n - 1$, with $n = m + 1$. $\blacksquare$

In the *indirect method* of proof, we assume the contrary of what we wish to prove, and then show that this assumption leads to a contradiction. The proofs of the next several theorems are indirect.

Theorem 1.5 If ab is odd, then a and b are both odd.

Proof: Suppose that a and b are not both odd. Then a or b is even. If a is even, then $a = 2m$, so $ab = (2m)b = 2(mb)$, so ab is even, contrary to hypothesis. Similarly, if b is even, then $b = 2m$, so $ab = a(2m) = 2(am)$, so ab is even, contrary to hypothesis. ■

Theorem 1.6 There are no integers m, n such that $4m + 1 = 4n - 1$.

Proof: Suppose that $4m + 1 = 4n - 1$. Then $4m + 2 = 4n$, so that $2m + 1 = 2n$. The left member of the last equation is odd, while the right member is even, an impossibility. ■

Theorem 1.7 If $ab = 4k - 1$, then $a = 4m - 1$ for some m or $b = 4n - 1$ for some n.

Proof: Since ab is odd by hypothesis, it follows from Theorem 1.5 that a and b are both odd. Therefore, by Theorem 1.4, we have $a = 4m \pm 1$ and $b = 4n \pm 1$. If $a = 4m + 1$ and $b = 4n + 1$, then $ab = (4m + 1)(4n + 1) = 16mn + 4m + 4n + 1 = 4(4mn + m + n) + 1$. By hypothesis, this is impossible. Therefore, by Theorem 1.6 we must have $a = 4m - 1$ or $b = 4n - 1$. ■

Section 1.2 Exercises

1. Referring to the sequences $\{u_n\}, \{v_n\}$ defined on page 5, prove by direct substitution that $v_{2n} = v_n^2 - 2Q^n$.

2. Prove that $u_{3n} = u_n(v_{2n} + Q^n)$.

3. Use indirect proof to show that if n^2 is odd, then n is odd.

4. Prove that there are no natural numbers a, b such that both $a + b$ and ab are odd.

1.3 Well-Ordering and Induction

Consider the usual ordering on N, namely, if $a, b \in N$, then $a < b$ if and only if $b - a > 0$. This yields $1 < 2 < 3 <$ *etc.* This ordering has the following properties: If a, b, c are distinct natural numbers, then (1) either $a < b$ or $b < a$, but not both; (2) if $a < b$ and $b < c$, then $a < c$. We also assume the following:

Well-Ordering Principle

N is well-ordered with respect to the usual order, that is, every non empty subset of N has a least element.

For example, if S is the subset of N consisting of even squares, then the least element of S is 4.

Although the Well-Ordering Principle may appear to be trivially true for N with the usual ordering, it is *not* true for Z, the set of all integers, with the usual ordering on Z. (There is no least negative integer.) The Well-Ordering Principle also fails on R^+, the set of all positive real numbers, with respect to the usual order. (There is no least positive real number.)

Using indirect proof and the Well-Ordering Principle, we prove the following useful theorem:

Theorem 1.8   Every strictly decreasing sequence of natural numbers is finite. That is, let $S = \{x_1, x_2, x_3,$ *etc.*$\}$ be a non empty subset of N such that $x_1 > x_2 > x_3 >$ *etc.* Then S is finite.

Proof: By hypothesis, S is non empty. Suppose that S is infinite. Since $x_1 > x_2 > x_3 >$ *etc.*, it follows that S has no least element. But this contradicts the Well-Ordering Principle. Therefore S is finite. ∎

An important consequence of the Well-Ordering Principle is

Theorem 1.9   **(Principle of Mathematical Induction)**

Let S be a subset of N with the two following properties:

(i) $1 \in S$;

(ii) If $n \in S$, then $n + 1 \in S$.

Then $S = N$.

Proof: Let T be the complement of S in N, that is, T is the set of all natural numbers that do *not* belong to S. It suffices to show that T is empty. If T is non empty, then by the Well-Ordering Principle, T has a least element, say n. Since $1 \in S$ by hypothesis, we have $1 \notin T$, so $n \neq 1$. Since 1 is the least natural number, we have $n > 1$, so $n - 1$ is a natural number. Since n is the least element of T, it follows that $n - 1 \in S$. But then (ii) implies that $n \in S$, an impossibility. Therefore T is empty, so $S = N$.   ∎

Remarks

The converse of Theorem 1.9 is also true. That is, assuming the Principle of Mathematical Induction on **N**, one can prove the validity of the Well-Ordering Principle.

Suppose that we wish to prove a statement about natural numbers. Let S be the subset of N for which the statement is true. We must show that $S = N$. If we use induction, we must show that (i) $1 \in S$; (ii) if $n \in S$, then also $n + 1 \in S$. The assertion $n \in S$ is called the *induction hypothesis*. Step (i) is called the *basis step*, while step (ii) is called the *induction step*.

We are now ready to use mathematical induction to prove equation (1.3) from page 4, which we have previously called a conjecture.   ∎

Theorem 1.10   Let S_n denote the sum of the first n natural numbers, that is, $S_n = 1 + 2 + 3 + \cdots + n$. Then $S_n = \frac{1}{2}n(n + 1)$.

Proof: (Induction on n)   To verify that condition (i) holds, we note that $\frac{1}{2}1(1 + 1) = \frac{1}{2}(2) = 1 = S_1$. To verify that condition (ii) holds, we must show that if $S_n = \frac{1}{2}n(n + 1)$, then $S_{n+1} = \frac{1}{2}(n + 1)(n + 2)$. Now

$$S_{n+1} = 1 + 2 + 3 + \cdots + n + (n + 1) = S_n + (n + 1)$$

By induction hypothesis, we have $S_n = \frac{1}{2}n(n + 1)$. Therefore

$$S_{n+1} = \frac{1}{2}n(n + 1) + (n + 1) = (n + 1)(\frac{1}{2}n + 1) = \frac{1}{2}(n + 1)(n + 2) \quad ∎$$

Induction may also be used to prove inequalities, as we demonstrate in Theorem 1.11.

Theorem 1.11 $2^n > n$ for every natural number n.

Proof: (Induction on n)

(i) $2^1 = 2$ and $2 > 1$, so $2^1 > 1$.

(ii) We must show that if $2^n > n$, then $2^{n+1} > n + 1$.

By induction hypothesis, we have

$$2^n > n$$

Multiplying this inequality by 2, we get $2^{n+1} > 2n$. We also know that $n \geq 1$, so $n + n \geq n + 1$, that is, $2n \geq n + 1$. A well-known property of inequalities states that If $a > b$ and $b \geq c$, then $a > c$. Therefore, we may conclude that $2^{n+1} > n + 1$. ■

Mathematical induction may also be used to prove statements that are true not for all n, but for all sufficiently large n. Such statements are true for all $n \geq n_0$, where n_0 is a certain constant. In such instances, step (i) in the inductive proof consists in verifying that the statement is true for $n = n_0$. As an example, consider the following, which is a variation of Theorem 1.11:

Theorem 1.12 If $n \geq 3$, then $2^{n-1} > n$.

Proof: (Induction on n.)

(i) $2^{3-1} = 2^2 = 4 > 3$, so the statement holds for $n = 3$.

(ii) Assuming that $2^{n-1} > n$, we must show that $2^n > n + 1$.

This part of the proof differs little from that of Theorem 1.11. By induction hypothesis, we have $2^{n-1} > n$. Multiplying this inequality by 2, we have $2^n > 2n$. But $2n = n + n > n + 1$ (since $n \geq 3$), so we have $2^n > n + 1$. ■

Mathematical induction can also be used to *define* a sequence of real numbers. Suppose that an initial term, a_1 , is defined. Suppose also that a_{n+1} is defined whenever a_n is defined. Then, by the Principle of Mathematical Induction, a_n is defined for all n. In the next few pages, we will present several examples of definition by mathematical induction.

Definition 1.1

Let $a_1 = 1$, and let $a_n = na_{n-1}$ for all $n \geq 2$. Then

$$a_2 = 2a_1 = 2 * 1 = 2$$

$$a_3 = 3a_2 = 3 * 2 * 1 = 6$$

$$a_4 = 4a_3 = 4 * 3 * 2 * 1 = 24$$

$$a_5 = 5a_4 = 5 * 4 * 3 * 2 * 1 = 120$$

The usual notation for a_n is $n!$ (called "n factorial"). This symbol is defined inductively by $1! = 1$, $n! = n((n-1)!)$ for $n \geq 2$. Note that $n!$ is the product of all the natural numbers from 1 to n. Our above calculations have shown that $2! = 2$, $3! = 6$, $4! = 24$, $5! = 120$. Note that $n!$ increases very rapidly with n. For example, $12! = 479,001,600$. If n is large, then $n!$ can be estimated by *Stirling's formula*, which states that

$$n! \approx \sqrt{2\pi} n^{n+\frac{1}{2}} e^{-n}$$

In general, $n!$ is the number of possible permutations, that is, rearrangements, of a set of n distinct objects. If you have n books that will all fit on the same shelf, then you can arrange them in $n!$ distinct ways.

It is sometimes more convenient to use an alternate form of the Principle of Mathematical Induction, known as Strong Induction. (We omit the proof that Theorem 1.9A is logically equivalent to Theorem 1.9.)

Theorem 1.9A

(Strong Induction)

Let **S** be a subset of **N** with the two following properties:

(i) $1 \in \mathbf{S}$;

(ii) If $m \in \mathbf{S}$ $\quad \forall m : 1 \leq m < n$, then $n \in \mathbf{S}$.

Then $\mathbf{S} = \mathbf{N}$.

Remarks

Strong induction may also be applied to prove statements that hold for all $n \geq n_0$, where n_0 is a certain constant. In such instances, 1 must be replaced by n_0 in (i) and (ii).

We now present an example of a theorem whose proof is by strong induction. If the integer $n \geq 2$, then n is called *composite* if there exist integers a, b such that $n = ab, 1 < a < n$, $1 < b < n$. Otherwise, n is called *prime*. For example, 6 is composite, since $6 = 2 * 3$, but 2 and 3 are both prime. We use strong induction to prove the following theorem:

Theorem 1.13

If the natural number $n \geq 2$, then n may be represented as a product of one or more primes.

Proof: If n is prime, which is the case for $n = 2$, then n is the product of one prime. If n is composite, then $n = ab$, where $2 \leq a < n$ and $2 \leq b < n$. By induction hypothesis, each of a, b is a product of primes. Therefore, so is n. ∎

Fibonacci (1170–1250)

Leonardo Pisano, also known as *Leonardo Bigollo* and *Fibonacci*, was born in Pisa, Italy. Pisa was then an independent city-state, competing with nearby Genoa and Florence. Fibonacci's father was involved in trade with North Africa. The father sent young Leonardo to the North African city of Bugia, on the Mediterranean coast of what is now Algeria. Bugia was a center where candles were made. There, Fibonacci learned the Hindu-Arabic numeral system, which at the time was largely unknown in Europe. Europeans were still using the clumsy Roman numeral system in their calculations.

When he returned to Pisa some years later, Fibonacci successfully introduced the Hindu-Arabic numeral system in Europe. He wrote several books, among them *Liber Abaci*, which appeared in 1202. In *Liber Abaci*, Fibonacci mentions a problem involving pairs of rabbits that leads to the inductively defined sequence given below:

$$1, 1, 2, 3, 5, 8, 13, 21, 34, 55, \cdots$$

These numbers are now known as *Fibonacci numbers*. They occur in nature and have many applications. For example, the surface of a pineapple consists of lozenge-shaped units, arranged in spirals about the trunk of the pineapple. Some of the spirals are clockwise, while others are counterclockwise. The ratio of spirals in these two directions is 8:13, two consecutive terms in the Fibonacci sequence. The phenomenon just described is just one of many botanical phenomena where the ratio of consecutive Fibonacci numbers occurs.

Fibonacci was respected as one of the best European mathematicians of his age. He was able to find an approximate solution to a cubic equation, quite an achievement at the time. He was presented to the Holy Roman Emperor Frederick Barbarossa when the latter visited Pisa in 1226. Fibonacci was awarded a salary by the city of Pisa in 1240 in recognition of his services in teaching and accounting. There is a larger-than-life statue of Fibonacci in Pisa today.

Definition 1.2

Let $F_1 = 1$, $F_2 = 1$, $F_n = F_{n-1} + F_{n-2}$ for all $n \geq 3$. F_n is called the nth Fibonacci number. Note that

$$F_3 = F_2 + F_1 = 1 + 1 = 2$$

$$F_4 = F_3 + F_2 = 2 + 1 = 3$$

$$F_5 = F_4 + F_3 = 3 + 2 = 5$$

$$F_6 = F_5 + F_4 = 5 + 3 = 8$$

$$\vdots$$

Fibonacci numbers occur in nature and have many interesting properties and many applications. Let the roots of the equation $x^2 = x + 1$ be $\alpha = \frac{1+\sqrt{5}}{2}$, $\beta = \frac{1-\sqrt{5}}{2}$. The constant α, whose approximate value is 1.618, is known as the *golden ratio*. Note that $\alpha + \beta = 1, \alpha - \beta = \sqrt{5}$, $\alpha\beta = -1$, $\alpha^2 = \alpha + 1$, $\beta^2 = \beta + 1$. The following theorem gives an explicit formula for F_n.

Theorem 1.14 $F_n = \frac{\alpha^n - \beta^n}{\alpha - \beta}$

Proof: (Strong induction on n) Since the Fibonacci sequence has not one but two initial terms, step (i) in the proof consists in verifying that the statement is true both for $n = 1$ and $n = 2$.

(i) $F_1 = \dfrac{\alpha^1 - \beta^1}{\alpha - \beta} = \dfrac{\alpha - \beta}{\alpha - \beta} = 1$

$F_2 = \dfrac{\alpha^2 - \beta^2}{\alpha - \beta} = \alpha + \beta = 1$

(ii) If $n \geq 3$, then by induction hypothesis, we have

$$F_{n-1} = \frac{\alpha^{n-1} - \beta^{n-1}}{\alpha - \beta}$$

$$F_{n-2} = \frac{\alpha^{n-2} - \beta^{n-2}}{\alpha - \beta}$$

Therefore

$$F_{n-1} + F_{n-2} = \frac{\alpha^{n-1} - \beta^{n-1} + \alpha^{n-2} - \beta^{n-2}}{\alpha - \beta}$$

$$= \frac{(\alpha^{n-1} + \alpha^{n-2}) - (\beta^{n-1} + \beta^{n-2})}{\alpha - \beta}$$

But $F_n = F_{n-1} + F_{n-2}$ by Definition 1.2. Furthermore, since $\alpha^2 = \alpha + 1$ and $\beta^2 = \beta + 1$, it follows that $\alpha^n = \alpha^{n-1} + \alpha^{n-2}$ and $\beta^n = \beta^{n-1} + \beta^{n-2}$. Substituting, we obtain our desired conclusion, namely,

$$F_n = \frac{\alpha^n - \beta^n}{\alpha - \beta}. \quad \blacksquare$$

The next theorem shows that the growth of F_n with n is essentially exponential.

Theorem 1.15 $\alpha^{n-2} \leq F_n \leq \alpha^{n-1}$ for all $n \geq 1$.

Proof: Exercise.

Yet another inductively defined sequence of natural numbers is the sequence of *Lucas numbers*, which we define below. Lucas numbers are related in many ways to Fibonacci numbers. Edouard Lucas (1842–1891) was a French mathematician who taught at the Lycee Charlemagne (a secondary school) in Paris.

Definition 1.3   **(Lucas Numbers)**

Let $L_1 = 1$, $L_2 = 3$, $L_n = L_{n-1} + L_{n-2}$ for $n \geq 3$. L_n is called the nth Lucas number. The next theorem is an explicit formula for L_n in terms of α and β.

Theorem 1.16 $L_n = \alpha^n + \beta^n$ for all $n \geq 1$.

Proof: Exercise

Section 1.3 Exercises

1. Compute $n!$ for $n = 6, 7, 8, 9, 10$.

2. Compute F_n for $n = 7, 8, 9, \cdots, 20$.

3. Compute L_n for $n = 3, 4, 5, \cdots, 20$.

Use mathematical induction to prove each of statements 4 through 26.

4. The sum of the first n odd natural numbers is n^2.

5. The sum of the first n squares is $n(n+1)(2n+1)/6$.

6. The sum of the first n odd squares is $n(4n^2 - 1)/3$.

7. The sum of the first n cubes is $n^2(n+1)^2/4$.

8. The sum of the first n odd cubes is $n^2(2n^2 - 1)$.

9. $3^n > 3n - 1$

10. $2^n > n^2$ if $n \geq 5$

11. Theorem 1.15

12. Theorem 1.16

13. $F_1 + F_2 + F_3 + \cdots + F_n = F_{n+2} - 1$

14. $F_2 + F_4 + F_6 + \cdots + F_{2n} = F_{2n+1} - 1$

15. $F_1 + F_3 + F_5 + \cdots + F_{2n-1} = F_{2n}$

16. $L_1 + L_2 + L_3 + \cdots + L_n = L_{n+2} - 3$

17. $L_2 + L_4 + L_6 + \cdots + L_{2n} = L_{2n+1} - 1$

18. $L_1 + L_3 + L_5 + \cdots + L_{2n-1} = L_{2n} - 2$

19. $\alpha^n = \alpha F_n + F_{n-1}$

20. $\beta^n = \beta F_n + F_{n-1}$

21. $F_{n+1} + F_{n-1} = L_n$ (Hint: $F_0 = 0$.)

22. $L_{n+1} + L_{n-1} = 5F_n$ (Hint: $L_0 = 2$.)

23. $F_{m+n} = F_{m+1}F_n + F_m F_{n-1}$ (Hint: Let m be fixed.)

24. $L_{m+n} = L_{m+1}F_n + L_m F_{n-1}$

25. $5F_{m+n} = L_{m+1}L_n + L_m L_{n-1}$

26. Let a "good" word have the following properties: (i) each letter is A or B, (ii) no two A's are consecutive. Let w_n be the number of "good" words with exactly n letters. Prove that $w_n = F_{n+2}$.

The identities given by Theorems 1.14 and 1.16, namely, $F_n = \frac{\alpha^n - \beta^n}{\alpha - \beta}$, $L_n = \alpha^n + \beta^n$, are known as the *Binet equations*. Use them to prove the following additional identities concerning Fibonacci and Lucas numbers.

27. $F_{2n} = F_n L_n$

28. $F_{2n+1} = F_{n+1}^2 + F_n^2$

29. $L_{2n} = L_n^2 - 2(-1)^n$

30. $L_{2n+1} = L_n L_{n+1} + (-1)^{n+1}$

31. $L_n^2 - 5F_n^2 = 4(-1)^n$

32. $F_{n-1} F_{n+1} - F_n^2 = (-1)^n$

33. $L_{n-1} L_{n+1} - L_n^2 = 5(-1)^{n+1}$

34. $F_{m+n} + (-1)^n F_{m-n} = F_m L_n$

35. $F_{m+n} - (-1)^n F_{m-n} = F_n L_m$

36. $2F_{m+n} = F_m L_n + F_n L_m$

37. $L_{m+n} + (-1)^n L_{m-n} = L_m L_n$

38. $L_{m+n} - (-1)^n L_{m-n} = 5F_m F_n$

39. $2L_{m+n} = L_m L_n + 5F_m F_n$

Section 1.3   Computer Exercises

40. Find the least n such that $n! > 10^{100}$.

41. Compute the first 100 Fibonacci and Lucas numbers.

42. Compute each of the ratios: (a) F_{n+1}/F_n; (b) L_{n+1}/L_n; (c) L_n/F_n for $1 \le n \le 50$. What seems to be happening to these ratios as n increases?

43. Compute $\alpha^n/\sqrt{5}$ to 8 decimals for $1 \le n \le 50$. What seems to be happening to this ratio as n increases?

44. Let $c_n = F_{n-1}/2^n$ and compute the sum $c_1 + c_2 + c_3 + \cdots + c_n$ for $1 \le n \le 50$. What seems to be happening to this sum as n increases?

1.4   Sigma Notation and Product Notation

In number theory and in other branches of mathematics, one frequently encounters sums of similar terms. A convenient abbreviation for the sum

$$a_1 + a_2 + a_3 + \cdots + a_n$$

is

$$\sum_{k=1}^{n} a_k$$

This symbol is read "the summation of a_k as k runs from 1 to n". The Greek letter Σ is used to indicate summation. The variable k is called the *running index* or *index of summation*, while 1 and n are called the *lower* and *upper limits of summation*, respectively. (The running index is sometimes i or j.)

The symbol a_k denotes the kth term in the sum, where $1 \leq k \leq n$. Often, a_k is a function of k.

For example, the sum of the first n odd natural numbers may be represented using sigma notation by

$$\sum_{k=1}^{n} (2k - 1)$$

Products of similar terms also arise in number theory and elsewhere. A convenient abbreviation for the product

$$a_1 a_2 a_3 \cdots a_n$$

is

$$\prod_{k=1}^{n} a_k$$

This symbol is read "the product of all a_k as k runs from 1 to n". The Greek letter Π is used to indicate taking a product. As before, k is the running index, and 1 and n are the lower and upper limits of the product. For example, if n is a natural number, then

$$\prod_{k=1}^{n} k = n!$$

Section 1.4   Exercises

1. Use sigma notation to write the *statement* of each of Exercises 4 through 18 of Section 1.3.

2. Evaluate $\sum_{k=1}^{10} \frac{1}{2^k}$. Present your result as an ordinary fraction.

3. Use product notation to represent each of the following products, and then use factorials, etc. to simplify your result: (a) product of the first n even natural numbers, (b) product of the first n odd natural numbers.

4. Evaluate the "telescoping" product:

$$\prod_{k=1}^{100} \frac{k}{k + 1}$$

1.5 Binomial Coefficients

The algebraic expression $x + y$ is called a *binomial*. If we expand $(x + y)^n$ for $n = 1, 2, 3$, etc., we obtain

$$
\begin{aligned}
(x+y)^1 &= & 1x + 1y \\
(x+y)^2 &= & 1x^2 + 2xy + 1y^2 \\
(x+y)^3 &= & 1x^3 + 3x^2y + 3xy^2 + 1y^3 \\
(x+y)^4 &= & 1x^4 + 4x^3y + 6x^2y^2 + 4xy^3 + 1y^4 \\
(x+y)^5 &= & 1x^5 + 5x^4y + 10x^3y^2 + 10x^2y^3 + 5xy^4 + 1y^5 \\
(x+y)^6 &= & 1x^6 + 6x^5y + 15x^4y^2 + 20x^3y^3 + 15x^2y^4 + 6xy^5 + 1y^6
\end{aligned}
$$

If we copy just the coefficients, we obtain the numerical array known as *Pascal's Triangle*, whose first 6 rows are

$$
\begin{array}{ccccccccccc}
 & & & & 1 & & 1 & & & & \\
 & & & 1 & & 2 & & 1 & & & \\
 & & 1 & & 3 & & 3 & & 1 & & \\
 & 1 & & 4 & & 6 & & 4 & & 1 & \\
 1 & & 5 & & 10 & & 10 & & 5 & & 1 \\
1 & & 6 & & 15 & & 20 & & 15 & & 6 & & 1
\end{array}
$$

Blaise Pascal (1623–1662)

Blaise Pascal, a French mathematician, was one of the founders of probability theory. There is evidence that Pascal's triangle was known earlier in Asia. The numbers that appear in Pascal's triangle are known as *binomial coefficients*. Pascal's father, Etienne, was also a mathematician. Pascal built a calculating machine and also wrote several works on religion.

In Section 1.3, we defined $n!$ for natural numbers, n. Now we extend this concept by defining $0! = 1$. Let k be an integer such that $0 \le k \le n$. We then define the *binomial coefficient* $\binom{n}{k}$ as follows:

Definition 1.4

$$
\binom{n}{k} = \frac{n!}{k!(n-k)!}
$$

The symbol $\binom{n}{k}$, which is read "n choose k", counts the number of distinct ways that a subset of k elements may be chosen, without regard to order, from a set of n distinct elements. Binomial coefficients occur in probability, statistics, abstract algebra, and combinatorics, as well as in number theory. Note that $\binom{n}{k} = 1$, whereas if $1 \leq k \leq n$, then

$$\binom{n}{k} = \frac{n(n-1)(n-2)\cdots(n-k+1)}{k(k-1)(k-2)1\cdots1} = \prod_{i=0}^{k-1} \frac{n-i}{k-i}$$

For example, we have

$$\binom{7}{3} = \frac{7*6*5}{3*2*1} = 35$$

$$\binom{12}{2} = \frac{12*11}{2*1} = 66$$

The following theorem gives several properties of binomial coefficients.

Theorem 1.17

Properties of Binomial Coefficients

Let k, n be integers such that $n \geq 1$ and $0 \leq k \leq n$. Then

$$B_1 : \binom{n}{n-k} = \binom{n}{k} \qquad \text{(Symmetry Property)}$$

$$B_2 : \binom{n}{n-k-1} + \binom{n}{k} = \binom{n+1}{k} \qquad \text{(Pascal's Identity)}$$

$$B_3 : (x+y)^n = \sum_{k=0}^{n} \binom{n}{k} x^{n-k} y^k \qquad \text{(Binomial Theorem)}$$

$$B_4 : \sum_{k=0}^{n} \binom{n}{k} = 2^n$$

$$B_5 : \sum_{k=0}^{n} (-1)^k \binom{n}{k} = 0$$

$$B_6 : \sum_{k \; odd} \binom{n}{k} = \sum_{k \; even} \binom{n}{k} = 2^{n-1}$$

$$B_7 : \sum_{k=0}^{n} \binom{n}{k}^2 = \binom{2n}{n}$$

Remarks

The symmetry property, B_1, says that Pascal's triangle is symmetric with respect to a vertical line drawn through the middle. B_1 is useful in evaluating $\binom{n}{k}$ when $k > \frac{n}{2}$. For example,

$$\binom{20}{18} = \binom{20}{2} = \frac{20 * 19}{2 * 1} = 190$$

Pascal's Identity (B_2) can be used to generate Pascal's triangle inductively, that is, the $(n+1)^{st}$ row can be obtained from the nth row.

We now prove $B_2, B_3,$ and B_6, leaving the proofs of the other parts of Theorem 1.17 as exercises.

Proof of B_2:

$$\binom{n}{k-1} + \binom{n}{k} = \frac{n!}{(k-1)!(n-k+1)!} + \frac{n!}{k!(n-k)!}$$

$$= \frac{n!}{(k-1)!(n-k)!}\left(\frac{1}{n-k+1} + \frac{1}{k}\right)$$

$$= \frac{n!}{(k-1)!(n-k)!}\left(\frac{n+1}{k(n+1-k)}\right)$$

$$= \frac{(n+1)!}{k!(n+1-k)!} = \binom{n+1}{k} \quad \blacksquare$$

Proof of B_3: (Induction on n)

$\binom{1}{0}x^1y^0 + \binom{1}{1}x^0y^1 = 1x + 1y = (x+y)^1$, so that B_3 holds when $n = 1$.

By induction hypothesis, we have

$$(x+y)^n = \sum_{k=0}^{n}\binom{n}{k}x^{n-k}y^k = x^n + \sum_{k=1}^{n}\binom{n}{k}x^{n-k}y^k = \sum_{j=0}^{n-1}x^{n-j}y^j + y^n$$

Now

$$(x+y)^{n+1} = (x+y)(x+y)^n = x(x+y)^n + y(x+y)^n$$

$$= x\left(x^n + \sum_{k=1}^{n}\binom{n}{k}x^{n-k}y^k\right) + y\left(\sum_{j=0}^{n-1}\binom{n}{j}x^{n-j}y^j + y^n\right)$$

$$= x^{n+1} + \sum_{k=1}^{n}\binom{n}{k}x^{n+1-k}y^k + \sum_{j=0}^{n-1}\binom{n}{j}x^{n-j}y^{j+1} + y^{n+1}$$

$$= x^{n+1} + \sum_{k=1}^{n}\binom{n}{k}x^{n+1-k}y^k + \sum_{k=1}^{n}\binom{n}{k-1}x^{n+1-k}y^k + y^{n+1}$$

$$= x^{n+1} + \sum_{k=1}^{n}\left(\binom{n}{k} + \binom{n}{k-1}\right)x^{n+1-k}y^k + y^{n+1}$$

$$= x^{n+1} + \sum_{k=1}^{n}\binom{n+1}{k}x^{n+1-k}y^k + y^{n+1}$$

$$= \sum_{k=0}^{n+1}\binom{n+1}{k}x^{n+1-k}y^k . \quad \blacksquare$$

Note that the proof of B_3 made use of B_2.

Proof of B_6:   Let

$$a = \sum_{k\ even}\binom{n}{k},$$

$$b = \sum_{k\ odd}\binom{n}{k}$$

Now

$$a + b = \sum_{k=0}^{n}\binom{n}{k} = 2^n \quad by\ B_4$$

$$a - b = \sum_{k=0}^{n}(-1)^k\binom{n}{k} = 0 \quad by\ B_5.$$

Solving for a, b, we obtain $a = b = 2^{n-1}$. $\blacksquare$

For example, in Row 4 of Pascal's triangle, we have $1 + 6 + 1 = 4 + 4 = 8 = 2^{4-1}$.

Section 1.5 Exercises

1. Evaluate each of the following binomial coefficients:

   (a) $\begin{pmatrix} 8 \\ 2 \end{pmatrix}$ (b) $\begin{pmatrix} 9 \\ 3 \end{pmatrix}$ (c) $\begin{pmatrix} 10 \\ 4 \end{pmatrix}$ (d) $\begin{pmatrix} 11 \\ 8 \end{pmatrix}$

   (e) $\begin{pmatrix} 9 \\ 0 \end{pmatrix}$ (f) $\begin{pmatrix} 12 \\ 12 \end{pmatrix}$ (g) $\begin{pmatrix} 100 \\ 98 \end{pmatrix}$

2. Use Pascal's Identity to generate the seventh and eighth rows of Pascal's triangle.

3. Simplify $\frac{n+k}{n} \binom{n}{k}$, where $1 \le k \le n$. (Your answer should contain no fractions.)

4. Prove the following parts of Theorem 1.17:
   (a) B_1; (b) B_4; (c) B_5; (d) B_7.

5. Use Pascal's Identity and mathematical induction to prove that $\binom{n}{k}$ is always an integer.

6. Prove that if $n \ge 2$, then $2^n < \binom{2n}{n} < 2^{2n}$.

Section 1.5 Computer Exercises

Write a computer program to do each of the following:

7. For each n such that $1 \le n \le 15$, and for each k such that $0 \le 2k \le n$, find the sum of all binomial coefficients $\binom{n-k}{k}$. Is there a pattern?

8. For each n such that $1 \le n \le 10$, evaluate

   (a) $\displaystyle\sum_{k=0}^{n} \binom{n}{k} F_k$

   (b) $\displaystyle\sum_{k=0}^{n} \binom{n}{k} L_k$

   Is there a pattern?

9. Find the first 10 values of n such that $\binom{2n}{n}$ is *not* divisible by 3, 5, or 7. (Ronald Graham, a well-known mathematician, has offered \$250 for a proof that there are infinitely many such n.)

1.6 Greatest Integer Function

Let us a second look at Theorem 1.17, part B_6, which says that

$$\sum_{k \ even} \binom{n}{k} = 2^{n-1}$$

Notice that we used a modified, somewhat vaguer form of sigma notation, that is, we did not specify the limits of summation. Why was this necessary? If k is even, then $k = 2j$ for some $j \geq 0$. Therefore, since $0 \leq k \leq n$, we have $0 \leq 2j \leq n$, so $0 \leq j \leq \frac{n}{2}$. We might be tempted to rewrite the preceding equation as

$$\sum_{j=0}^{\frac{n}{2}} \binom{n}{k} = 2^{n-1}$$

but this would only be valid when $\frac{n}{2}$ is an integer, that is, when n is even. Upon reflection, we see that j, the index of summation, can be as large as the *greatest integer* that is less than or equal to $\frac{n}{2}$. This leads us to define what is known as the *greatest integer function*.

Definition 1.5 Greatest Integer Function

If x is a real number, let n be the unique integer such that $n \leq x < n+1$. Then we say that n is the *integer part* of x, and we write

$$[x] = n.$$

For example, $[4.6] = 4$, $[\sqrt{5}] = 2$, $[\pi] = 3$, $[\frac{1}{2}] = 0$, $[-4.6] = -5$. Note that Definition 1.5 implies $x = [x] + h$, where $0 \leq h < 1$. The quantity $h = x - [x]$ is called the *fractional part* of x. The following theorem lists some properties of the greatest integer function.

Theorem 1.18 Properties of the Greatest Integer Function

I_1: $x - 1 < [x] \leq x < [x] + 1$ for all real x.

I_2: $[x] = x$ if and only if x is an integer.

I_3: $[x + n] = [x] + n$ for all real x and all integers, n.

I_4: If $x \leq y$, then $[x] \leq [y]$.

I_5: $[x] + [y] \leq [x + y] \leq [x] + [y] + 1$ for all real x, y.

I_6: If $x \geq 0$ and $y \leq 0$, then $[x][y] \leq [xy]$.

I_7:   If a, b, q, r are non-negative integers such that $a = qb + r$, with $0 \leq r < b$, then $q = [\frac{a}{b}]$ and $r = a - b[\frac{a}{b}]$.

We will prove I_1, I_4, I_5, and I_6. The proofs of I_2, I_3, and I_7 are left as exercises.

Proof of I_1: By Definition 1.5, $x = [x] + h$, where $0 \leq h < 1$. Therefore $x - 1 < x - h$, that is, $x - 1 < [x]$. Also, $h \geq 0$ implies $[x] \leq x$. Finally, since $x - [x] = h < 1$, we have $x < [x] + 1$.   ∎

Proof of I_4: It suffices to prove that if $[x] > [y]$, then $x > y$. Let $y = [y] + k$, where $0 \leq k < 1$. Since $[x]$ and $[y]$ are integers, and $[x] > [y]$ by hypothesis, it follows that $[x] \geq [y] + 1$, so $[x] \geq y - k + 1 > y$. Now $x \geq [x]$ by I_1, so $x > y$.   ∎

Proof of I_5: Let $x = [x] + h, y = [y] + k$ with $0 \leq h < 1, 0 \leq k < 1$. Thus we have $x + y = [x] + [y] + (h + k)$, with $0 \leq h + k < 2$. Therefore, by virtue of I_3 and I_2, we have $[x + y] = [x] + [y] + [h + k]$. If $0 \leq h + k < 1$, then $[h + k] = 0$, so $[x + y] = [x] + [y]$. If $1 \leq h + k < 2$, then $[h + k] = 1$, so $[x + y] = [x] + [y] + 1$.   ∎

Proof of I_6: Using the same notation as in the proof of I_5, we have $xy = ([x] + h)([y] + k) = [x][y] + h[y] + k[x] + hk$. Furthermore, since $x \geq 0$ and $y \geq 0$ by hypothesis, it follows from I_4 and I_2 that $[x] \geq 0$ and $[y] \geq 0$. Therefore $xy \geq [x][y]$. Now I_4 implies $[xy] \geq [[x][y]]$. But I_2 implies $[[x][y]] = [x][y]$. Therefore $[xy] \geq [x][y]$.   ∎

Now that the greatest integer function is at our disposal, we may remove the ambiguity in the statement of Theorem 1.17, part B_6, by using $[\frac{n}{2}]$ as the upper limit of summation. This yields

$$\sum_{j=0}^{[\frac{n}{2}]} \binom{n}{2j} = 2^{n-1}$$

The second identity in Theorem 1.17, part B_6, said that

$$\sum_{k \ odd} \binom{n}{k} = 2^{n-1}$$

If k is odd, then $k = 2j + 1$ for some $j \geq 0$. Since $0 \leq k \leq n$, we have $2j + 1 \leq n$, which implies $j \leq \frac{1}{2}(n - 1)$, hence $j \leq [\frac{1}{2}(n - 1)]$. Therefore we write

$$\sum_{j=0}^{[\frac{n-1}{2}]} \binom{n}{2j + 1} = 2^{n-1}$$

We conclude this section with a theorem regarding the greatest integer function that will be useful later on.

Theorem 1.19 | If x is real and n is a natural number, then

$$\left[\frac{[x]}{n}\right] = \left[\frac{x}{n}\right]$$

Proof: Exercise.

Section 1.6 Exercises

1. Evaluate (a) $\left[\frac{23}{3}\right]$ (b) $\left[\sqrt{24}\right]$ (c) $[10\pi]$ (d) $[-10\pi]$.

2. Prove the following parts of Theorem 1.18: I_2, I_3, I_7.

3. Prove Theorem 1.19.

Section 1.6 Computer Exercises

Write a computer program to do each of the following:

4. For each n such that $1 \leq n \leq 10$, evaluate

$$\frac{1}{2^{n-1}} \sum_{k=0}^{[(n-1)/2]} \binom{n}{2k+1} 5^k$$

$$\frac{1}{2^{n-1}} \sum_{k=0}^{[n/2]} \binom{n}{2k} 5^k$$

Is there a pattern?

5. For each n such that $1 \leq n \leq 10$, evaluate

$$\sum_{k=-[(n+1)/5]}^{[n/5]} (-1)^k \binom{n}{\frac{n-5k}{2}}$$

$$\sum_{k=-[(n+1)/5]}^{[n/5]} (-1)^k \frac{n}{n-k} \binom{n}{\frac{n-5k}{2}}$$

Is there a pattern?

Review Exercises

1. Use indirect proof to show that if m and n are integers whose product is even, then at least one of m, n is even.

2. Use indirect proof to show that if m and n are integers whose sum is odd, then m and n have opposite parity, that is, one is odd while the other is even.

3. Use induction to prove that $\sum_{k=1}^{n} 2^k = 2^{n+1} - 2$.

4. Let F_n denote the nth Fibonacci number. Use induction to prove the identity:

$$\sum_{k=1}^{n} F_{3k} = \frac{F_{3n+2} - 1}{2}$$

5. Use induction to prove that $2^n > 3n$ if $n \geq 4$.

6. Use the Binet identities to prove that $F_{3n} = F_n(L_{2n} + (-1)^n)$.

7. Evaluate $\sum_{k=1}^{5} \frac{1}{k}$. Present your result as an ordinary fraction.

8. Evaluate $\prod_{k=1}^{5} \frac{2^k - 1}{2^k}$. Present your result as an ordinary fraction.

9. Evaluate each of the following:

   (a) $\binom{12}{3}$

   (b) $\binom{13}{1}$

   (c) $\binom{14}{0}$

   (d) $\binom{15}{14}$

10. Evaluate each of the following:

    (a) $[8\pi]$

    (b) $[.7]$

    (c) $[\sqrt{7}]$

Chapter 2

Divisibility

"Divide and conquer"
   —From Julius Caesar

2.1 Introduction

In number theory, certain questions arise concerning the divisors of a given natural number. Recall that a natural number is called *perfect* if it equals the sum of its own proper divisors. Can one find a formula that yields all perfect numbers? If F_m and F_n are Fibonacci numbers, what conditions on m and n guarantee that F_m divides F_n?

In this chapter, we develop the divisibility properties of natural numbers, thereby taking the first steps toward answering the questions just raised. Of particular importance are the concepts of *greatest common divisor* and *least common multiple* of a pair of natural numbers. *Note that lowercase letters refer to natural numbers, unless otherwise stated.*

2.2 Divisibility, Greatest Common Divisor, Euclid's Algorithm

Definition 2.1   Divisibility

We say that a *divides* b, and we write $a|b$ if $b = ka$ for some k.

For example, $3|12$, since $12 = 4*3$; also $8|40$, since $40 = 5*8$. If $a|b$, we might also say "a is a divisor of b", "a is a factor of b", or "b is a multiple of a". If a does *not* divide b, then we write $a \nmid b$. For example, $3 \nmid 14$; also, $8 \nmid 42$.

Definition 2.2   Proper Divisor

We say that a is a *proper divisor* of b if $a|b$ and $a < b$.

For example, 3 is a proper divisor of 6, but 6 is not a proper divisor of 6. Note that for any b, all divisors of b are proper except for b itself.

Definition 2.3   Nontrivial Divisor

We say that a is a *nontrivial divisor* of b if $a|b$ and $1 < a < b$.

For example, the nontrivial divisors of 6 are 2 and 3. Note that 1 is a proper divisor of every larger integer, but 1 has no nontrivial divisors. The following theorem states some properties of divisibility.

Theorem 2.1   Properties of Divisibility

$\mathbf{D_1}$: $1|a$ and $a|a$ for all a.

$\mathbf{D_2}$: If $a|b$, then $a \leq b$.

$\mathbf{D_3}$: If $a|b$ and $b|a$, then $a = b$.

$\mathbf{D_4}$: If $a|b$ and $b|c$, then $a|c$.

$\mathbf{D_5}$: For all c, $a|b$ if and only if $ac|bc$.

$\mathbf{D_6}$: If $a|b$ and $c|d$, then $ac|bd$.

$\mathbf{D_7}$: If $a|b$ and $a|c$, then $a|(bx + cy)$ for all x, y.

We prove parts D_2, D_4, and D_7, leaving the proofs of the other parts of Theorem 2.1 as exercises.

Proof of D_2: If $a|b$, then by Definition 2.2, $b = ka$ for some $k \geq 1$, so $ka \geq a$, that is, $b \geq a$. ∎

Proof of D_4: If $a|b$ and $b|c$, then $b = ka$ and $c = jb$ for some j, k. Therefore $c = j(ka) = (jk)a$, so $a|c$. ∎

Proof of D_7: If $a|b$ and $a|c$, then $b = ka$ and $c = ja$ for some j, k. Therefore $bx + cy = (ka)x + (ja)y = (kx + jy)a$, so $a|(bx + cy)$. ∎

Remarks

In order to prove that $m = n$, it is sometimes convenient to use D_3, that is, to prove that $m|n$ and $n|m$.

∎

Further Remarks

If a, b, c refer to nonzero integers rather than to natural numbers, then Theorem 2.1 remains valid except for parts D_2, D_3 and D_5, which must be adjusted as follows: If $a|b$, then $|a| \leq |b|$; if $a|b$ and $b|a$, then $a = \pm b$; if $c \neq 0$, then $a|b$ if and only if $ac|bc$.

∎

When 23 is divided by 5, the quotient is 4 and the remainder is 3. Since we prefer to deal with integers, instead of writing $\frac{23}{5} = 4\frac{3}{5}$, we write $23 = 4(5) + 3$. Whenever one natural number is divided by another, the remainder is strictly less than the divisor. The following theorem explains why this is so.

Theorem 2.2

Division Algorithm

If a and b are integers such that $a \geq b > 0$, then there exist unique integers q, r such that $a = qb + r$ and $0 \leq r < b$.

Proof: If $b|a$, then $q = a/b$ and $r = 0$. Having disposed of this case, let us assume that $b < a$ and $b \nmid a$. Let S be the set of all natural numbers of the form $x_n = a - nb$, with $n \geq 1$. By hypothesis, we know that $x_1 = a - b > 0$, so $x_1 \in S$. Therefore S is nonempty. By the Well-Ordering Principle, S has a least element, x_q. Let $x_q = a - qb = r$. If $r = b$, then $a = (q + 1)b$, so $b|a$, contrary to hypothesis. If $r > b$, then $x_{q+1} = a - (q+1)b > 0$, so $x_{q+1} \in S$, yet $x_{q+1} < x_q$. This contradicts the definition of x_q. Therefore $0 < r < b$. Finally, suppose that $a = q_1 b + r_1 = q_2 b + r_2$, with $r_1 > r_2$. But then $b|(r_1 - r_2)$, so $b \le r_1 - r_2$ by D_2, hence $b < r_1$, an impossibility. Therefore r is unique, from which it follows easily that q is also unique. ■

Remarks

Note that when a is divided by b, then q is the quotient and r is the remainder, that is, $q = [a/b]$ and $r = a - b[a/b]$. For example, if $a = 50$ and $b = 13$, then $q = [50/13] = 3$ and $r = 50 - 13[50/13] = 50 - 13 * 3 = 50 - 39 = 11$.

■

Definition 2.4   Common Divisor

If $c|a$ and $c|b$, then we say that c is a *common divisor* of a and b.

For example, the common divisors of 36 and 60 are 1, 2, 3, 4, 6, and 12. Also, the common divisors of 20 and 30 are 1, 2, 5, and 10. Note that 1 is a common divisor of every pair a, b.

Definition 2.5   Greatest Common Divisor

Without loss of generality, suppose that a and b are given natural numbers, with $a \ge b$. Suppose that d is an integer such that (i) $d|a$ and $d|b$; (ii) if $c|a$ and $c|b$, then $c|d$. Then we say that d is the *greatest common divisor* of a and b. We write $d = GCD(a, b)$ or simply $d = (a, b)$.

We have seen that 12 is the greatest common divisor of 60 and 36, so we write $(60, 36) = 12$. Also, 10 is the greatest common divisor of 30 and 20, so we write $(30, 20) = 10$.

Theorem 2.3   Properties of the Greatest Common Divisor

Let $d = (a, b)$. Then the following are true:

$\mathbf{G_1}$: If $c|a$ and $c|b$, then $c \le d$.

$\mathbf{G_2}$: $(a, b) = b$ if and only if $b|a$.

$\mathbf{G_3}$: $1 \le d \le Min\{a, b\}$.

$\mathbf{G_4}$: If $c|a$ and $c|b$, then $(\frac{a}{c}, \frac{b}{c}) = \frac{d}{c}$.

$\mathbf{G_5}$: $(\frac{a}{d}, \frac{b}{d}) = 1$.

$\mathbf{G_6}$: $(an, bn) = dn$ for all n.

We prove parts G_4 and G_6, leaving the proofs of the other parts of Theorem 2.3 as exercises.

Proof of G_4: Since $c|a$ and $c|b$ by hypothesis, we know that $\frac{a}{c}$ and $\frac{b}{c}$ are integers. By Definition 2.5, $\frac{d}{c}$ is an integer. Now $d|a$, that is, $c(\frac{d}{c})|c(\frac{a}{c})$. Therefore, by Theorem 2.1, part D_5, we have $\frac{d}{c}|\frac{a}{c}$. Similarly, $\frac{d}{c}|\frac{b}{c}$. We have shown that $\frac{d}{c}$ is a common divisor of $\frac{a}{c}$ and $\frac{b}{c}$. It remains to show that $\frac{d}{c}$ is their *greatest* common divisor. If $k|\frac{a}{c}$ and $k|\frac{b}{c}$, then Theorem 2.1, part D_5 implies $kc|a$ and $kc|b$. Therefore, by Definition 2.5, $kc|d$. Again, applying Theorem 2.1, part D_5, we have $k|\frac{d}{c}$. Therefore $(\frac{a}{c}, \frac{b}{c}) = \frac{d}{c}$. ∎

Proof of G_6: Since $n|an$ and $n|bn$, it follows from Definition 2.5 that $n|(an, bn)$. Let $(an, bn) = dn$. Therefore it follows from G_4 that $(\frac{an}{n}, \frac{bn}{n}) = \frac{dn}{n}$, that is $(a, b) = d$. Thus we have $(an, bn) = (a, b)n$. ∎

> ### Euclid (330–270 B.C.)
> Euclid was a leading scholar at the *Museum of Alexandria* in Egypt, an ancient equivalent of today's universities. His book *Elements*, written about 300 B.C., is the oldest known and surviving mathematical textbook. When asked to make mathematics simpler, he reputedly said, *"There is no royal road to geometry."* On another occasion, when a student asked how he would benefit by learning mathematics, Euclid reputedly told his servant to give the student a few coins.

Euclid's Algorithm, which is discussed below, is a method for computing (a, b) via a finite sequence of divisions.

Greatest Common Divisor via Euclid's Algorithm

We wish to compute $d = (a, b)$, where a, b are given natural numbers with $a \geq b$. We start by letting a be the *dividend*, D, and letting b be the *divisor*, t. Next, perform the

Key Step: Divide D by t, obtaining a quotient, q, and a remainder, r.

If $r > 0$, then replace D by t, replace t by r, and repeat the Key Step. If $r = 0$, then $d = t$ (the last divisor).

Example 2.1 | Find (87,27).

Solution: $87 = 3(27)+6$; $27 = 4(6)+3$; $6=2(3)+0$

Result: $(87,27) = 3$.

Example 2.2   Find (418,165).

Solution: $418 = 2(165) + 88; \quad 165 = 1(88) + 77; \quad 88 = 1(77) + 11;$
$77 = 7(11) + 0$

Result: $(418,165) = 11.$

Next, we explain why the Euclidean Algorithm works. Consider the sequence of equations that it produces. Let $a = r_{-1}, b = r_0$.

The first iteration yields: $\qquad\qquad\qquad r_{-1} = q_1 r_0 + r_1, \, 0 \le r_1 < r_0.$

If $r_1 > 0$, then the second iteration yields: $\qquad r_0 = q_2 r_1 + r_2, \, 0 \le r_2 < r_1.$

If $r_2 > 0$, then the second iteration yields: $\qquad r_1 = q_3 r_2 + r_3, \, 0 \le r_3 < r_2.$

If j or more iterations are needed, then the jth iteration yields

$$r_{j-2} = q_j r_{j-1} + r_j, 0 \le r_j < r_{j-1}.$$

Note that the sequence: $r_{-1}, r_0, r_1, r_2, \cdots$ is strictly decreasing. Therefore, by Theorem 1.8, this sequence must be finite. This means that for some $k \ge 0$, we must have $r_{k+1} = 0$, hence

$$r_{k-1} = q_k r_k$$

To recapitulate, Euclid's Algorithm yields the following sequence of equations:

$$
\begin{array}{lll}
(1) & r_{-1} = q_1 r_0 + r_1 & 0 < r_1 < r_0 \\
(2) & r_0 = q_2 r_1 + r_2 & 0 < r_2 < r_1 \\
(3) & r_1 = q_3 r_2 + r_3 & 0 < r_3 < r_2 \\
& \vdots & \\
(j) & r_{j-2} = q_j r_{j-1} + r_j, & 0 < r_j < r_{j-1} \qquad (2.1)\\
& \vdots & \\
(k-1) & r_{k-3} = q_{k-1} r_{k-2} + r_{k-1}, & 0 < r_1 < r_0 \\
(k) & r_{k-2} = q_k r_{k-1} + r_k, & 0 < r_k < r_{k-1} \\
(k+1) & r_{k-1} = q_{k+1} r_k &
\end{array}
$$

If $d = (a, b)$, we claim that $d = r_k$. We will prove this by showing that $r_k | d$ and $d | r_k$. By Equation (2.1), form $(k + 1)$, we have $r_k | r_{k-1}$. Therefore, by form (k) and Theorem 2.1, part D_7, we have $r_k|r_{k-2}$. Since $r_k|r_{k-1}$ and $r_k|r_{k-2}$, it follows from form (k-1) and Theorem 2.1, part B_7 that $r_k|r_{k-3}$. Continuing in like fashion, we eventually obtain $r_k|r_0$ and $r_k|r_{-1}$, that is, $r_k|b$ and $r_k|a$. Therefore, $r_k|(a,b)$, that is, $r_k|d$.

On the other hand, form (1) implies $d|r_1$. Since $d|r_0$ and $d|r_1$, form (2) implies $d|r_2$. Since $d|r_1$ and $d|r_2$, form (3) implies $d|r_3$. Continuing in like fashion, we ultimately see that form (k) implies $d|r_k$. Therefore $d = r_k$.

Note that the number of iterations needed to execute Euclid's Algorithm varies with the input. In Example 1 above, 3 iterations were needed to determine that (87,27)=3. In Example 2 above, 4 iterations were needed to determine that (418,165) = 11. The following theorem, credited to Lamé, gives a relation between the size of the input (a and b) and the number of iterations, n , needed to compute (a,b) using Euclid's Algorithm. This proof makes use of properties of the Fibonacci numbers.

Theorem 2.4

(Lamé) Let $\alpha = \frac{1+\sqrt{5}}{2}$. If a and b are integers such that $a > b > 0$ and n is the number of iterations needed to compute (a,b) using Euclid's Algorithm, then $n \leq 1 + \log_\alpha b$.

Proof: Again, we let $a = r_{-1}, b = r_0$ and refer to Equations (1) through $(k+1)$ above. We will use strong induction on j to prove that $r_{k-j} \geq F_{j+2}$ for all j such that $0 \leq j \leq k$. Since $r_k \geq 1$ and $F_2 = 1$, we have $r_k \geq F_2$. Since $q_{k+1} \geq 2$ (otherwise we would have $r_{k-1} = r_k$), it follows from Equation $(k+1)$ that $r_{k-1} \geq 2 * 1$. Since $F_3 = 2$, we have $r_{k-1} \geq F_3$. In general, we have

$$(k-j) \qquad r_{k-j} = q_{k-j+2} r_{k-j+1} + r_{k-j+2}.$$

Since $q_{k-j+2} \geq 1$, we have $r_{k-j} \geq r_{k-j+1} + r_{j-k+2}$. By induction hypothesis, we have $r_{j-k+1} \geq F_{j+1}$ and $r_{j-k+2} \geq F_j$. Therefore $r_{k-j} \geq F_{j+1} + F_j$, that is, $r_{k-j} \geq F_{j+2}$. Letting $j = k$, we get $r_0 \geq F_{k+2}$, that is, $b \geq F_{k+2}$. If n is the total number of iterations, then $n = k + 1$, so $b \geq F_{n+1}$. By Theorem 1.15, $F_n \geq \alpha^{n-1}$. Therefore $\alpha^{n-1} \leq b$, so $n \leq 1 + \log_\alpha b$. ∎

Remarks

It can be shown that the time needed to compute (a,b) using Euclid's Algorithm is proportional to $(\log_\alpha b)^3$. Since $\log_\alpha b$ is very small compared to b, this implies that Euclid's Algorithm is *computationally feasible*, that is, suitable even for very large inputs. For example, finding the greatest common divisor of two 200-digit numbers takes only about 8 times long as the same computation for two 100-digit numbers.

The following theorem, whose proof is omitted, offers an alternate procedure for evaluating (a, b). (see [11])

Theorem 2.5

$$(a, b) = 2 \sum_{i=1}^{b-1} [\frac{ai}{b}] + a + b - ab$$

The following theorem states an important property of the greatest common divisor that we will need in Chapter 4.

Theorem 2.6

Linear Form of the Greatest Common Divisor

If a and b are natural numbers, then (a, b) is the least natural number of the form $ax + by$, where x and y are integers. Furthermore, if the natural number $t = au + bv$ for integers u, v, then $(a, b)|t$.

Proof: Let S be the set of all natural numbers of the form $ax + by$, where x and y are integers. Since $a = a(1) + b(0)$, we have $a \in S$, so S is non empty. By the Well-Ordering Principle, S has a least element, d, where $d = am + bn$ for some integers m, n. By the Division Algorithm (Theorem 2.2), there exist integers q, r such that $a = qd + r$, where $0 \le r < d$. Now $r = a - qd = a - q(am + bn) = a(1 - qm) + b(-qn)$. If $r \ne 0$, then $r \in S$, yet $r < d$. This contradicts the definition of d. Therefore $r = 0$, so $d|a$. Similarly, $d|b$. If $c|a$ and $c|b$, then by Theorem 2.1, part D_7, $c|(am + bn)$, that is, $c|d$. Therefore $d = (a, b)$. Finally, if $t = au + bv$, then by Theorem 2.1, part D_7, we have $d|t$. ∎

The following theorem gives a method for finding integers x, y such that $(a, b) = ax + by$.

Theorem 2.7

Let $a \ge b > 0$. If $a = r_{-1}$ and $b = r_o$, let q_j and r_j denote respectively the quotient and remainder that are generated by the jth iteration of Euclid's Algorithm. Define sequences $\{s_j\}$ and $\{t_j\}$ as follows:

$$s_{-1} = 1 , \ s_0 = 0 , \ s_j = s_{j-2} - q_j s_{j-1} \quad \text{for} \quad j \ge 1$$
$$t_{-1} = 0 , \ t_0 = 1 , \ t_j = t_{j-2} - q_j t_{j-1} \quad \text{for} \quad j \ge 1$$

Then $r_j = as_j + bt_j$ for all j; in particular, $(a, b) = r_k = as_k + bt_k$.

Proof: (Strong induction on j)   Exercise.

Example 2.3   Find x, y such that $(87, 27) = 87x + 27y$.

Solution: Recall from Example 2.1 (page 32) that

$$87 = 3(27) + 6 \; ; \; 27 = 4(6) + 3 \; ; \; 6 = 2(3) \quad so \quad (87, 27) = 3$$

In order to compute x and y, we set up a table:

j	q_j	s_j	t_j
-1		1	0
0		0	1
1	3	1	-3
2	4	-4	13

Note that $s_1 = s_{-1} - q_1 s_0 = 1 - 0(3) = 1$; $t_1 = t_{-1} - q_1 t_0 = 0 - 1(3) = -3$.
    Also $s_2 = s_0 - q_2 s_1 = 0 - 1(4) = -4$; $t_2 = t_0 - q_2 t_1 = 1 - 4(-3) = 13$.
We obtain x and y from the bottom row of the table, so that we have

Result: $(87,27) = 3 = 87(-4)+27(13)$,
that is, $x = -4$, $y = 13$.

Check: $87(-4) + 27(13) = -348 + 351 = 3$.

Example 2.4   Find x, y such that $(418, 165) = 418x + 165y$.

Solution: Recall from Example 2.2 (page 32) that
$418 = 2(165) + 88 \; ; \; 165 = 1(88) + 77 \; ; \; 88 = 1(77) + 11; \; 77 = 7(11)$.
In order to compute x and y, we set up a table:

j	q_j	s_j	t_j
-1		1	0
0		0	1
1	2	1	-2
2	1	-1	3
3	1	2	-5

Result: $(418,165) = 11 = 418(2) + 165(-5)$,
so $x = 2$, $y = -5$.

Check: $418(2) + 165(-5) = 836 - 825 = 11$.

Remarks

In setting up the table, be sure to *omit* the last quotient in Euclid's Algorithm, that is, the quotient that yields the zero remainder. Note also that x and y are of opposite sign.

The next theorem is a consequence of Theorem 2.6.

Theorem 2.8   Let a, b be natural numbers. If there exist integers x, y such that $ax + by = 1$, then $(a, b) = 1$.

Proof: By hypothesis and Theorem 2.5, $(a, b)|1$. Therefore $(a, b) = 1$. ■

Definition 2.6   **Relatively Prime**

If a and b are natural numbers such that $(a, b) = 1$, then we say that a and b are *relatively prime*.

We can now prove the following:

Theorem 2.9   If a, b, m are natural numbers such that $(a, m) = (b, m) = 1$, then $(ab, m) = 1$.

Proof: By hypothesis and by Theorem 2.8, there exist integers u, v, x, y such that

$$1 = au + mv; \quad also \quad 1 = bx + my$$

Therefore

$$1 = (au + mv)(bx + my) = ab(ux) + m(auy + bvx + mvy)$$

Now Theorem 2.8 implies $(ab, m) = 1$. ■

Generalizing Theorem 2.9, we have

Theorem 2.10   Let $A = \prod_{i=1}^{n} a_i$ where $n \geq 2$. If $(a_i, m) = 1$ for all i, then $(A, m) = 1$.

Proof: Exercise (induction on n).
The following three theorems, whose proofs are left as exercises, will be useful later on.

Theorem 2.11   If $a > 1$ and $n \geq 1$, then $(a - 1)|(a^n - 1)$.

Proof: Exercise.

Theorem 2.12   If $ab = c^n$ and $(a, b) = 1$, then $a = x^n$ and $b = y^n$ for x, y such that $(x, y) = 1$.

Proof: Exercise.

Theorem 2.13   If $a \geq 2$, then $(a^m - 1, a^n - 1) = a^{(m,n)} - 1$.

Proof: Exercise.

The next theorem, known as *Euclid's Lemma*, is another consequence of Theorem 2.6. In Chapter 3, Euclid's Lemma helps establish an important property of prime numbers.

Theorem 2.14   **Euclid's Lemma**
If $a|bc$ and $(a, b) = 1$, then $a|c$.

Proof: Since $(a, b) = 1$ by hypothesis, Theorem 2.6 implies $ax + by = 1$ for some integers x, y. Therefore $c(ax + by) = c(1)$, that is, $cax + cby = c$. Now $a|a$ by Theorem 2.1, part D_1, and $a|bc$ by hypothesis. Therefore, by Theorem 2.1, part D_7, $a|(cax + cby)$, that is, $a|c$.   ■

The concept of greatest common divisor can be extended to sets of three or more natural numbers. Let $a_1, a_2, a_3, \ldots, a_n$ be natural numbers. If c is a natural number such that $c|a_i$ for all i, then we say that c is a *common divisor* of the a_i. For example, 5 is a common divisor of 20, 30, and 40. Also, 6 is a common divisor of 36, 54, and 72.

If d is a common divisor of the a_i, and if for every common divisor, c, of the a_i, we have $c|d$, then we say that d is the *greatest common divisor* of the a_i, and we write

$$d = (a_1, a_2, a_3, \ldots, a_n)$$

For example, 10 is the greatest common divisor of 20, 30, and 40, so we write $(20,30,40) = 10$. Also, 18 is the greatest common divisor of 36, 54, and 72, so we write $(36,54,72) = 18$. The following theorem may be used to compute (a_1, a_2, a_3):

Theorem 2.15
$$(a_1, a_2, a_3) = ((a_1, a_2), a_3)$$

Proof: Exercise.

For example, $(12,18,32) = ((12,18),32) = (6,32) = 2$. Also $(36,84,105) = ((36,84),105) = (12,105) = 3$. Theorem 2.15 may be generalized as follows:

Theorem 2.16   If $n \geq 3$, then

$$(a_1, a_2, a_3, \ldots, a_{n-1}, a_n) = ((a_1, a_2, a_3, \ldots, a_{n-1}), a_n)$$

Proof: Exercise (induction on n).

Note that Theorem 2.6 could be restated as:

Theorem 2.6A   If a_1 and a_2 are natural numbers, then there exist integers x_1, x_2 such that $(a_1, a_2) = a_1 x_1 + a_2 x_2$.

This theorem extends easily to sets of n natural numbers, where $n \geq 3$:

Theorem 2.17   Let $a_1, a_2, a_3, \ldots, a_n$ be natural numbers, where $n \geq 2$. Then there exist integers $x_1, x_2, x_3, \ldots, x_n$ such that

$$(a_1, a_2, a_3, \ldots, a_n) = a_1 x_1 + a_2 x_2 + a_3 x_3 + \ldots + a_n x_n$$

Proof: Exercise.

The following definition is useful when dealing with a set of three or more natural numbers:

Definition 2.7   **Pairwise Relatively Prime**

Let $a_1, a_2, a_3, \ldots, a_n$ be natural numbers, where $n \geq 3$. We say that the a_i are *pairwise relatively prime* if $(a_i, a_j) = 1$ for all $i \neq j$.

For example, 6, 55, 91 are pairwise relatively prime, since $(6, 55) = (6, 91) = (55, 91) = 1$.

Section 2.2   Exercises

1. Find all the divisors of each integer from 20 to 30.

2. Prove parts D_1, D_3, D_5, D_6 of Theorem 2.1.

3. Prove that if $a|1$, then $a = 1$.

4. Prove that if $a|b$ and $n \geq 1$, then $a^n|b^n$.

5. Prove that if $n|(a-1)$ and $n|(b-1)$, then $n|(ab-1)$.

6. Prove that if n is odd, then $8|(n^2 - 1)$.

7. Find all common divisors of each pair of integers:
   (a) 12, 30   (b) 12, 16   (c) 8, 21   (d) 10, 20   (e) 36, 63.

8. Prove parts G_1, G_2, G_3, G_5 of Theorem 2.5.

9. Prove that $(n, 1) = 1$ for all n.

10. Prove that $(n+1, n) = 1$ for all n.

11. Prove that $(2n+1, 2n-1) = 1$ for all n.

12. Prove that for all n, (a) $(F_{n+1}, F_n) = 1$; (b) $(F_{n+2}, F_n) = 1$.

13. Prove that $(a, b) = 1$ if and only if $(a+b, ab) = 1$.

14. For each pair m, n, find (m, n) and find integers x, y such that $(m, n) = mx + ny$: (a) 30, 12 (b) 121, 29 (c) 95, 75 (d) 273, 231 (e) 299, 104 (f) 273, 221 (g) 621, 126 (h) 402, 102 (i) 15493, 9593.

15. Prove each of the following Theorems: (a) 2.8 (b) 2.11 (c) 2.12 (d) 2.13 (e) 2.14 (f) 2.15 (g) 2.16 (h) 2.17

16. Prove that if n is odd and $a \geq 0$, then $(a+1)|(a^n + 1)$.

17. Prove that $2|F_n$ if and only if $3|n$. (Hint: Use identity 23 from p. 15 and induction on n.)

18. Prove that $F_{nk+r} = \sum_{j=0}^{k} \binom{k}{j} F_{n-1}^{k-j} F_n^j F_{r+j}$. (Hint: You will need Theorem 1.14 and the results of Section 1.3, Exercises 19 and 20.)

19. Prove that $F_n|F_{kn}$.

20. Prove that $(F_m, F_n) = F_{(m,n)}$.

Section 2.2 Computer Exercises

21. Write a computer program to do each of the following:

22. Given natural numbers a, b, find (a, b) and find integers x, y such that $(a, b) = ax + by$.

23. Given natural numbers $a_1, a_2, a_3, \ldots, a_n$, find $(a_1, a_2, a_3, \ldots, a_n)$ and find integers $x_1, x_2, x_3, \ldots, x_n$ such that
$(a_1, a_2, a_3, \ldots, a_n) = a_1 x_1 + a_2 x_2 + a_3 x_3 + \ldots + a_n x_n$.

2.3 Least Common Multiple

So far, we have considered divisors, common divisors, and greatest common divisors. Recall that if a divides b, then b is a multiple of a. We now consider the analogous concepts of common multiple and least common multiple.

Definition 2.8   Common Multiple

If $a|n$ and $b|n$, then we say that n is a *common multiple* of a and b.

For example, the common multiples of 8 and 12 are 24, 48, 72, etc. Also, the common multiples of 9 and 15 are 45, 90, 135, etc. Since $a|ab$ and $b|ab$, ab is a common multiple of a and b.

Definition 2.9   Least Common Multiple

Given a and b, suppose that m has the following properties:
(i) $a|m$ and $b|m$; (ii) if $a|n$ and $b|n$, then $m|n$. Then we say that m is the *least common multiple* of a and b. We write $m = LCM(a, b)$, or simply $m = [a, b]$.

For example, the least common multiple of 8 and 12 is 24, so we write $[8,12]=24$. Also, the least common multiple of 9 and 15 is 45, so we write $[9,15] = 45$. The following theorem, which is analogous to Theorem 2.3, lists some properties of the least common multiple of two natural numbers.

Theorem 2.18   Properties of the Least Common Multiple

Let $m = [a, b]$.

$\mathbf{L}_1$: If $a|n$ and $b|n$, then $m \leq n$.

$\mathbf{L}_2$: $[a, b] = a$ if and only if $b|a$.

$\mathbf{L}_3$: $\text{Max}\{a, b\} \leq m \leq ab$.

L$_4$: If $c|a$ and $c|b$, then $[\frac{a}{c} \cdot \frac{b}{c}] = \frac{m}{c}$.

L$_5$: $[an, bn] = [a, b]n$ for all n.

L$_6$: If $(a, b) = 1$, then $[a, b] = ab$.

L$_7$: $[a, b] = ab/(a, b)$.

We prove parts L_4 through L_7, leaving the proofs of the other parts of Theorem 2.18 as exercises.

Proof of L_4: Since $c|a$ and $c|b$ by hypothesis, it follows that $\frac{a}{c}$ and $\frac{b}{c}$ are integers. Let $j = [\frac{a}{c}, \frac{b}{c}]$, so that $\frac{a}{c}|j$ and $\frac{b}{c}|j$. We will show that $j = \frac{m}{c}$. Now Theorem 2.1, Part D_5 implies that $a|cj$ and $b|cj$. Therefore, by Definition 2.9, we have $m|cj$. Since $c|a$ and $a|m$, it follows that $c|m$, so $\frac{m}{c}$ is an integer. Now Theorem 2.1, part D_5 implies $\frac{m}{c}|j$. Now Definition 2.9 implies $a|m$ and $b|m$; therefore Theorem 2.1, part D_5 implies $\frac{a}{c}|\frac{m}{c}$ and $\frac{b}{c}|\frac{m}{c}$. This in turn implies $j|\frac{m}{c}$. By Theorem 2.1, part D_3, we have $j = \frac{m}{c}$. ∎

Proof of L_5: If $n|r$ and $n|s$, then L_4 implies $[\frac{r}{n}, \frac{s}{n}] = \frac{[r,s]}{n}$, so that $[r, s] = n[\frac{r}{n}, \frac{s}{n}]$. Letting $r = an, s = bn$, we obtain $[an, bn] = n[a, b]$. ∎

Proof of L_6: Since $m = [a, b]$, we have $a|m, b|m$, so $m = au = bv$ for some integers u, v. Now L_3 implies $m \leq ab$, so $bv \leq ab$, hence $v \leq a$. Now $a|bv$ and by hypothesis, we have $(a, b) = 1$. Therefore Theorem 2.13 implies $a|v$, so $a \leq v$. Thus $v = a$ and $m = ab$. ∎

Proof of L_7: Since $d = (a, b)$, by Theorem 2.3, part G_4, we have $(\frac{a}{d}, \frac{b}{d}) = 1$. Therefore L_6 implies $[\frac{a}{d}, \frac{b}{d}] = (\frac{a}{d})(\frac{b}{d})$. Applying L_5 with $n = d$, we get $[a, b] = ab/(a, b)$. ∎

In order to compute $[a, b]$, we first use Euclid's Algorithm to find (a, b), and then use Theorem 2.18, part L_7.

Example 2.5   Find [87,27].

Solution: Recall from Example 2.1 that **(87,27)** = 3. Therefore

$$[\textbf{87,27}] = \textbf{87*27}/\textbf{(87,27)}$$

$$= \textbf{87*27}/\textbf{3}$$

$$= \textbf{29*27}$$

$$= \textbf{783}.$$

Example 2.6	Find $[418,165]$.

Solution: Recall from Example 2.2 that $(418,165) = 11$. Therefore

$$[418,165] = 418*165/(418,165)$$

$$= 418*165/11$$

$$= 38*165$$

$$= 6270.$$

The concept of least common multiple can be extended to sets of three or more natural numbers. Suppose that $a_1, a_2, a_4, \ldots, a_n$ are natural numbers such that $a_i | c$ for each index, i. Then we say that c is a *common multiple* of the a_i. For example, 144 is a common multiple of 6, 8, and 9. Also, 60 is a common multiple of 6, 10, and 15.

If m is a common multiple of the a_i, and if for every common multiple, c, of the a_i, we have $m | c$, then we say that m is the least common multiple of the a_i, and we write

$$m = [a_1, a_2, a_3, \ldots, a_n]$$

For example, the least common multiple of 6, 8, and 9 is 72, so we write $[6, 8, 9] = 72$. Also, the least common multiple of 6, 10, and 15 is 30, so we write $[6, 10, 15] = 30$.

In order to compute $[a_1, a_2, a_3]$, we use the following theorem:

Theorem 2.19	$[a_1, a_2, a_3] = [[a_1, a_2], a_3]$.

Proof: Exercise.

For example, $[6, 8, 9] = [[6, 8], 9] = [24, 9] = 72$.

Section 2.3 Exercises

1. (a) Prove parts L_1, L_2, L_3 of Theorem 2.17. (b) Prove Theorem 2.18.

2. Suppose that $(a, b) = 1$, $a | m$, $b | m$. Prove that $ab | m$.

3. Prove the following generalization of Theorem 2.19: If $n \geq 3$, then $[a_1, a_2, a_3, \ldots, a_n] = [[a_1, a_2, a_3, \ldots, a_{n-1}], a_n]$.

4. For each pair of integers m, n of Section 2.2, Exercise 14, find $[m, n]$.

5. Find the greatest common divisor and the least common multiple of each of the following sets of three integers:

   (a) 30, 42, 70

   (b) 54, 90, 240

   (c) 296, 444, 555

   (d) 21, 28, 36

   (e) 20, 30, 60

   (f) 32, 72, 84

   (g) 90, 120, 135

   (h) 114, 133, 171

   (i) 45, 81, 87

6. Prove the following generalization of Theorem 2.18, part L_6: If n natural numbers are pairwise relatively prime, then their least common multiple is their product.

7. Is it always true that $[a, b, c] = abc/(a, b, c)$? (Explain.)

2.4 Representations of Integers

We usually represent integers in the base 10 or decimal system. We now examine the decimal system and present several divisibility tests. We further discuss representation of natural numbers to the base m, where the natural number $m \geq 2$. Of particular interest is the case $m = 2$, that is, the binary number system.

Decimal Representation of Integers

Recall that

$$4372 = 4000 + 300 + 70 + 2 = 4(10^3) + 3(10^2) + 7(10^1) + 2(10^0)$$

In the decimal system, each natural number is represented as a sum of multiples of powers of 10. Specifically, if n is the natural number whose decimal representation is $a_r a_{r-1} a_{r-2} \cdots a_2 a_1 a_0$ where $0 \leq a_i \leq 9$ for all i and $a_r > 0$, then

$$n = \sum_{i=0}^{r} a_i 10^i$$

The use of 10 as the base of our most common number system is arbitrary. It may be related to the use of fingers for counting in times past. The a_i that appear above are called the *decimal digits* of n. We now present several divisiblity tests for integers in base 10. We prove the first two tests, leaving proofs of the other tests as exercises.

Theorem 2.20 n is even if and only if its last decimal digit is even.

Proof: Let $n = \sum_{i=0}^{r} a_i 10^i = a_0 + \sum_{i=1}^{r} a_i 10^i$. Since $2|10^i$ for all $i \geq 1$, it follows that $2| \sum_{i=1}^{r} a_i 10^i$. Therefore $2|n$ if and only if $2|a_0$. ∎

Before we discuss the next divisibility test, we introduce the concept of the sum of the decimal digits of a natural number. If the natural number n has the decimal representation, $n = \sum_{i=0}^{r} a_i 10^i$, we define $t_{10}(n) = \sum_{i=0}^{r} a_i$. For example, $t_{10}(4372) = 4 + 3 + 7 + 2 = 16$.

Theorem 2.21 $3|n$ if and only if $3|t_{10}(n)$.

Proof: Let

$$n = \sum_{i=0}^{r} a_i 10^i$$

$$= \sum_{i=0}^{r} a_i (1 + 10^i - 1)$$

$$= \sum_{i=0}^{r} a_i + \sum_{i=0}^{r} (10^i - 1)$$

$$= t_{10}(n) + \sum_{i=0}^{r} (10^i - 1)$$

Now $3|(10 - 1)$, and by Theorem 2.11, $(10 - 1)|(10^i - 1)$ for all $i \geq 1$. Therefore $3| \sum_{i=0}^{r} a_r (10^i - 1)$. It now follows that $3|n$ if and only if $3|t_{10}(n)$. ∎

Remarks

In plain language, Theorem 2.21 says that a natural number is divisible by 3 if and only if the sum of its digits is divisible by 3. For example, $t_{10}(78162) = 7 + 8 + 1 + 6 + 2 = 24$. Since $3|24$, it follows from Theorem 2.20 that $3|78162$.

∎

The next theorem generalizes Theorem 2.20.

Theorem 2.22 If n is a natural number, and $m \geq 1$, let $k_m(n)$ be the integer consisting of the m rightmost decimal digits of n. Then $2^m | n$ if and only if $2^m | k_m(n)$.

Proof: Exercise.

For example, looking at 537148, we see that $2^1 | 8$ and $2^2 | 48$, but $2^3 \nmid 148$. Therefore $2^2 | 537148$, but $2^3 \nmid 537148$.

A similar theorem holds regarding divisibility by powers of 5, namely,

Theorem 2.23 If n is a natural number, and $m \geq 1$, let $k_m(n)$ be the integer consisting of the m rightmost decimal digits of n. Then $5^m | n$ if and only if $5^m | k_m(n)$.

Proof: Exercise.

For example, looking at 97450, we see that $5^1 | 0$ and $5^2 | 50$, but $5^3 \nmid 450$. Therefore $5^2 | 97450$, but $5^3 \nmid 97450$.

The next theorem, which is reminiscent of Theorem 2.21, is a test for divisibilty by 9.

Theorem 2.24 $9 | n$ if and only if $9 | t_{10}(n)$.

Proof: Exercise.

Remarks

In plain language, Theorem 2.24 says that a natural number is divisible by 9 if and only if the sum of its digits is divisible by 9. For example, $t_{10}(62451) = 6 + 2 + 4 + 5 + 1 = 18$. Since $9 | 18$, it follows from Theorem 2.24 that $9 | 62451$.

■

The following is a test for divisibility by 11:

Theorem 2.25 If $n = \sum_{i=0}^{r} a_i 10^i$, then $11 | n$ if and only if

$$11 | \left(\sum_{i=0}^{[r/2]} a_{2i} - \sum_{i=1}^{[r/2]} a_{2i-1} \right)$$

Proof: Exercise.

Remarks

In plain language, Theorem 2.25 says that a natural number is divisible by 11 if and only if the difference between the sums of its alternate digits is divisible by 11. For example, looking at 90816, we see that $(9 + 8 + 6) - (0 + 1) = 23 - 1 = 22$. Now $11 | 22$, so $11 | 90816$.

■

More generally, let the integer $m \geq 2$. Then the natural number n can be represented to the base m as

$$n = \sum_{i=0}^{r} a_i m^i$$

where $0 \leq a_i \leq m - 1$ for all i and $a_r > 0$. We write

$$n = (a_r a_{r-1} a_{r-2} \ldots a_2 a_1 a_0)_m$$

The a_i are called the m-ary digits of n.

Binary Representation of Integers

If $m = 2$, then

$$n = \sum_{i=0}^{r} a_i 2^i$$

where $a_r = 1$ and $a_i \in \{0, 1\}$ for all i such that $0 \leq i \leq r - 1$. The a_i are called the *binary digits* or *bits* of n. Note that r is not fixed, but varies with n.

For example,

$$372 = 256 + 64 + 32 + 16 + 4$$

$$= 1(2^8) + 0(2^7) + 1(2^6) + 1(2^5) + 1(2^4) + 0(2^3) + 1(2^2) + 0(2^1) + 0(2^0)$$

Therefore the binary representation of 372 is 101110100. When more than one base is used to represent a given natural number, it is customary to indicate the appropriate base at the lower right in order to avoid confusion. Therefore, we write

$$372_{10} = 101110100_2$$

The binary representation of 372 shown above was obtained by subtracting the highest power of 2 from 372 and then repeating this process with the remainder as far as possible. That is,

$$372 - 256 = 116$$

$$116 - 64 = 52$$

$$52 - 32 = 20$$

$$20 - 16 = 4$$

$$4 - 4 = 0$$

An alternate method of decimal-to-binary conversion requires successive divisions by 2 and preservation of the remainders. For example, to convert 372_{10} to binary, we proceed as follows:

$$372 = 2(186) + 0$$

$$186 = 2(93) + 0$$

$$93 = 2(46) + 1$$

$$46 = 2(23) + 0$$

$$23 = 2(11) + 1$$

$$11 = 2(5) + 1$$

$$5 = 2(2) + 1$$

$$2 = 2(1) + 0$$

$$1 = 2(0) + 1$$

Taking the remainders *in the reverse order*, we get $372_{10} = 101110100_2$.

To convert from binary to decimal, one multiplies each 1 in the binary representation by the appropriate power of 2 and then adds those powers of 2. For example,

$$11001_2 = 1(2^4) + 1(2^3) + 0(2^2) + 0(2^1) + 1(2^0)$$

$$= 16 + 8 + 1$$

$$= 25_{10}$$

The number of bits in the binary representation of n is $1 + [\log_2 n]$, while the number of digits in the decimal representation of n is $1 + [\log_{10} n]$.

If $m > 10$, then the base m system requires the use of $m - 10$ additional numerals. For example, the *hexadecimal* system uses the base 16. In this system, the numbers from 10 to 16 may be represented by the letters from u through z, namely,

ten	u
eleven	v
twelve	w
thirteen	x
fourteen	y
fifteen	z

In order to convert from decimal to hexadecimal, one performs successive divisions by 16 and preserves the remainders. For example, to convert 76542_{10} to hexadecimal, one proceeds as follows:

$$76542 = 16(4783) + 14$$
$$4783 = 16(298) + 15$$
$$298 = 16(18) + 10$$
$$18 = 1(16) + 2$$
$$1 = 0(16) + 1$$

Result: $76542_{10} = 12uzx_{16}$

To convert from hexadecimal to decimal, one multiplies each hexadecimal digit by the appropriate power of 16 and then adds those products. For example,

$$12uzx_{16} = 1(16^4) + 2(16^3) + 10(16^2) + 15(16^1) + 14(16^0)$$
$$= 1(65536) + 2(4096) + 10(256) + 15(16) + 14$$
$$= 65536 + 8192 + 2560 + 240 + 14$$
$$= 76542_{10}$$

Section 2.4   Exercises

1. Prove the following Theorems:

   (a) 2.21

   (b) 2.22

   (c) 2.23

   (d) 2.24

2. Test the integer 12167320 for divisibility by

   (a) powers of 2

   (b) powers of 5

   (c) 3

   (d) 11

3. Convert each of the following decimal representations to binary:

   (a) 13

   (b) 36

   (c) 127

   (d) 100

   (e) 1001

4. Convert each of the following binary representations to decimal:

   (a) 1010

   (b) 10000

   (c) 1100001

   (d) 10101

   (e) 111000

5. Convert each of the following decimal representations to hexadecimal:

   (a) 12345

   (b) 9679

   (c) 30000

   (d) 8192

   (e) 777

6. Convert each of the following hexadecimal representations to decimal:

   (a) 1234

   (b) $wv9$

   (c) $6x78z$

   (d) $3y56$

   (e) $9u2$

7. Let r be the sum of the bits in the binary representation of n. Prove that the number of odd coefficients in row n of Pascal's triangle is 2^r.

Section 2.4 Computer Exercises

8. Write computer programs to convert the representation of an integer

   (a) from decimal to binary

   (b) from binary to decimal

   (c) from decimal to hexadecimal

   (d) from hexadecimal to decimal.

9. Let $r(n)$ be the number of zeroes at the end of the binary representation of $\dbinom{2n}{n}$. Let $s(n)$ be the number of 1's in the binary representation of n. Compute $r(n)$ and $s(n)$ for each n such that $1 \le n \le 15$. Care to conjecture?

Review Exercises

1. For each pair m, n of integers given below, do the following:

   (i) Use the Euclidean Algorithm to find (m, n).

   (ii) Find $[m, n]$.

   (iii) Find integers x, y such that $(m, n) = mx + ny$.

   (a) 396, 315

   (b) 2002, 847

   (c) 2431, 1378

2. For each set of three integers m, n, r, find (m, n, r) and $[m, n, r]$.

   (a) 60, 72, 88

   (b) 9, 10, 11

   (c) 20, 36, 42

3. Convert each of the following integers from decimal to binary form:

   (a) 125

   (b) 1000

   (c) 77

   (d) 231

   (e) 58

4. Convert each of the following integers from binary to decimal form:

   (a) 10101

   (b) 1100011

(c) 1110000

(d) 11010011

(e) 100001

5. Find the simplest expression, in terms of n , of a natural number whose binary representation consists of n 1's.

6. Find the simplest expression, in terms of n , of a natural number whose binary representation consists of n 1's, alternating with 0's.

7. Let a, b, c be natural numbers. Prove that if $a|b, b|c$, and $c|a$, then $a = b = c$.

8. Prove that if $3|F_n$, then $4|n$. Hint: Use Identity 23 from p. 15 and induction on n.

9. Prove that if n is even, then $(n + 1, n^2 + 1) = 1$.

10. Prove that

$$(F_n, L_n) = \begin{cases} 2 & \text{if } 3|n \\ 1 & \text{otherwise} \end{cases}$$

(Hint: Use Identity 31 from p. 16 and the result of Exercise 17 from p. 39.)

Chapter 3

Primes

"The study of the distribution of primes should be considered as one of the most important chapters of mathematical science."

—Edmund Landau (1877–1938)

3.1 Introduction

Most integers can be represented as products of smaller integers. For example, $91 = 7 * 13$. An integer that has such a representation is called *composite*. Thus 91 is composite. On the other hand, an integer that has no such representation is called *prime*. For example, both 7 and 13 are prime. The number 1 is called a *unit*. Thus every integer greater than 1 is either prime or composite.

Primes are the building blocks of the multiplicative structure of the natural numbers. We shall see that if the natural number $n > 1$, then n can be represented as a product of one or more primes in a way that is essentially unique. As a consequence, many questions concerning n can be reduced to questions concerning the prime factors of n.

3.2 Primes, Prime Counting Function, Prime Number Theorem

Definition 3.1   Composite Number

A natural number n , is said to be *composite* if it has a nontrivial divisor, d , such that $1 < d < n$. If so, then $n = dc$, where $c = n/d$.

Definition 3.2   Prime Number

A natural number, p , is said to be *prime* if $p > 1$ and p has no nontrivial divisor. (In other words, p has no divisors other than 1 and p.)

The primes below 100 are: 2, 3, 5, 7, 11, 13, 17, 19, 23, 29, 31, 37, 41, 43, 47, 53, 59, 61, 67, 71, 73, 79, 83, 89, 97. Table 1 in Appendix B lists the primes below 1,000.

Our first theorem about primes is the following:

Theorem 3.1   If m is a natural number and p is a prime, then

$$(m, p) = \begin{cases} p & \text{if} \quad p | m \\ 1 & \text{if} \quad p \nmid m \end{cases}$$

Proof: Let $d = (m, p)$. Since $d | p$ and p is prime, we know that either $d = 1$ or $d = p$. By Theorem 2.3, part G_2, $d = p$ if and only if $p | m$. Therefore, if $p \nmid m$, then $d = 1$. ∎

The following theorem establishes one of the most important properties of primes.

Theorem 3.2   If p is prime and $p | ab$, then $p | a$ or $p | b$.

Proof: It suffices to show that if p is prime, $p | ab$, and $p \nmid a$, then $p | b$. If p is prime and $p \nmid a$, then Theorem 3.1 implies $(a, p) = 1$. Since $p = ab$ by hypothesis, Euclid's Lemma (Theorem 2.13) implies $p | b$. ∎

Remarks

If n is composite and $n|ab$, then it need not follow that $n|a$ or $n|b$. For example, $6|(2*3)$, yet $6 \nmid 2$ and $6 \nmid 3$. In fact, an alternate way to *define* primes would be to say that p is prime if whenever $p|ab$, then $p|a$ or $p|b$.

∎

In the next theorem, whose proof is left as an exercise, the result of Theorem 3.2 is extended to products of two or more factors.

Theorem 3.3

If p is prime, $r \geq 2$, and $p| \prod_{i=1}^{r} a_i$, then $p|a_i$ for some i.

Proof: Exercise (induction on r).

Theorem 3.4

If $n > 1$, then n has a prime factor, p.

Proof: If n is prime, then the conclusion follows from the fact that $n|n$. If n is composite, let p be the least nontrivial divisor of n. We will show that p is prime. If p is composite, then p has a nontrivial divisor, d. Now $d|p$ and $p|n$, so $d|n$, yet $d < p$. This contradicts the definition of p. Therefore p is prime. ∎

For example, the least prime divisor of 111 is 3; the least prime divisor of 323 is 17; the least prime divisor of 1001 is 7.

The following theorem leads to a test of primality.

Theorem 3.5

If n is composite, and if p is the least prime factor of n, then $p \leq \sqrt{n}$.

Proof: By hypothesis, $n = pm$, with $m > 1$. By Theorem 3.4, m has a prime factor, q, so $q \leq m$. If $m < p$, then $q < p$. Since $m|n$, we have $q|n$, yet $q < p$. This contradicts the definition of p. Therefore $p \leq m$, so $p^2 \leq pm$, that is $p^2 \leq n$, hence $p \leq \sqrt{n}$. ∎

Suppose we wish to determine whether 10001 is prime or composite. If 10001 is composite, then according to Theorem 3.5, 10001 has a prime factor, p, such that $p \leq \sqrt{10001}$, that is $p \leq 100$. Therefore, we try to divide 10001 by each of the primes below 100, in order. Eventually, we find that $73|10001$, and $10001 = 73 * 137$. This procedure can be generalized as follows:

Test of Primality by Trial Division

If $j \geq 1$, let p_j denote the jth prime, so that $p_1 = 2, p_2 = 3, p_3 = 5$, etc. Given $n > 1$, let k be the largest integer such that $p_k \leq \sqrt{n}$. If $p_i | n$ for some i such that $1 \leq i \leq k$, then n is composite (and p_i is a factor of n). If no such p_i divides n, then n is prime.

For example, suppose we wish to determine whether 103 is prime or composite. Now $p_4 = 7 < \sqrt{103} < 11 = p_5$, so $k = 4$. We now try to divide 103 by each of the first 4 primes, that is, by 2, 3, 5, and 7. Since 103 is not divisible by 2, 3, 5, or 7, we conclude that 103 is prime.

Testing the natural number n for primality by trial division requires that we know all the primes, p, such that $p \leq [\sqrt{n}]$. We shall see that if n is sufficiently large, then there are too many such primes, so that trial division would require too much time and space (computer memory). Nevertheless, trial division is an effective way to find all the *small* prime factors of n, if any.

Although primes grow more scarce among large integers, they never run out, as Euclid proved about 300 B.C.

Theorem 3.6   (**Euclid**)   There exist infinitely many primes.

Proof: Suppose that $p_1, p_2, p_3, \dots, p_n$ were the only primes. Let M be the product of these primes. Since $M + 1 > 1$, by Theorem 3.4, there exists a prime, q, such that $q|(M+1)$. Since $(M+1, M) = 1$, we know that $q \nmid M$. Therefore $q \neq p_i$ for all i such that $1 \leq i \leq n$, contrary to hypothesis. ■

The distribution of primes seems highly irregular. There is no convenient formula that yields the precise value of p_n, the nth prime, in terms of n. The prime that follows 197 is 199, but then the next prime is 211. If p and q are odd primes with $p < q$, then $q - p \geq 2$. If $q - p = 2$, then we say that p and q are a pair of *twin primes*. Some examples of twin prime pairs are 5 and 7, 197 and 199, 9929 and 9931. The existence of infinitely many twin prime pairs is suspected, but has not been proven. On the other hand, the following is known:

Theorem 3.7   There exist arbitrarily large gaps between consecutive primes.

Proof: We will show that for any $n \geq 2$, there exist consecutive primes p, q such that $q - p \geq n$. For each k such that $2 \leq k \leq n$, we have $k | (n! + k)$. Therefore, $n! + 2, n! + 3, n! + 4, \ldots, n! + n$ is a sequence of $n - 1$ consecutive composite numbers. Let p be the greatest prime such that $p \leq n! + 1$. (Such a p exists, since $2 < 2! + 1 \leq n! + 1$.) Let q be the least prime such that $q \geq n! + n + 1$. Theorem 3.6 implies that such a prime, q, exists. Now p and q are consecutive primes, and $q - p \geq n$. ∎

As a consequence of the Prime Number Theorem, which follows shortly, it can be shown that in the vicinity of n, the average distance between consecutive primes is approximately $\log n$ (the natural logarithm of n). For example, let us choose an interval of length 200, centered at $n = 1000$. By referring to Table 1 in Appendix B, the reader can verify that there are 28 primes in this interval, and that the average distance between consecutive primes is 6.8. On the other hand, $\log 1000$ is approximately 6.9.

Definition 3.3   **Prime Counting Function**

If x is a positive real number, then the *prime counting function*, denoted $\pi(x)$ counts the number of primes, p, such that $p \leq x$.

For example, since the primes up to 10 are 2, 3, 5, 7 , we have $\pi(10) = 4$. Also, since the primes up to 30 are 2, 3, 5, 7, 11, 13, 17, 19, 23, 29, we have $\pi(30) = 10$. Theorem 3.5 implies that $\pi(x)$ tends to infinity as x tends to infinity. Table 3.1 below shows how $\pi(10^n)$ grows with n.

If x is a large positive integer, we would like to be able to *estimate* $\pi(x)$ without actually listing all primes p such that $p \leq x$. What does it mean

Table 3.1

Table of the Prime Counting Function

n	$\pi(10^n)$	n	$\pi(10^n)$
1	4	10	455051511
2	25	11	4118054813
3	168	12	37607912018
4	1229	13	346065536839
5	9592	14	3204941750802
6	78498	15	29844570422669
7	664579	16	279238341033925
8	5761455	20	2220819602560918840
9	50847534		

to say that one function provides a good estimate of another function? An answer is given by the following definition:

Definition 3.4

Let $f(x)$ and $g(x)$ be functions of the real variable x that are defined for $x > 0$. We say that $f(x)$ is *asymptotic* to $g(x)$, and we write $f(x) \sim g(x)$, if

$$\lim_{x \to \infty} \frac{f(x)}{g(x)} = 1$$

If this is so, then it seems reasnable to use $g(x)$ to estimate $f(x)$ for large x. Now we are ready to state one of the most important theorems in number theory:

Theorem 3.8

Prime Number Theorem

$$\pi(x) \sim \frac{x}{\log x}$$

That is

$$\lim_{x \to \infty} \frac{\pi(x)}{x/\log x} = 1$$

The Prime Number Theorem, which suggests that we use $x/\log x$ in order to estimate $\pi(x)$, was first conjectured independently by Legendre and Gauss in the 1790s. The first complete proofs, based on complex analysis, were given independently by Hadamard and De la Vallee Poussin in 1896. A proof using only real analysis was given by Selberg and Erdos in 1949. We omit the proof of the Prime Number Theorem, which is beyond the scope of this text.

Let $f(x) = \pi(x) \log x/x$. The convergence of $f(x)$ to its limiting value of 1 is slow. For example, $f(10^{20}) = 1.023$. This means that relative error of 2.3 % results when $\pi(10^{20})$ is estimated by $10^{20}/\log 10^{20}$. A better approximation to $\pi(x)$ is given by the so-called logarithmic integral:

$$li(x) = \int_2^x \frac{dt}{\log t}$$

Special methods exist to verify the primality of natural numbers t such that $t + 1$ is an easily factored number. An example thereof is provided by the *Mersenne numbers*. (Marin Mersenne 1588–1648 was a French monk who did much to stimulate mathematical research.) Let $M_n = 2^n - 1$. M_n is called the nth Mersenne number. Mersenne numbers are discussed in Chapter 5. At present, the largest verified prime is $M_{25964951} = 2^{25964951} - 1$, an integer with over 7,000,000 decimal digits.

Consider the problem of finding a function from the non-negative integers to the natural numbers that evaluates only primes. Let us try the polynomial $f(x) = 2x^2 + 29$. Then $f(0) = 29, f(1) = 31, f(2) = 37, \ldots$ $f(28) = 1597$ (all primes). But $f(29) = 1711 = 29 * 59$ (a composite). In fact, $f(29k)$ is divisible by and greater than 29 for all $k \geq 1$, and is therefore composite.

Similarly, if $g(x) = x^2 + x + 41$, then $g(n)$ is prime for $0 \leq n \leq 39$, but $g(40) = 1681 = 41^2$ (a composite). In fact, $g(41k - 1)$ is divisible by and greater than 41 for all $k \geq 1$ and is therefore composite. The following theorem states that every polynomial with positive lead coefficient must evaluate infinitely many composite numbers.

Theorem 3.9 Let $f(x)$ be a polynomial of degree $r > 1$ with integer coefficients and positive lead coefficient, that is,

$$f(x) = \sum_{k=0}^{r} a_k x^k$$

with $a_r > 0$. Then there are infinitely many n such that $f(n)$ is composite.

Proof: Without loss of generality, we may assume that all $a_k \geq 0$. If $f(t)$ is composite for all t, then we are done. Otherwise, there exists t such that $f(t) = p$, a prime. Let $m \geq 1$. Now

$$f(t + mp) = \sum_{k=0}^{r} a_k (t + mp)^k$$

$$= \sum_{k=0}^{r} a_k \left(t^k + \sum_{j=1}^{k} \binom{k}{j} t^{k-j} (mp)^j \right)$$

$$= \sum_{k=0}^{r} a_k t^k + \sum_{k=0}^{r} a_k \left(\sum_{j=1}^{k} \binom{k}{j} t^{k-j} (mp)^j \right)$$

$$= f(t) + \sum_{k=0}^{r} a_k \left(\sum_{j=1}^{k} \binom{k}{j} t^{k-j} (mp)^j \right)$$

Therefore $f(t + mp)$ is divisible by, but greater than p, hence composite. Since m is arbitrary, we are done. ∎

Section 3.2   Exercises

1. Prove that if $a \geq 2, n \geq 2$, and $a^n - 1$ is prime, then $a = 2$ and n is prime.

2. Prove that if $a \geq 2, n \geq 2$, and $a^n + 1$ is prime, then $2|a$ and $n = 2^k$ for some $k \geq 1$.

3. Prove the following generalization of Theorem 3.5: If n has k prime factors, of which p is the least, then $p \leq n^{1/k}$.

4. Use induction on r to prove Theorem 3.3.

5. With very little computation, find a prime factor of $10^{10} + 1$.

6. Using Table 1 in Appendix B, find the 10 smallest primes of the form $n^2 + 1$.

7. Using Table 1 in Appendix B, find the 10 smallest primes of the form $n^2 + n + 1$.

8. Using Table 1 in Appendix B, find the 10 smallest primes, p, such that $2p + 1$ is also prime. (Such primes are called *Sophie Germain* primes.) Sophie Germain (1776–1830) was an outstanding woman mathematician.

9. Prove that if p is a prime such that $p^k | a^n$ but $p^{k+1} \nmid a^n$, then $n | k$.

10. Prove that there exist arbitrarily long sequences of consecutive odd composite numbers.

11. Use the prime number theorem to estimate the number of primes below 10^{10}. Now look up the actual number from Table 3.1 and compute the relative error in the estimate.

12. Suppose that a prime greater than 10^{100} is needed in an application to cryptology. If odd numbers slightly greater than 10^{100} are chosen at random until a prime is found, how many such odd numbers should one expect to test for primality?

Section 3.2   Computer Exercises

13. For a given natural number, n, let $S_n = \{2, 3, 4, \ldots, q\}$ be the set of all primes p such that $p \leq \sqrt{n}$. Suppose that S_n has k elements, that is, $k = \pi(\sqrt{n})$. A counting argument shows that $\pi(n)$ may be computed from the formula:

$$\pi(n) = k + n - 1 - \sum [\frac{n}{p}] + \sum [\frac{n}{p_1 p_2}] - \sum [\frac{n}{p_1 p_2 p_3}] + \ldots$$

(Each sum is taken over all products of one or more primes from S_n.) For example, if $n = 10$, then $S_{10} = \{2, 3\}$, so $k = 2$ and

$$\pi(10) = 2 + 10 - 1 - ([\frac{10}{2}] + [\frac{10}{3}]) + [\frac{10}{6}]$$

$$= 11 - (5 + 3) + 1$$

$$= 4$$

Write a computer program to compute $\pi(10^6)$ using the above formula.

14. Write a computer program to find the least odd number n such that n is the least of 10 consecutive odd composite numbers.

15. Write a computer program to test natural numbers for primality by trial division.

3.3 Sieve of Eratosthenes, Canonical Factorization, Fundamental Theorem of Arithmetic

Let us now consider the problem of finding all the primes in a given interval. A method for so doing was given by Eratosthenes (276–195 B.C.) This method is known as the *Sieve of Eratosthenes*.

Sieve of Eratosthenes

Given natural numbers $a < b$, we wish to find all primes p such that $a \le p \le b$. By Theorem 3.5, if n is a composite number in this interval, then n has a prime factor not exceeding $\sqrt{b}$. Suppose that the primes not exceeding $\sqrt{b}$ are $p_1 = 2, p_2 = 3, p_3 = 5, \dots, p_k$. First, we write down all the integers from a to b. Then, for each i such that $1 \le i \le k$, we eliminate all nontrivial multiples of p_i. That is to say, we cross out all the nontrivial multiples of 2, then all the nontrivial multiples of 3, etc. After we have eliminated all the nontrivial multiples of p_k, the remaining integers, if any, are all prime, since we have "sieved out" all the composites.

For example, let us use the sieve to find all primes, if any, between 110 and 130. We can simplify our task by writing down only the odd integers in this interval, namely,

$$111, 113, 115, 117, 119, 121, 123, 125, 127, 129$$

since all even numbers in this range are composite. Since $11 < \sqrt{130} < 13$, any odd composite number in our interval must be divisible by an odd prime not exceeding 11, that is, by 3, 5, 7, or 11.

Now

$$3|111, \ 3|117, \ 3|123, \ 3|129$$
$$5|115, \ 5|125$$
$$7|119$$
$$11|121$$

The remaining integers, namely, 113 and 127, are prime.

In 332 B.C., Alexander the Great (of Macedonia) conquered Egypt. Shortly afterwards, he founded the city of Alexandria on Egypt's Mediterranean coast. The city grew rapidly and was home to the Museum, which was an ancient equivalent of today's universities. Scholars in residence wrote books and gave lectures, living comfortably on a government stipend. One such scholar was Euclid (330–270 B.C.). Another was Eratosthenes.

Eratosthenes (276–195 B.C.)

Eratosthenes was born in Cyrene, in what is now Libya. He became head librarian at the Museum in Alexandria. One of the foremost scholars of his time, he wrote works on geography, mathematics, literary criticism, grammar, and poetry. Eratosthenes devised an experiment to measure the circumference of the earth. He also served as tutor to the two sons of the Pharoah, Ptolemy II.

Recall Theorem 1.13. If the natural number $n \geq 2$, then n may be represented as the product of one or more primes. For example,

$$2646 = 2 * 1323$$

$$= 2 * 3 * 441$$

$$= 2 * 3^2 * 147$$

$$= 2 * 3^3 * 49$$

$$= 2^1 3^3 7^2$$

The rightmost expression is the *canonical factorization* of 2646.

Definition 3.5   **Canonical Factorization**

If $n \geq 2$, then we write n as a product of primes:

$$n = p_1^{e_1} p_2^{e_2} p_3^{e_3} \cdots p_r^{e_r}$$

That is

$$n = \prod_{i=1}^{r} p_i^{e_i}$$

where the p_i are distinct primes such that $p_1 < p_2 < p_3 < \cdots < p_r$ and each exponent $e_i \geq 1$. This is called the *canonical factorization* of n.

For example, the canonical factorization of 202,400 is

$$202,400 = 2^5 5^2 11^1 23^1$$

Before discussing how to obtain the canonical factorization of a given natural number, we need to introduce some more notation. We are concerned with the highest power of a prime that divides a given integer. If p is a prime such that $p^k | n$ but $p^{k+1} \nmid n$, then we write

$$p^k || n \quad or \quad o_p(n) = k$$

For example, $3^4|162$, but $3^5 \nmid 162$, so we write $3^4||162$. Also, $2^3|40$, but $2^4 \nmid 40$, so we write $2^3||40$. The two preceding examples show that $o_3(162) = 4$ and $o_2(40) = 3$.

Determining the Canonical Factorization of a Natural Number

Given $n > 1$, divide n by each of the primes in increasing order, that is, by 2, 3, 5, 7, etc. until p_1, the smallest prime factor of n, is found. (Recall that if n is not divisible by any prime less than $\sqrt{n}$, then n itself is prime.)

Determine the exponent e_1 such that $p_1^{e_1}||n$. Let $n_1 = n/p_1^{e_1}$. Find p_2, the smallest prime factor of n_1. Determine the exponent e_2 such that $p_2^{e_2}||n_1$. Let $n_2 = n_1/p_2^{e_2}$. Find p_3, the smallest prime factor of n_2. Determine the exponent e_3 such that $p_3^{e_3}||n_2$. Let $n_3 = n_2/p_3^{e_3}$, etc. Ultimately, for some $r \geq 1$, we get $n_{r+1} = 1$. Then our result is

$$n = \prod_{i=1}^{r} p_i^{e_i}$$

For example, let us obtain the canonical factorization of 1386. $1386 = 2*693$. Now $2 \nmid 693$, but $693 = 3*231 = 3^2*77$. Now $3 \nmid 77$, and $5 \nmid 77$, but $77 = 7*11$. Since 11 is prime, we are done. Therefore $1386 = 2^1 3^2 7^1 11^1$.

The *Fundamental Theorem of Arithmetic*, which follows, states that if $n > 1$, then the canonical factorization of n is unique.

Theorem 3.10

Fundamental Theorem of Arithmetic
If $n > 1$, then the canonical factorization of n is unique.

Proof: Let n have a canonical factorization:

$$n = \prod_{i=1}^{r} p_i^{e_i}$$

where $p_1 < p_2 < \ldots < p_r$. First, we show that the set of prime factors of n, namely $\{p_1, p_2, \ldots, p_r\}$ is uniquely determined. Suppose that q is a prime such that $q|n$. By the hypothesis and by Theorem 3.3, we must have $q|p_i$ for some index, i. However, this implies $q = p_i$.

Next we show that the set of exponents, namely, $\{e_1, e_2, \ldots, e_r\}$, is uniquely determined. Suppose that $n = p_i^{e_i}c = p_i^{f_i}d$, where $p_i \nmid cd$, and, without loss of generality, $e_i \leq f_i$. Therefore $c = p_i^{f_i-e_i}d$. If $e_i < f_i$, then $p_i|c$ is an impossibility. Therefore $e_i = f_i$. Since i is arbitrary, we are done. ∎

Remarks

Although the Fundamental Theorem of Arithmetic may seem "obvious," at least for natural numbers, it is noteworthy that number systems exist in which the Fundamental Theorem fails to hold. In such number systems, the factorization of certain numbers as products of primes may fail to be unique. The study of such number systems is a topic in *algebraic number theory*.

■

For example, consider $Z[\sqrt{-5}]$, the set of all numbers $a + b\sqrt{-5}$ where $a, b \in Z$. We define multiplication in $Z[\sqrt{-5}]$ by

$$(a + b\sqrt{-5})(c + d\sqrt{-5}) = (ac - 5bd) + (ad + bc)\sqrt{-5}$$

In this system, we have

$$6 = 2 * 3$$

Also,

$$6 = (1 + \sqrt{-5})(1 - \sqrt{-5})$$

It can be shown that in $Z[\sqrt{-5}]$, each of the factors 2, 3, $1 + \sqrt{-5}$, and $1 - \sqrt{-5}$ is prime. Therefore, in $Z[\sqrt{-5}]$, the factorization of 6 as a product of primes is *not* unique.

We saw earlier, in Theorem 3.6, that infinitely many primes exist. By adapting the argument used to prove Theorem 3.6, and by using the following generalization of Theorem 1.7, we can prove the existence of infinitely many primes of the form $4k - 1$.

Theorem 3.11   If $\prod_{i=1}^{r} a_i = 4k - 1$, then $a_i = 4m - 1$ for some factor, a_i.

Proof: Exercise (induction on r).

Theorem 3.12   There exist infinitely many primes of the form $4k - 1$.

Proof: Suppose that $q_1, q_2, q_3, \ldots, q_n$ are the only primes of the form $4k - 1$. Let $m = -1 + 4\prod_{i=1}^{n} q_i$. Since m has the form $4k - 1$, it follows from Theorem 3.11 that m has a prime factor, p, of the form $4k - 1$. But then $p = q_i$ for some index, i. Now $p|m$ and $p|(m + 1)$, we have $p|1$, an impossibility. ■

> ### Remarks
>
> The first few primes of the form $4k - 1$ are 3, 7, 11, 19, 23. It is also true that there are infinitely many primes of the form $4k + 1$, such as 5, 13, 17, 29, 37, but this cannot be proven in the same manner as Theorem 3.12. We furnish a proof of this assertion at the end of Chapter 4.
>
> Next, we state two important theorems regarding primes. The first theorem, which is due to Lejeune Dirichlet, greatly generalizes Theorem 3.12.
>
> ∎

Theorem 3.13

Dirichlet's Theorem

If $(a, b) = 1$, then there are infinitely many primes of the form $an + b$.

For example, there are infinitely many primes whose decimal representations end in 3, such as 3, 13, 23, 43, 53, 73, 83, 103, 113, etc., since these are the primes of the form $10n + 3$, and $(10, 3) = 1$. We omit the proof of Dirichlet's Theorem, which requires analytic methods that are beyond the scope of this text.

Theorem 3.14

Bertrand's Postulate

For all $n \geq 2$, there is a prime, p, such that $n < p < 2n$.

We omit the proof of Bertrand's Postulate, which is rather lengthy.

For example, if $n = 5$, then $p = 7$. Note that in terms of the prime counting function, Theorem 3.14 says that $\pi(2n) - \pi(n) \geq 1$ if $n \geq 2$.

We conclude this chapter by presenting a theorem that allows us to obtain the canonical factorization of $n!$ without computing $n!$. First, a preliminary result is needed.

Theorem 3.15

If d and n are natural numbers, then

$$[\frac{n}{d}] - [\frac{n-1}{d}] = \begin{cases} 1 & \text{if} \quad d \mid n \\ 0 & \text{if} \quad d \nmid n \end{cases}$$

Proof: Exercise.

Theorem 3.16   For any natural number, n , we have

$$n! = \prod p^{\sum_{k \geq 1}[n/p^k]}$$

the product being taken over all primes.

Remarks

Although the product that appears in the statement of Theorem 3.16 is taken over all primes, only those finitely many primes p such that $p \leq n$ will carry a positive exponent. Therefore, we could rewrite the statement of Theorem 3.16 as

$$n! = \prod_{p \leq n} p^{\sum_{k \geq 1}[n/p^k]}$$

Furthermore, if $p^f \leq n < p^{f+1}$, then $[n/p^k] = 0$ for all $k > f$. Therefore every sum that occurs as an exponent is actually finite.

■

Proof: (Induction on n)   Theorem 3.16 is trivially true for $n = 1$. By induction hypothesis, we have

$$(n-1)! = \prod p^{\sum_{k \geq 1}[(n-1)/p^k]}$$

We wish to prove that

$$n! = \prod p^{\sum_{k \geq 1}[n/p^k]}$$

Since $n = n!/(n-1)!$, it suffices to show that

$$n = \prod p^{\sum_{k \geq 1}\{[n/p^k]-[(n-1)/p^k]\}}$$

In other words, it suffices to show that if $p^j || n$, then

$$j = \sum_{k \geq 1} \left\{ [n/p^k] - [(n-1)/p^k] \right\}$$

that is,

$$j = \sum_{k=1}^{j} \left\{ [n/p^k] - [(n-1)/p^k] \right\}.$$

The last equation is valid because $p^k | n$ for $1 \leq k \leq j$, so that each of the j summands in the right member of the last equation is 1, by Theorem 3.15.   ■

For example, $10! = 2^a 3^b 5^c 7^d$, where $a = [\frac{10}{2}] + [\frac{10}{4}] + [\frac{10}{8}] = 5 + 2 + 1 = 8$; $b = [\frac{10}{3}] + [\frac{10}{9}] = 3 + 1 = 4$; $c = [\frac{10}{5}] = 2$; $d = [\frac{10}{7}] = 1$. That is, $10! = 2^8 3^4 5^2 7^1$.

The following theorems, whose proofs are left as exercises, will be useful in Chapter 4.

Theorem 3.17   If p is prime and $0 < k < p$, then $p | \binom{p}{k}$.

Proof: Exercise.

Theorem 3.18   If $(a, b) = 1$ and $ab = c^n$, then there exist x, y such that $a = x^n$, $y = b^n$, and $(x, y) = 1$.

Proof: Exercise.

There are many open questions concerning primes. For example, one may ask whether there exist infinitely many primes of any of the following forms:

1. $n^2 + 1$
2. $n^2 + n + 1$
3. $2^n - 1$
4. $2^n + 1$
5. Fibonacci numbers.

A celebrated open problem concerning primes is the following:

Goldbach's Conjecture   If $n \geq 4$, then there are distinct primes p, q such that $p + q = 2n$.

In addition, it is not known whether there is always a prime between two consecutive squares. It is also not known whether for all $n \geq 4$, there is a prime, p, such that $n < p < n + \sqrt{n}$.

Section 3.3   Exercises

1. Using the sieve of Eratosthenes, find all primes between 220 and 250.

2. Without using Dirichlet's Theorem, prove that there are infinitely many primes of the form $6k - 1$.

3. Find the canonical factorization of each of the following integers:

   (a) 299

   (b) 480

   (c) 777

   (d) 399

   (e) 145

   (f) 221

   (g) 72

   (h) 2450

   (i) 5005

   (j) 191

   (k) 896

   (l) 529

   (m) 104

   (n) 171

4. Prove that if $n \geq 9$ and if $n - 2$ and $n + 2$ are both prime, then $3|n$.

5. Prove

   (a) Theorem 3.15

   (b) Theorem 3.18

   (c) Theorem 3.19

6. Prove that if p and q are distinct primes such that $pq|n^2$, then $pq|n$.

7. Find all integers n with the property that $p|n$ for every prime, p, such that $p \leq \sqrt{n}$.

8. Prove that if m and k are natural numbers such that $m^{k-1} = k > 1$, then $m = k = 2$. (Use calculus.)

9. Prove that if $1 < m < n$ and $m^n = n^m$, then $m = 2$ and $n = 4$.

   (Hint: Show that m, n have the same prime factors and use the result of Exercise 8.)

10. Let $f(x)$ be a monic polynomial with integer coefficients, that is, a polynomial whose lead coefficient is 1. Prove that if r is a rational root of $f(x)$, that is, if $f(r) = 0$, then r is an integer.

11. Use the result of Exercise 10 to prove that $\sqrt{2}$ is irrational.

12. Prove that if p_n is the nth prime and $n \geq 5$, then $p_n^2 < p_1 p_2 p_3 \ldots p_{n-1}$.

13. Prove that if $m! = a^n$ and $a > 1$, then $n = 1$.

14. Theorem 3.16 could be restated as follows: If p is prime and $p^j||n!$, then $j = \sum_{k \geq 1}[n/p^k]$. Prove the following alternative formula: Let $n = \sum_{i=0}^{r} a_i p^i$, where $0 \leq a_i \leq p-1$ for all i, and $a_r > 0$. (This is the representation of n to the base p.) Let $t_p(n) = \sum_{i=0}^{r} a_i$. In other words, $t_p(n)$ is the sum of the digits of n to the base p. Then $j = (n - t_p(n))/(p-1)$. For example, if $3^j||22!$, then according to Theorem 3.16, we have $j = [22/3] + [22/9] = 7 + 2 = 9$. If we use the alternative formula, first we note that $22_{10} = 211_3$, since $22 = 2(3^2) + 1(3^1) + 1(3^0)$. Thus $j = \frac{1}{2}\{22 - (2 + 1 + 1)\} = \frac{1}{2}(18) = 9$.

15. Using the alternative formula from the preceding exercise, prove that if $p^j|| \begin{pmatrix} n \\ k \end{pmatrix}$, then

$$j = \frac{t_p(k) + t_p(n-k) - tg_p(n)}{p-1}$$

16. Prove that if $2^j|| \begin{pmatrix} 2n \\ n \end{pmatrix}$, then j is the sum of the binary digits of n, that is, $j = t_2(n)$. Conclude that $\begin{pmatrix} 2n \\ n \end{pmatrix}$ is even for all $n \geq 1$.

17. Prove that the number of odd terms in the nth row of Pascal's triangle is $2^{t_2(n)}$.

18. Find the value of k such that $3^k||100!$.

19. Find the number of terminal zeroes in the decimal representation of 1000!.

20. Prove that

$$[2, 3, 4, \ldots, n] = \prod_{p \leq n} p^{[\log_p n]}$$

21. Prove that if $n \geq 2$, then $\frac{1}{2} + \frac{1}{3} + \frac{1}{4} + \ldots + \frac{1}{n}$ is not an integer.

22. Let $p_1 = 2, p_2 = 3, p_3 = 5$, etc. Find all natural numbers m such that there exists $n \geq 2$ such that

$$m! = p_1^{2^{n-1}} p_2^{2^{n-2}} p_3^{2^{n-3}} \cdots p_{n-1}^2 p_n$$

23. Prove that $(1 + 2 + 3 + \cdots + n)|n!$ if and only if $n + 1$ is composite.

Section 3.3 Computer Exercises

24. Write a computer program that, given a large integer, n, will attempt to at least partially factor n by using trial division to find all prime factors of n that are less than 100.

25. Write a computer program that uses the Sieve of Eratosthenes to find all primes between A and B, where $A < B \leq 10,000$.

26. Write a computer program that, given an integer n such that $4 \le n \le 100$, will find all representations of $2n$ as a sum of two distinct primes.

27. Write a computer program to find all n such that $1 \le n \le 100$ and $\dbinom{2n}{n}$ is not divisible by 3, 5, or 7. (Use the alternative formula from Exercise 14 in order to avoid needless computation.)

28. The following problem is related to Goldbach's Conjecture.

   If $n \ge 4$, let $g(n)$ be the number of pairs of noncomposite integers p, q such that $1 \le p < q \le 2n - 1$ and $p + q = 2n$.

   (a) Prove that $g(n) \le \pi(2n) - \pi(n)$.

   (b) Prove that if $g(n) = \pi(2n) - \pi(n)$, then $3 | n$.

   (c) Write a computer program to find all $n \le 200$ such that
   $$g(n) = \pi(2n) - \pi(n).$$

Review Exercises

1. Use the formula for $\pi(n)$ given on p. 57 to verify that $\pi(100) = 25$.

2. Find the canonical factorization of each of the following:

   (a) 111

   (b) 128

   (c) 2002

   (d) 735

   (e) 899

3. Prove that there are infintely many primes of the form $6k - 1$.

   Hint: Review the proof of Theorem 3.12, and note that if the prime $p \ge 5$, then $p = 6k \pm 1$ for some positive integer k.

4. Use trial division to verify that 1003 is prime.

5. Use the Sieve of Eratosthenes to find all primes between 500 and 550.

6. Find the number of terminal zeroes in the decimal representation of 100!.

7. Is it true that there are infinitely many primes whose decimal representations end in the digit 1? (Explain.)

8. List the first 10 primes whose last decimal digit is 1.

9. List all representations of 50 as a sum of two distinct primes.

10. Recall that the prime p is called a *Sophie Germain prime* if $2p + 1$ is also prime. Using Table 1 in Appendix B, find all Sophie Germain primes p such that $p < 500$.

Chapter 4

Congruences

4.1 Introduction

Many questions in number theory concern the nature of the remainder when one natural number is divided by another. An integer is *odd* if division by 2 leaves a remainder of 1; it is *even* if division by 2 leaves a remainder of 0. We saw by way of Theorem 3.11 that there exist infinitely many primes of the form $4k - 1$. Now $4k - 1 = 4(k - 1) + 3$, so these are the primes which when divided by 4 leave a remainder of 3. Toward the end of this chapter we will prove that there are also infinitely many primes of the form $4k + 1$.

If a and b are integers that leave the same remainder when divided by the natural number m, then we say that a is *congruent* to b (mod m), and we write

$$a \equiv b \pmod{m}$$

Such a statement is called a *congruence*. For example, $11 \equiv 3 \pmod{2}$, since 11 and 3 both leave remainder 1 when divided by 2.

The theory of congruences was invented by the great German mathematician Carl Friedrich Gauss (1777–1855). We shall see that many questions involving divisibility become more tractable when phrased in terms of congruences.

4.2 Congruences and Equivalence Relations

We begin by giving the definition and properties of *congruences*.

Let a, b, c, and d denote integers, while m and n denote natural numbers.

Definition 4.1   **Congruence** (mod m)

If $m|(a-b)$, then we say a is *congruent* to b (mod m) and we write

$$a \equiv b \quad (\text{mod } m)$$

We call m the *modulus* of the congruence.

For example, $17 \equiv 1 \pmod 8$, since $17-1 = 16 = 2(8)$; $21 \equiv 0 \pmod 7$, since $21-0 = 21 = 3(7)$; and $13 \equiv -2 \pmod 5$, since $13-(-2) = 15 = 3(5)$. If $m \nmid (a-b)$, then we say a is *not congruent* to b (mod m) and we write

$$a \not\equiv b \quad (\text{mod } m)$$

For example, $17 \not\equiv 1 \pmod 6$, since $17-1 = 16$; and $6 \nmid 16$; $21 \not\equiv 0 \pmod 8$, since $21 - 0 = 21$ and $8 \nmid 21$; and $13 \not\equiv -2 \pmod 7$, since $13 - (-2) = 15$ and $7 \nmid 15$.

Theorem 4.1   **Properties of Congruence**

C_1: If $a \equiv b \pmod m$ and $c \equiv d \pmod m$, then $a \pm c \equiv b \pm d \pmod m$.

C_2: If $a \equiv b \pmod m$, then $ac \equiv bc \pmod m$.

C_3: If $a \equiv b \pmod m$ and $c \geq 1$, then $ac \equiv bc \pmod{mc}$.

C_4: If $a \equiv b \pmod m$ and $c \equiv d \pmod m$, then $ac \equiv bd \pmod m$.

C_5: If $a \equiv b \pmod m$, then $a^n \equiv b^n \pmod m$ for all $n \geq 1$.

C_6: $a \equiv 0 \pmod m$ if and only if $m|a$.

C_7: If $ac \equiv bc \pmod m$ and $(c,m) = 1$, then $a \equiv b \pmod m$.

C_8: If $ac \equiv bc \pmod m$ and $(c,m) = n$, then $a \equiv b \pmod{m/n}$.

C_9: If $a \equiv b \pmod m$ and $n|m$, then $a \equiv b \pmod n$.

We prove several of these properties, leaving the proofs of the remaining properties as exercises.

Proof of C_2: By hypothesis and Definition 4.1, $m|(a-b)$.
Now $(a-b)|c(a-b)$, so Theorem 2.1, part D_4 implies that $m|c(a-b)$; that is, $m|(ac-bc)$. Therefore, $ac \equiv bc \pmod m$. ∎

Proof of C_7: If $ac \equiv bc \pmod{m}$, then $m|(ac - bc)$; that is, $m|c(a - b)$. By hypothesis, $(c, m) = 1$. Therefore, Euclid's Lemma (Theorem 2.13) implies that $m|(a - b)$, so $a \equiv b \pmod{m}$. ■

Proof of C_8: As in the proof of C_7, $m|c(a - b)$. Since $(c, m) = n$ by hypothesis, we know that $n|c$ and $n|m$, so m/n and c/n are integers. Now $n(m/n)|n(c/n)(a - b)$. Theorem 2.1, part D_5 implies that $(m/n)|c/n(a - b)$, but Theorem 2.4, part G_5 implies that $(m/n, c/n) = 1$. Therefore, Euclid's Lemma (Theorem 2.13) implies that $m/n|(a - b)$; that is, $a \equiv b \pmod{m/n}$. ■

Note that arithmetic operations on congruences are similar to, but not identical with, arithmetic operations on equations. Parts C_1 and C_4 of Theorem 4.1 say that congruences (mod m) may be added, subtracted, and multiplied. Part C_8, which generalizes part C_7, tells what happens when a common factor is cancelled from a congruence (mod m); namely, the resulting congruence may have a reduced modulus. For example,

$$110 \equiv 50 \pmod{15}$$

That is,

$$10 \cdot 11 \equiv 10 \cdot 5 \pmod{15}$$

Now $(10, 15) = 5$. Therefore, if we cancel 10 from both sides of the last congruence, we get

$$11 \equiv 5 \pmod{15/5}$$

That is,

$$11 \equiv 5 \pmod{3}$$

Congruence (mod m) is a special case of a more general mathematical concept known as an *equivalence relation*. We therefore digress from our discussion of congruences to study equivalence relations and the related concept of equivalence classes.

Equivalence Relations

Equivalence relations, an important topic in mathematics, belong to the branch of mathematics known as *set theory*. A set, S, is a collection of objects known as *elements*. If a is an element of set S, we write $a \in S$. For example, if $S = \{1, 2\}$, (read "S is the set whose elements are 1 and 2"), then $1 \in S$ and $2 \in S$. The symbol (a, b) denotes the *ordered pair* whose first element is a and whose second element is b. Ordered pairs can be defined in terms of sets, but we omit the details. Order counts, that is, $(a, b) = (c, d)$ if and only if $a = c$ and $b = d$.

Definition 4.2   Equivalence Relation

If S is a non-empty set and R is a set of ordered pairs of elements from S, we say that R is an *equivalence relation* on S if all three of the following properties hold:

1. (Reflexive Property) For all a in S, $(a, a) \in R$.

2. (Symmetric Property) For all a and b in S, if $(a, b) \in R$, then also $(b, a) \in R$.

3. (Transitive Property) For all a, b, and c in S, if $(a, b) \in R$ and $(b, c) \in R$, then also $(a, c) \in R$.

Example 4.1

Let $S = Z$ (the set of all integers). Define R on Z as follows: If $a, b \in Z$, then $(a, b) \in R$ if and only if $a - b = 2n$ for some $n \in Z$. (n is not fixed, but varies with a and b.) Let us verify that R is an equivalence relation on Z.

1. (Reflexivity) For all $a \in Z$, $a - a = 0 = 2(0) \rightarrow (a, a) \in R \rightarrow R$ is reflexive.

2. (Symmetry) If $(a, b) \in R$, then $a - b = 2n$. Multiplying by -1, we get $-(a - b) = -2n$, that is $b - a = 2(-n)$. Therefore $(b, a) \in R$, so R is symmetric.

3. (Transitivity) If $(a, b) \in R$ and $(b, c) \in R$, then $a - b = 2m$ and $b - c = 2n$ for some integers m, n. Adding these equations, we have $(a - b) + (b - c) = 2m + 2n$, that is, $a - c = 2(m + n)$. Therefore $(a, c) \in R$, so R is transitive.

Since R is reflexive, symmetric, and transitive, it follows that R is an equivalence relation on Z.

Example 4.2

T is the set of all plane triangles. R is the relation of similarity. (Two triangles are said to be similar if all three pairs of corresponding sides are in the same proportion.) In less precise language, two triangles are said to be similar if they have the same shape but not necessarily the same size. R is an equivalence relation on T.

Example 4.3

W is the set of all English words. R is the relation of beginning with the same letter. R is an equivalence relation on W.

Example 4.4

$P[x]$ is the set of all polynomials in x of degree ≥ 1 with integer coefficents. R is the relation of having the same degree. (That is, if $f(x) \in P[x]$ and $g(x) \in P[x]$, then $(f(x), g(x)) \in R$ if and only if $f(x)$ and $g(x)$ have the same degree.) R is an equivalence relation on $P[x]$.

We now give an example of a relation that is not an eqivalence relation. Let a relation R be defined on N by aRb if $a = b^m$ for some $m \geq 1$. Now for all $a \in N$, we have $a = a^1$, so $(a, a) \in R$. Therefore R is reflexive. Also, if $(a, b) \in R$ and $(b, c) \in R$, then $a = b^m$ and $b = c^n$, so $a = (c^n)^m = c^{nm}$. Therefore $(a, c) \in R$, so R is transitive. However if $(a, b) \in R$, then $a = b^m$. It does not follow that $b = a^n$, so we cannot say that $(b, a) \in R$. Therefore, R is not symmetric, so R is *not* an equivalence relation on N.

Suppose that S and T are sets such that each element of S is also an element of T. Then we say that S is a *subset* of T, and we write $S \subset T$. For example, if $S = \{1, 2\}$, and $T = \{1, 2, 3\}$, then S is a subset of T. Sets A and B are said to be *equal* (write $A = B$) if $A \subset B$ and $B \subset A$. In this case A and B are two different names for the same set.

Associated with any equivalence relation R on set S are certain subsets of S known as *equivalence classes*, which are defined below.

Definition 4.3   **Equivalence Class**

If R is an equivalence relation on set S, and if a belongs to S, we define the *equivalence class* of a under R, denoted $[a]_R$, as the set of all b in S such that $(b, a) \in R$. That is,

$$[a]_R = \{b \in S : (b, a) \in R\}$$

Example 4.5

$S = Z$, $(a, b) \in R$ if and only if $a - b = 2n$.

$$[0]_R = \{b \in Z : b - 0 = 2n\} = \{b \in Z : b = 2n\} = \{0, \pm2, \pm4, \pm6, \cdots\}$$

$$[1]_R = \{b \in Z : b - 1 = 2n\} = \{b \in Z : b = 2n + 1\} = \{\pm1, \pm3, \pm5, \cdots\}$$

That is, $[0]_R$ is the set of all even integers, while $[1]_R$ is the set of all odd integers.

Example 4.6

If t is an equilateral triangle, then $[t]_R$ is the set of all equilateral triangles (see example 4.2).

| Example 4.7 | Referring to example 4.3, $[mathematics]_R$ is the set of all English words that start with m. |

| Example 4.8 | $[3x^2 + 4x + 5]_R$ is the set of all polynomials in x of degree 2, with integer coefficients. |

An equivalence class may have many different names, as is shown by the following theorem.

| Theorem 4.2 | Let a and b belong to set S. Let R be an equivalence relation on S. Then aRb if and only if $[a]_R = [b]_R$. |

Proof: First we show that if $(a, b) \in R$, then $[a]_R = [b]_R$. Suppose that $c \in [a]_R$. Then $(c, a) \in R$. Since $(a, b) \in R$ by hypothesis and R is transitive, it follows that $(c, b) \in R$, so $c \in [b]_R$. Therefore $[a]_R \subset [b]_R$. Similarly, $[b]_R \subset [a]_R$. Therefore $[a]_R = [b]_R$. Conversely, suppose that $[a]_R = [b]_R$. Since R is reflexive, we have $(a, a) \in R$, so $a \in [a]_R$. Therefore $a \in [b]_R$, so $(a, b) \in R$. ∎

Referring to Example 4.1, we have

$$[0]_R = [2]_R = [-2]_R = [4]_R = [-4]_R = \cdots$$

also

$$[1]_R = [-1]_R = [3]_R = [-3][_R = \cdots$$

Referring to Example 4.3, we have

$$[\text{mathematics}]_R = [\text{milk}]_R = [\text{monarch}]_R = \text{etc.}$$

If we look at all the equivalence classes that result from an equivalence relation R on set S, we may notice that (i) each equivalence class is a subset of S, and (ii) each element of S belongs to exactly one equivalence class. A collection of subsets of S with property (ii) is called a *partition* of S. The definition of partition of a set is given below.

Definition 4.4   Partition of a Set

A *partition* of a nonempty set is a collection of one or more subsets of S such that each element of S belongs to precisely one subset.

For example, if $S = \{1, 2, 3, 4, 5, 6, 7, 8, 9, 10\}$, one partition of S consists of the subsets $S_1 = \{1, 4, 7, 10\}$, $S_2 = \{2, 5, 8\}$, and $S_3 = \{3, 6, 9\}$.

One reason that equivalence relations are important is that if R is an equivalence class on set S, then the set of all equivalence classes under R forms a partition of S. In the example above, the subsets that form the partition are equivalence classes under the equivalence relation $(a, b) \in R$ if $3|(a - b)$. Conversely, a partition of S may be used to define an equivalence relation on R as follows: if $a, b \in S$, then $(a, b) \in R$ if and only if a and b belong to the same subset in the partition.

For example, if $S = \{1, 2, 3, 4, 5\}$, let $S_1 = \{1, 4\}$, $S_2 = \{2, 5\}$, $S_3 = \{3\}$. The corresponding equivalence relation consists of the following ordered pairs:

$$R = \{(1, 1), (1, 4), (4, 1), (4, 4), (2, 2), (2, 5), (5, 2), (5, 5), (3, 3)\}$$

Section 4.2 Exercises

1. Prove each of the following parts of Theorem 4.1.

   (a) C_1

   (b) C_3

   (c) C_4

   (d) C_5

   (e) C_6

   (f) C_9

2. Reduce each of the following congruences as much as possible by cancellation.

   (a) $650 \equiv 350 \pmod{75}$

   (b) $320 \equiv 32 \pmod{12}$

   (c) $215 \equiv 5 \pmod{21}$

3. Determine which of the following relations on $\mathbf{N}$, if any, is an equivalence relation. Justify your conclusion.

   (a) aRb if $a|b$.

   (b) aRb if $\frac{1}{2}(a + b)$ is an integer

   (c) aRb if $a + b$ is prime

   (d) aRb if $ab = k^2$ for some integer k

   (e) aRb if $a + b$ is even or composite.

4. Given that parallelism is an equivalence relation on the set of lines in the xy-plane, what is the equivalence class of

   (a) the x-axis?

   (b) the y-axis?

   (c) the line: $y = x$?

   (Your answer should be a geometric description.)

5. Under the equivalence relation of Example 3 in this section, how many distinct equivalence classes are there?

6. In the xy-plane, say that two points are equivalent if they are equidistant from the origin. What is the equivalence class of the point $(3, 4)$? (Describe it geometrically.)

7. Prove that if R is an equivalence relation on set S, then

   (a) no equivalence class is empty;

   (b) any two distinct equivalence classes are disjoint, and

   (c) the union of all the equivalence classes under R is S.

8. Let a relation, R , be defined on Z as follows: If $a, b \in Z$, then $(a, b) \in R$ if, and only if $3a + 4b = 7n$ for some $n \in Z$.

   (a) Prove that R is an equivalence relation on Z.

   (b) Find 5 elements of $[4]_R$.

9. Let a relation, R , be defined on Z as follows: If $a, b \in Z$, then $(a, b) \in R$ if, and only $a^2 - b^2 = 5n$ for some $n \in Z$.

   (a) Prove that R is an equivalence relation on Z.

   (b) How many distinct equivalence classes are there?

10. Let a relation, R , be defined on Z as follows: If $a, b \in Z$, then $(a, b) \in R$ if, and only $\frac{a}{b} = 3^n$ for some $n \in Z$.

    (a) Prove that R is an equivalence relation on Z.

    (b) Find 5 elements of $[4]_R$.

11. d Let the set $S = \{1, 2, 3, 4, 5, 6\}$ be partitioned into subsets as follows:

$$S_1 = \{1, 2, 3\}, \ S_2 = \{4, 5\}, \ S_6 = \{6\}$$

List all the ordered pairs in the corresponding equivalence relation.

Returning to the subject of congruences, we see by way of the following theorem that for every natural number m, congruence (mod m) is an equivalence relation on **Z**.

Theorem 4.3   Let m be a natural number. Then congruence (mod m) is an equivalence relation on Z.

Proof: Since $m|0$, we have $m|(a-a)$ for all a in Z, so $a \equiv a \pmod{m}$ for all a in Z. Therefore, congruence $\pmod{m}$ is reflexive. If $a \equiv b \pmod{m}$, then $m|(a-b)$, so $a-b = km$ for some k. This implies that $b-a = (-k)m$, so $m|(b-a)$; hence $b \equiv a \pmod{m}$. Therefore, congruence $\pmod{m}$ is symmetric. For all a, b, and c in Z, if $a \equiv b \pmod{m}$ and $b \equiv c \pmod{m}$, then $m|(a-b)$ and $m|(b-c)$. Therefore, $m|[(a-b)+(b-c)]$; that is, $m|(a-c)$. Therefore, $a \equiv c \pmod{m}$, so congruence $\pmod{m}$ is transitive. Since congruence $\pmod{m}$ is reflexive, symmetric, and transitive, it is an equivalence relation on Z. ∎

Let us look at the equivalence classes under congruence $\pmod{m}$. There are m such classes, namely, $[0]_m$, $[1]_m$, $[2]_m$, ... , $[m-1]_m$. Now $[0]_m = \{0, \pm m, \pm 2m, \pm 3m, ...\}$. In other words, $[0]_m = \{km : k \text{ is in } Z\}$. More generally, if $0 \le r \le m-1$, then $[r]_m = \{km + r : k \text{ is in } Z\}$. That is to say, $[r]_m$ consists of those integers that are r more than a multiple of m. Another way to put this would be to say that $[r]_m$ consists of those integers which leave a remainder r when divided by m.

For example, when $m = 2$, the two resulting equivalence classes should be familiar to you.

$$[0]_2 = \{0, \pm 2, \pm 4, \pm 6, ...\} = \text{ the set of all even integers}$$
$$[1]_2 = \{\pm 1, \pm 3, \pm 5, \pm 7, ...\} = \text{ the set of all odd integers}$$

Again, when $m = 3$, the three resulting equivalence classes are

$$[0]_3 = \{..., -6, -3, 0, 3, 6, ...\}$$
$$[1]_3 = \{..., -5, -2, 1, 4, 7, ...\}$$
$$[2]_3 = \{..., -4, -1, 2, 5, 8, ...\}$$

4.3 Linear Congruences

For all natural numbers m, b, we know by the Division Algorithm (Theorem 2.2) that there exist q and r such that $b = qm + r$ and $0 \le r \le m-1$. This fact also holds for arbitrary integers b. We now introduce the concept of least positive residues.

Definition 4.5   **Least Positive Residue** $(\bmod\ m)$

If m is a natural number and b is an integer, let $b = qm + r$, where $0 \le r \le m-1$. We say that r is the *least positive residue of b* $\pmod{m}$.

For example, the least positive residue of 35 $\pmod 8$ is 3, since $35 = 4(8) + 3$; the least positive residue of 35 $\pmod 7$ is 0, since $35 = 5(7) + 0$; and the least positive residue of -13 $\pmod{10}$ is 7, since $-13 = (-2)10 + 7$.

The set of least positive residues (mod m) is the set $\{0, 1, 2, ..., m-1\}$. For example, the set of least positive residues (mod 4) is $\{0, 1, 2, 3\}$, and the set of least positive residues (mod 7) is $\{0, 1, 2, 3, 4, 5, 6\}$.

Note that $a \equiv b$ (mod m) if and only if a and b have the same least positive residue (mod m).

Congruence may be used to prove results regarding divisibility without the necessity of calculating large numbers. As a first example, we prove that $11|(233^5 + 1)$. Since $233 > 11$, we begin by finding the least positive residue of 233 (mod 11). Since

$$233 = 21(11) + 2$$

we have

$$233 \equiv 2 \text{ (mod 11)}$$

Therefore, $233^5 \equiv 2^5 = 32 \equiv 10$ (mod 11), so $233^5 + 1 \equiv 10 + 1 = 11 \equiv 0$ (mod 11); that is, $11|(233^5 + 1)$.

Note that whenever a number greater than or equal to a modulus is generated, it should be replaced by its least positive residue. (This rule does not apply to exponents, however.)

As a second example, we prove that $31|(15^{10} - 1)$

$$
\begin{aligned}
15^2 &= 225 \equiv 8 \text{ (mod 31)} \\
15^4 &= (15^2)^2 \equiv 8^2 = 64 \equiv 2 \text{ (mod 31)} \\
15^5 &= 15(15^4) \equiv 15(2) = 30 \text{ (mod 31)} \\
15^{10} &= (15^5)^2 \equiv 30^2 = 900 \equiv 1 \text{ (mod 31)}
\end{aligned}
$$

We have $15^{10} \equiv 1$ (mod 31), so $31|(15^{10} - 1)$.

We shall often need to compute the least positive residue of a^m (mod n), where the integers a, m, and n are all greater than 1. One could obtain the desired result by successively computing the least positive residues (mod n) of $a^2, a^3, a^4, ..., a^m$, but this would require $m-1$ multiplications. A more efficient procedure essentially makes use of the binary representation of the exponent m. Note that any even power of a given base can be computed by squaring, since $a^{2k} = (a^k)^2$. Also, any odd power of a given base can be computed by multiplying the base by a square, since $a^{2k+1} = a(a^{2k})$.

Suppose, for example, that we wish to compute the least positive residue of 7^{19} (mod 31). Now

$$
\begin{aligned}
19 &= 2(9) + 1 = 18 + 1 \\
18 &= 2(9) \\
9 &= 2(4) + 1 = 8 + 1 \\
8 &= 2(4) \\
4 &= 2(2) \\
2 &= 2(1)
\end{aligned}
$$

Therefore we need compute only 7^2, 7^4, 7^8, 7^9, 7^{18}, and 7^{19} (mod 31):

$$
\begin{aligned}
7^2 &= (7^1)^2 = 49 \equiv 18 \ (\text{mod } 31) \\
7^4 &= (7^2)^2 \equiv 18^2 = 324 \equiv 14 \ (\text{mod } 31) \\
7^8 &= (7^4)^2 \equiv 14^2 = 196 \equiv 10 \ (\text{mod } 31) \\
7^9 &= 7(7^8) \equiv 7(10) \equiv 70 \equiv 8 \ (\text{mod } 31) \\
7^{18} &= (7^9)^2 \equiv 8^2 = 64 \equiv 2 \ (\text{mod } 31) \\
7^{19} &= 7(7^{18}) \equiv 7(2) = 14 \ (\text{mod } 31)
\end{aligned}
$$

In general, the number of squaring operations needed to compute a^m in this manner is $[\log_2 m]$. For example, in order to compute the least positive residue of 7^{19} (mod 31), the number of squarings was $[\log_2 19] = 4$.

In elementary algebra, one learns how to solve linear equations, that is, equations of the type $ax = b$, where $a \neq 0$ and x is unknown. The corresponding problem in number theory involves linear congruences, which are defined below.

Definition 4.6   **Linear Congruence**

If a, b, and m are integers such that $a \not\equiv 0$ (mod m) and x is unknown, then the congruence

$$ ax \equiv b \ (\text{mod } m) $$

is called a *linear congruence (mod m)*.

Before we discuss how to solve a linear congruence, we must discuss what is meant by a solution. Suppose, for example, that we wish to solve the linear congruence

$$ 3x \equiv 1 \ (\text{mod } 5) \tag{4.1} $$

If we consider a solution to be a number, then we are faced with a clumsy situation, since there are infinitely many numbers that are solutions, namely, $x = 2, 7, 12, 17,$ etc., as well as $x = -3, -8, -13, -18,$ etc. These solutions all have the form $x = 5k + 2$, where k is an integer. Therefore, it makes sense to consider the solution of Congruence (4.1) to be the congruence

$$ x \equiv 2 \ (\text{mod } 5) \tag{4.2} $$

We also could have said that the solution to Congruence (4.1) is the congruence

$$ x \equiv 7 \ (\text{mod } 5) \tag{4.3} $$

but it is more convenient to use least positive residues. In general, the solution of a linear congruence

$$ax \equiv b \ (\text{mod } m) \tag{4.4}$$

if it exists, will have the form

$$x \equiv t \ (\text{mod } m) \tag{4.5}$$

where t is the least positive residue (mod m); that is, $0 \leq t \leq m - 1$. Any solution not in this form can, and usually should, be reduced to this form.

For example, the congruence

$$x \equiv 47 \ (\text{mod } 20)$$

should be reduced to

$$x \equiv 7 \ (\text{mod } 20)$$

Let us look at some additional examples. The congruence

$$3x \equiv 7 \ (\text{mod } 13) \tag{4.6}$$

has the solution

$$x \equiv 11 \ (\text{mod } 13) \tag{4.7}$$

since $3(11) = 33 = 2(13) + 7$. The congruence

$$5x \equiv 2 \ (\text{mod } 37) \tag{4.8}$$

has the solution

$$x \equiv 30 \ (\text{mod } 37) \tag{4.9}$$

since $5(30) = 150 = 4(37) + 2$. The congruence

$$3x \equiv 2 \ (\text{mod } 6) \tag{4.10}$$

has no solution.

We are about to begin the study of linear congruences, that is, polynomial congruences of degree 1. In Section 4.6 we will see how to obtain all solutions, if any, of the congruence

$$f(x) \equiv 0 \ (\text{mod } p^n) \tag{4.11}$$

where $f(x)$ is a polynomial of degree at least 2, p is prime, and $n \geq 2$, given the solutions of the congruence

$$f(x) \equiv 0 \ (\mathrm{mod} \ p) \tag{4.12}$$

In Chapter 7 we will study quadratic congruences, that is, polynomial congruences of degree 2.

Before we commence, some further comments are in order. Let $f(x)$ be a polynomial of degree at least 1 with integer coefficients. Any solution to the congruence

$$f(x) \equiv 0 \ (\mathrm{mod} \ m) \tag{4.13}$$

will have the form of Conguence (4.5), namely,

$$x \equiv t \ (\mathrm{mod} \ m)$$

Therefore, it is always theoretically possible to obtain all solutions of Congruence 4.11 by trial and error, that is, by the successive evaluation of $f(0), f(1), f(2), ..., f(m-1) \ (\mathrm{mod} \ m)$. For example, suppose we wish to solve the congruence

$$x^3 + x + 2 \equiv 0 \ (\mathrm{mod} \ 3) \tag{4.14}$$

Here $f(x) = x^3 + x + 2$. Now $f(0) = 2 \equiv 2 \ (\mathrm{mod} \ 3)$, $f(1) = 4 \equiv 1 \ (\mathrm{mod} \ 3)$, $f(2) = 12 \equiv 0 \ (\mathrm{mod} \ 3)$. Therefore, the unique solution of (4.14) is

$$x \equiv 2 \ (\mathrm{mod} \ 3) \tag{4.15}$$

The trial-and-error method for solving polynomial congruences works well for small m but is impractical for large m.

We shall often have the need to solve a linear congruence. The two following theorems tell us when a linear congruence admits solutions and, if so, how many.

Theorem 4.4

The congruence $ax \equiv b \ (\mathrm{mod} \ m)$ has at least one solution if and only if $(a, m) | b$.

Proof: Let $d = (a, m)$. If $ax \equiv b \ (\mathrm{mod} \ m)$, then $ax - b = km$ for some k, so $ax - km = b$. Since $d | a$ and $d | m$, we must have $d | b$; that is, $b = jd$ for some j. On the other hand, we know by Theorem 2.4 that there exist u and v such that $d = au + mv$, so $b = j(au + mv) = a(ju) + m(jv)$. Letting $x = ju$, we have $ax + m(jv) = b$, so $ax + m(jv) \equiv b \ (\mathrm{mod} \ m)$. This reduces to $ax \equiv b \ (\mathrm{mod} \ m)$. ∎

For example, we stated earlier that Congruence (4.10), namely,

$$3x \equiv 2 \pmod{6}$$

has no solution. This is true because $(3,6) = 3$, and $3 \nmid 2$.

Theorem 4.5 Let $d = (a, m)$. If $d|b$, then Congruence (4.4),

$$ax \equiv b \pmod{m}$$

has exactly d solutions (mod m).

─────────────────

Proof: By Theorem 4.4, we know that Congruence (4.4) has no solution unless $d|b$. Suppose that $x \equiv y \pmod{m}$ and $x \equiv z \pmod{m}$ are solutions of Congruence (4.4), where y and z are least positive residues (mod m). Then $ay \equiv az \equiv b \pmod{m}$. Now Theorem 4.1, part C_8, implies that $y \equiv z$ (mod m/d); that is, $y = z + k(m/d)$ for some k. By the Division Algorithm, (Theorem 2.2), $k = qd + r$, with $0 \leq r \leq d - 1$, so $y = z + (qd + r)(m/d) = z + qm + r(m/d)$. Therefore,

$$y \equiv z + r(m/d) \pmod{m} \qquad (4.16)$$

Letting r assume all possible values, we obtain at most d solutions of Congruence (4.4) that are distinct, that is, incongruent (mod m). We now show that these d solutions are all distinct (mod m). If $z + i(m/d) \equiv z + j(m/d)$ (mod m), then $i(m/d) \equiv j(m/d) \pmod{m}$, so that $m| [i(m/d) + j(m/d)]$, that is, $m|(i - j)(m/d)$. This implies that $d|(i - j)$. However, this is impossible if $0 \leq i < j \leq d - 1$. ■

For example, the congruence

$$8x \equiv 12 \pmod{20} \qquad (4.17)$$

has four solutions (mod 20), since $(8, 20) = 4$ and $4|12$. The congruence

$$9x \equiv 15 \pmod{30} \qquad (4.18)$$

has three solutions (mod 30), since $(9, 30) = 3$ and $3|15$. The congruence

$$9x \equiv 15 \pmod{27} \qquad (4.19)$$

has no solution, since $(9, 27) = 9$ but $9 \nmid 15$.

Suppose that $(a, m) = 1$, so that the congruence $ax \equiv b \pmod{m}$ has a unique solution. How do we actually find it without resorting to trial and error? The answer is found in the proof of Theorem 4.4. Using the Euclidean Algorithm to compute (a, m), we find integers u and v such that $au + mv = 1$. This yields the congruence

$$au \equiv 1 \pmod{m} \tag{4.20}$$

Then multiplication by b produces the congruence

$$a(ub) \equiv b \pmod{m} \tag{4.21}$$

so our solution to Congruence (4.4) is

$$x \equiv bu \pmod{m} \tag{4.22}$$

For example, suppose that we wish to solve the congruence

$$7x \equiv 2 \pmod{37} \tag{4.23}$$

First, we compute $(37, 7)$. We have

$$37 = 5(7) + 2$$
$$7 = 3(2) + 1$$

Therefore, $(37, 7) = 1$, and Congruence (4.23) has a unique solution (mod 37). Next, we use Theorem 2.6 to find integers u, v such that $(37, 7) = 7u + 37v$. We construct a 4-column table, with headings j, q_j, s_j, t_j as follows:

j	q_j	s_j	t_j
-1		1	0
0		0	1
1	5	1	-5
2	3	-3	16

Here $u = 16$ (the entry in the bottom right) and $b = 2$, so our solution of Congruence (4.23) is given by

$$x \equiv 16(2) = 32 \pmod{37} \tag{4.24}$$

Check: $7(32) = 224 = 6(37) + 2$, so $7(32) \equiv 2 \pmod{37}$.

As a second example, suppose that we wish to solve the congruence

$$11x \equiv 4 \pmod{25} \tag{4.25}$$

First, we compute (25,11). We have

$$25 = 2(11) + 3$$
$$11 = 3(3) + 2$$
$$3 = 1(2) + 1$$

Therefore, $(25, 11) = 1$, and Congruence (4.25) has a unique solution (mod 25). Now we use Theorem 2.6 to find integers u, v such that $(25,11) = 11u+25v$.

j	q_j	s_j	t_j
-1		1	0
0		0	1
1	2	1	-2
2	3	-3	7
3	1	4	-9

Here $u = -9$ and $b = 4$, so our solution of Congruence (4.25) is given by

$$x \equiv 4(-9) = -36 \ (\text{mod } 25) \tag{4.26}$$

Reducing to a least positive residue (mod 25), we obtain

$$x \equiv 14 \ (\text{mod } 25) \tag{4.27}$$

Check: $11(14) = 154 = 6(25) + 4$, so $11(14) \equiv 4 \ (\text{mod } 25)$.

Now suppose that $(a, m) = d$, $d|b$, and $d > 1$. We wish to find all d solutions of Congruence (4.4). By Theorem 4.1, part C_8 we know that

$$(a/d)x \equiv b/d \ (\text{mod } m/d) \tag{4.28}$$

Furthermore, by Theorem 2.4, part G_5 and Theorem 4.4, Congruence (4.27) has a unique solution (mod m/d), namely,

$$x \equiv t \ (\text{mod } m/d) \tag{4.29}$$

Let $x_k \equiv t + (k - 1)(m/d) \ (\text{mod } m)$, where $k = 1, 2, 3, ..., d$. Then each x_k is a solution of Congruence (4.4), and all the x_k are distinct (mod m).

For example, let us solve the congruence

$$8x \equiv 6 \ (\text{mod } 10) \tag{4.30}$$

Since $(8, 10) = 2$ and $2|6$, we expect two solutions (mod 10). Now Congruence (4.30) reduces to the congruence

$$4x \equiv 3 \ (\text{mod } 5) \tag{4.31}$$

whose unique solution is

$$x \equiv 2 \pmod{5} \tag{4.32}$$

Therefore, our desired solutions (mod 10) are

$$x_1 \equiv 2 + 0(5) \equiv 2 \pmod{10} \tag{4.33}$$

$$x_2 \equiv 2 + 1(5) \equiv 7 \pmod{10} \tag{4.34}$$

As a second example, let us solve the congruence

$$9x \equiv 12 \pmod{21} \tag{4.35}$$

Since $(9, 21) = 2$ and $3|12$, we expect three solutions (mod 21). Now Congruence (4.35) reduces to the congruence

$$3x \equiv 4 \pmod{7} \tag{4.36}$$

whose unique solution is

$$x \equiv 6 \pmod{7} \tag{4.37}$$

Therefore, our desired solutions of Congruence (4.35) are

$$x_1 \equiv 6 + 0(7) \equiv 6 \pmod{21} \tag{4.38}$$

$$x_2 \equiv 6 + 1(7) \equiv 13 \pmod{21} \tag{4.39}$$

$$x_3 \equiv 6 + 2(7) \equiv 20 \pmod{21} \tag{4.40}$$

To summarize, given a linear congruence of the form (4.4), namely,

$$ax \equiv b \pmod{m}$$

where $(a, m) = d$, $d|b$, and $d > 1$, we first reduce Congruence (4.4) to Congruence (4.28), namely,

$$(a/d)x \equiv b/d \pmod{m/d}$$

Then we obtain the unique solution of Congruence (4.28), and finally, we use this solution to obtain all d solutions of Congruence (4.4).

Section 4.3   Exercises

1. Use congruences to verify each of the following divisibility statements.

   (a) $13|(145^6 + 1)$

   (b) $17|(1855^4 + 1)$

(c) $233 \mid (2^{29} - 1)$

(d) $223 \mid (2^{37} - 1)$

(e) $431 \mid (2^{43} - 1)$

(f) $167 \mid (2^{83} - 1)$

2. Use trial and error to find all solutions, if any, to each of the following congruences.

(a) $x^3 + x + 1 \equiv 0 \pmod 3$

(b) $x^3 - 2x + 3 \equiv 0 \pmod 5$

(c) $x^2 - 2 \equiv 0 \pmod 7$

(d) $x^4 + 2x^3 + x - 2 \equiv 0 \pmod 7$

3. Find all solutions, if any, to each of the following linear congruences.

(a) $3x \equiv 1 \pmod{23}$

(b) $7x \equiv 1 \pmod{47}$

(c) $5x \equiv 2 \pmod{210}$

(d) $6x \equiv 7 \pmod{25}$

(e) $8x \equiv 5 \pmod{31}$

(f) $8x \equiv 12 \pmod{32}$

(g) $12x \equiv 6 \pmod{21}$

(h) $2x \equiv 2 \pmod{16}$

(i) $20x \equiv 8 \pmod{24}$

4. If p is prime, prove that $(a + b)^p \equiv a^p + b^p \pmod p$.

5. If x is odd and $n \geq 1$, prove that $x^{2^n} \equiv 1 \pmod{2^{n+2}}$.

6. If m and n are natural numbers, prove that $m^{2n} \equiv 1 \pmod{m + 1}$.

7. If p is an odd prime and $2 \leq k \leq p - 1$, prove that $\binom{p+1}{k} \equiv 0 \pmod p$.

8. If p is an odd prime and $1 \leq k \leq p - 2$, prove that $\binom{p-1}{k} \equiv (-1)^k \pmod p$.

9. If $\prod_{i=1}^{r} a_i \equiv 2 \pmod 4$, prove the following:

(a) There exists j such that $1 \leq j \leq r$ and $a_j \equiv 2 \pmod 4$.

(b) a_i is odd for all $i \neq j$.

Section 4.3 Computer Exercises

10. Write a computer program that will perform modular exponentiation, that is, when given natural numbers a, m, and n, will compute the least positive residue of a^m (mod n).

11. Let p be an odd prime. Let $0 < a < p$. Let $f(x) = x^2 - a$. Write a computer program to find all solutions, if any, of the congruence $f(x) = 0$ (mod p) by trial and error, that is, by evaluating $f(1)$, $f(2)$, $f(3)$, ..., $f(p-1)$.

12. Using the program from the preceding exercise, for each prime p from 31 to 47, find the number of integers a such that $0 < a < p$ and the congruence $f(x) \equiv 0$. (mod p) has solutions. Is there a pattern?

4.4 Linear Diophantine Equations and the Chinese Remainder Theorem

An equation in two or more unknown *integers* is called a *Diophantine equation*, in honor of Diophantus of Alexandria, who lived about 250 A.D. We shall now apply linear congruences to the solution to linear Diophantine equations in two unknowns.

Consider the equation

$$ax + by = c \tag{4.41}$$

where a, b, and c are given integers and x and y are unknown. Let $d = (a, b)$. Since $d|a$ and $d|b$, if Equation (4.41) has a solution, then $d|c$. If $d > 1$, then equation (41) may be simplified by dividing out d. Therefore, we assume henceforth that $d = 1$. We convert Equation (4.41) to the congruence

$$ax \equiv c \pmod{b} \tag{4.42}$$

Since $d = 1$, $d|c$, so Congruence (4.42) has a unique solution (mod b), namely,

$$x \equiv x_0 \pmod{b} \tag{4.43}$$

with $0 \leq x_0 < b$. Converting Congruence (4.43) to an equation, we get

$$x = x_0 + bt \tag{4.44}$$

where t is an arbitrary integer. Let $y_0 = (c - ax_0)/b$. Then $y = (c - ax)/b = [c - a(x_0 + bt)]/b = (c - ax_0)/b - at = y_0 - at$. Therefore, the complete solution of Equation (4.41) is given (in parametric form) by the equations:

$$x = x_0 + bt \tag{4.45}$$

$$y = y_0 - at \tag{4.46}$$

where the parameter t is an arbitrary integer.

For example, let us solve the Diophantine equation

$$6x + 11y = 41 \tag{4.47}$$

We convert Equation (4.47) to a congruence (mod 11), namely,

$$6x \equiv 41 \;(\mathrm{mod}\; 11) \tag{4.48}$$

which reduces first to

$$6x \equiv 8 \;(\mathrm{mod}\; 11) \tag{4.49}$$

and then further to

$$3x \equiv 4 \;(\mathrm{mod}\; 11) \tag{4.50}$$

The unique solution of Congruence (4.50) is given by

$$x \equiv 5 \;(\mathrm{mod}\; 11) \tag{4.51}$$

which we convert to the equation

$$x = 5 + 11t \tag{4.52}$$

Note that $x_0 = 5$. Now $y_0 = [41 - 6(5)]/11 = 1$, so we have

$$y = 1 - 6t \tag{4.53}$$

The complete solution to equation (4.47) is given by equations (4.52) and (4.53), where t denotes an arbitrary integer.

Next, we consider a "story" problem that leads to a linear Diophantine equation in two unknowns. Old McDonald (the farmer of reknown) sold chickens at \$5 each and geese at \$8 each. He collected a total of \$99. Assuming that he sold at least one bird of each kind, how many of each kind did he sell?

To solve this problem, let x be the number of chickens sold, and let y be the number of geese sold. Since the total amount of money collected is the sum of the value of the sold chickens and the value of the sold geese, we obtain the equation

$$5x + 8y = 99 \tag{4.54}$$

where we further specify that $x \geq 1$ and $y \geq 1$. To solve equation (4.54), we first reduce it to a congruence (mod 8), namely,

$$5x \equiv 99 \;(\mathrm{mod}\; 8) \tag{4.55}$$

which then reduces to

$$5x \equiv 3 \;(\mathrm{mod}\; 8) \tag{4.56}$$

The unique solution of Congruence (4.56) is

$$x \equiv 7 \pmod{8} \tag{4.57}$$

so $x_0 = 7$, and we convert Congruence (4.57) to the equation

$$x = 7 + 8t \tag{4.58}$$

Now $y_0 = [99 - 5(7)]/8 = 8$, so we obtain the equation

$$y = 8 - 5t \tag{4.59}$$

Since $x \geq 1$, equation (4.58) implies that $t \geq 0$; since $y \geq 1$, equation (4.59) implies that $t \leq 1$. Therefore, $t = 0$ or 1, so our problem has two solutions:

$$x = 7, \quad y = 8 \tag{4.60}$$

$$x = 15, \quad y = 3 \tag{4.61}$$

Suppose that we wish to solve a system of simultaneous linear congruences in one unknown, such as

$$\begin{cases} x \equiv 3 \pmod{5} \\ x \equiv 2 \pmod{6} \\ x \equiv 4 \pmod{7} \end{cases} \tag{4.62}$$

Problems of this type were considered by the Chinese mathematician Sun Tzu in the first century A.D. The following theorem, known as the *Chinese Remainder Theorem*, guarantees the existence of a solution and tells us how to find it, provided that the moduli of the various given linear congruences are pairwise relatively prime.

Theorem 4.6

Chinese Remainder Theorem

Let m_1, m_2, m_3, ... , m_r (where $r \geq 2$) be natural numbers that are pairwise relatively prime and whose product is M. Then the system of r simultaneous linear congruences,

$$\begin{cases} x \equiv a_1 \pmod{m_1} \\ x \equiv a_2 \pmod{m_2} \\ x \equiv a_3 \pmod{m_3} \\ \qquad \vdots \\ x \equiv a_r \pmod{m_r} \end{cases} \tag{4.63}$$

has a unique solution (mod M).

Proof: For each index i such that $1 \leq i \leq r$, let $M_i = M/m_i$. By hypothesis, $(m_i, m_j) = 1$ whenever $i \neq j$. Therefore, Theorem 2.9 implies that $(M_i, m_i) = 1$ for all i. It follows from Theorem 4.5 that there exists x_i such that

$$M_i x_i \equiv 1 \;(\text{mod } m_i) \tag{4.64}$$

Now let

$$x = \sum_{i=1}^{r} M_i x_i a_i \tag{4.65}$$

Since $M_j \equiv 0 \;(\text{mod } m_i)$ for all $j \neq i$, this implies that

$$x \equiv M_i x_i a_i \equiv a_i \;(\text{mod } m_i) \tag{4.66}$$

for all i; that is, x is a solution of System (4.63).

If z is any other solution of System (4.63), then $z \equiv a_i \;(\text{mod } m_i)$ for all i, so $z - x \equiv 0 \;(\text{mod } m_i)$; that is, $m_i | (z - x)$ for all i.

Therefore, $[m_1, m_2, m_3, ..., m_r] | (z - x)$. Since the m_i are pairwise relatively prime by hypothesis, it follows from the result of Exercise 6 in Section 2.3 that $[m_1, m_2, m_3, ..., m_r] = M$. Therefore, $M | (z-x)$, that is, $z \equiv x \;(\text{mod } M)$. ∎

For example, let us solve System (4.62) of linear congruences, namely,

$$\begin{cases} x \equiv 3 \;(\text{mod } 5) \\ x \equiv 2 \;(\text{mod } 6) \\ x \equiv 4 \;(\text{mod } 7) \end{cases}$$

We are given $m_1 = 5$, $m_2 = 6$, and $m_3 = 7$, so $M = m_1 m_2 m_3 = 5 \cdot 6 \cdot 7 = 210$. Also, $M_1 = M/m_1 = 210/5 = 42$, $M_2 = M/m_2 = 210/6 = 35$, and $M_3 = M/m_3 = 210/7 = 30$. Now

$$M_1 x_1 \equiv 1 \quad (\text{mod } m_1) \;;\; M_2 x_2 \equiv 1 \quad (\text{mod } m_2) \;;\; M_3 x_3 \equiv 1 \quad (\text{mod } m_3)$$
$$42 x_1 \equiv 1 \quad (\text{mod } 5) \;\;;\;\; 35 x_2 \equiv 1 \quad (\text{mod } 6) \;\;;\;\; 30 x_3 \equiv 1 \quad (\text{mod } 7)$$
$$2 x_1 \equiv 1 \quad (\text{mod } 5) \;\;;\;\; 5 x_2 \equiv 1 \quad (\text{mod } 6) \;\;;\;\; 2 x_3 \equiv 1 \quad (\text{mod } 7)$$
$$x_1 \equiv 3 \quad (\text{mod } 5) \;\;;\;\; x_2 \equiv 5 \quad (\text{mod } 6) \;\;;\;\; x_3 \equiv 4 \quad (\text{mod } 7)$$

Finally, $x \equiv M_1 x_1 a_1 + M_2 x_2 a_2 + M_3 x_3 a_3 \;(\text{mod } M)$, that is,

$$x \equiv (42)(3)(3) + (35)(5)(2) + (30)(4)(4)$$

$$\equiv 378 + 350 + 480$$

$$\equiv 1208 \;(\text{mod } 210)$$

Reducing, we obtain

$$x \equiv 158 \pmod{210} \tag{4.67}$$

Check: $158 = 31(5) + 3 = 26(6) + 2 = 22(7) + 4$.

Section 4.4 Exercises

1. Use congruences to find all positive integer solutions, if any, to each of the following equations:

   (a) $2x + 5y = 47$

   (b) $7x + 4y = 59$

   (c) $8x + 5y = 46$

   (d) $4x + 9y = 50$

   (e) $13x + 10y = 499$

   (f) $12x + 17y = 299$

2. Solve each of the following systems of simultaneous congruences:

   $(a) \begin{cases} x \equiv 1 \pmod{4} \\ x \equiv 2 \pmod{7} \end{cases}$
   $(b) \begin{cases} x \equiv 2 \pmod{5} \\ x \equiv 5 \pmod{8} \end{cases}$
   $(c) \begin{cases} x \equiv 2 \pmod{3} \\ x \equiv 3 \pmod{4} \end{cases}$

   $(d) \begin{cases} x \equiv 2 \pmod{5} \\ x \equiv 3 \pmod{7} \\ x \equiv 1 \pmod{8} \end{cases}$
   $(e) \begin{cases} x \equiv 3 \pmod{4} \\ x \equiv 5 \pmod{7} \\ x \equiv 2 \pmod{9} \end{cases}$
   $(f) \begin{cases} x \equiv 1 \pmod{5} \\ x \equiv 3 \pmod{6} \\ x \equiv 2 \pmod{7} \end{cases}$

   $(g) \begin{cases} x \equiv 2 \pmod{4} \\ x \equiv 4 \pmod{5} \\ x \equiv 3 \pmod{7} \end{cases}$
   $(h) \begin{cases} x \equiv 4 \pmod{7} \\ x \equiv 6 \pmod{11} \\ x \equiv 9 \pmod{13} \end{cases}$
   $(i) \begin{cases} x \equiv 1 \pmod{9} \\ x \equiv 2 \pmod{10} \\ x \equiv 3 \pmod{11} \end{cases}$

3. A sack of gold coins is stolen by a gang of nine thieves. If each thief gets an equal share of the coins, then two coins remain. If one thief is caught before the coins are divided so that each of the others gets an equal share, then one coin remains. If two thieves are caught before the coins are divided so that each of the seven others gets an equal share, then no coins remain. Find the smallest possible number of coins in the sack.

4. Old McDonald collected $75 at the market by selling chickens and geese. He got $4 for each chicken and $7 for each goose. How many of each did he sell?

5. Prove that if $m_1, m_2, m_3, ..., m_r$ are odd natural numbers that are pairwise relatively prime, and whose product is M, and if $x \equiv \frac{1}{2}(m_i - 1) \pmod{m_i}$ for each i, then $x \equiv \frac{1}{2}(M - 1) \pmod{M}$.

6. Find an odd integer k such that $k(2^n) - 1$ is composite for all $n \geq 1$.

7. An alternate approach to solving System (4.63) is as follows: Using the notation of Theorem 4.6, for each i such that $1 \leq i \leq r$, multiply the congruence $x \equiv a_i \pmod{m_i}$ by M_i to obtain $M_i x \equiv M_i a_i \pmod{M}$. Now add these r congruences to obtain a single congruence:

$$(\sum_{i=1}^{r} M_i)x \equiv \sum_{i=1}^{r} M_i a_i \pmod{M}$$

It is easily verified that $(\sum_{i=1}^{r} M_i, M) = 1$, so that the last congruence has a unique solution $\pmod{M}$, which is the solution of System (4.63). Using this approach, solve each system of congruences from Exercise 2.

Section 4.4   Computer Exercises

8. Write a computer program to solve systems of three simultaneous linear congruences with moduli that are pairwise relatively prime.

Write a computer program to solve the Diophantine equation $ax + by = c$, where a, b, and c are given and $(a, b) = 1$.

4.5 Polynomial Congruences $(\bmod\ p^n)$

Let $f(x)$ be a polynomial of degree $r \geq 1$ with integer coefficients. Let $m \geq 2$. There are various methods of determining all solutions, if any, of Congruence (4.13), namely,

$$f(x) \equiv 0 \pmod{m}$$

depending on the values of r and m. In Section 4.3 we saw how to solve Congruence (4.13) by trial and error for arbitrary r and m. We also saw how to solve linear congruences (the case $r = 1$) by using the Euclidean Algorithm. In Chapter 7 we will consider quadratic congruences (the case $r = 2$). In this section, where p denotes a prime and n denotes any natural number, we will learn how to obtain the solutions of the congruence

$$f(x) \equiv 0 \pmod{p^{n+1}} \tag{4.68}$$

from the solutions of the congruence

$$f(x) \equiv 0 \pmod{p^n}. \tag{4.69}$$

First of all, note that if Congruence (4.68) has a solution, then by Theorem 4.1, part C_9, so does every congruence of the type

$$f(x) \equiv 0 \pmod{p^k} \tag{4.70}$$

where $1 \leq k \leq n$. In particular, suppose that Congruence (4.69) has only the solutions $x \equiv x_1, x_2, x_3, \ldots, x_m \pmod{p^n}$. If y is a solution of Congruence

(4.68), then y is also a solution of Congruence (4.69). Therefore, there exists t such that $0 \leq t \leq p - 1$ and $y = x_i + tp^n$. The following theorem tells us under what conditions such a t exists and how many such t's exist.

Theorem 4.7

Let $f(x)$ be a polynomial with integer coefficients and degree $r \geq 1$. Let p be prime, $n \geq 1$. Let y be a solution of Congruence (4.68),

$$f(x) \equiv 0 \ (\text{mod } p^{n+1})$$

Then $y = x_i + tp^n \ (\text{mod } p^{n+1})$, where $0 \leq x_i \leq p^n$, and x_i is a solution of Congruence (4.69), namely,

$$f(x) \equiv 0 \ (\text{mod } p^n)$$

such that $0 \leq t \leq p - 1$, and t satisfies the congruence

$$tf'(x_i) \equiv \frac{-f(x_i)}{p^n} \ (\text{mod } p) \tag{4.71}$$

Furthermore, if h is the number of solutions of Congruence (4.71), then

$$h = \begin{cases} 1 \text{ if } p \nmid f'(x_i) \\ 0 \text{ if } p | f'(x_i) \text{ and } p^{n+1} \nmid f'(x_i) \\ p \text{ if } p | f'(x_i) \text{ and } p^{n+1} | f'(x_i) \end{cases}$$

Proof: Let $x_1, x_2, x_3, \ldots, x_m$ be the only solutions of Congruence (4.69). If $f(y) \equiv 0 \ (\text{mod } p^{n+1})$, then also $f(y) \equiv 0 \ (\text{mod } p^n)$, so there exists i such that $1 \leq i \leq m$ and $y \equiv x_i \ (\text{mod } p^n)$. Therefore, $y = x_i + tp^n$ for some t such that $0 \leq t \leq p - 1$. Using a Taylor series expansion, we get

$$f(y) = f(x_i + tp^n) = f(x_i) + tp^n f'(x_i) + \tfrac{1}{2} t^2 p^{2n} f''(x_i) + \ldots \tag{4.72}$$

If we convert this equation to a congruence (mod p^{n+1}) and reduce (mod p^{n+1}), then since $2n \geq n + 1$, we get

$$f(y) \equiv f(x_i) + tp^n f'(x_i) \quad (\text{mod } p^{n+1}) \tag{4.73}$$

Furthermore, since $f(y) \equiv 0 \ (\text{mod } p^{n+1})$ by hypothesis, we obtain

$$tp^n f'(x_i) \equiv -f(x_i) \quad (\text{mod } p^{n+1}) \tag{4.74}$$

Since $f(x_i) \equiv 0 \ (\text{mod } p^n)$, we have $p^n | f(x_i)$. Dividing Congruence (74) by p^n and applying Theorem 4.1, part C_8, we have

$$tf'(x_i) \equiv \frac{-f(x_i)}{p^n} \quad (\text{mod } p) \tag{4.75}$$

Let h be the number of solutions of Congruence (4.75). If $p \nmid f'(x_i)$, then Theorem 4.5 implies that $h = 1$. If $p|f'(x_i)$ and $p^{n+1} \nmid f(x_i)$, then Theorem 4.4 implies that $h = 0$. If $p|f'(x_i)$ and $p^{n+1}|f(x_i)$, then Theorem 4.5 implies that $h = p$, and the solutions of Congruence (4.75) are $t = 0, 1, 2, .., p - 1$.

∎

For example, suppose that we wish to solve the congruence

$$x^3 + 2x + 2 \equiv 0 \ (\text{mod } 49) \tag{4.76}$$

According to Theorem 4.7, we must first solve the congruence

$$x^3 + 2x + 2 \equiv 0 \ (\text{mod } 7) \tag{4.77}$$

By trial and error, the only solutions of Congruence (4.77) are $x_1 \equiv 2 \ (\text{mod } 7)$ and $x_2 \equiv 3 \ (\text{mod } 7)$. In this problem, $f(x) = x^3 + 2x + 2$, $f'(x) = 3x^2 + 2$, $p = 7$, and $n = 1$. If we let $i = 1$, then Congruence (4.75) becomes

$$tf'(2) \equiv \frac{-f(2)}{7} \ (\text{mod } 7) \tag{4.78}$$

Therefore we have

$$14t \equiv -2 \quad (\text{mod } 7) \quad \rightarrow \quad 0t \equiv 5 \quad (\text{mod } 7) \tag{4.79}$$

an impossibility. Therefore there is no solution of Congruence (4.76) corresponding to $x_1 = 2$.

If we let $i = 2$, then Congruence (4.75) becomes

$$tf'(3) \equiv \frac{-f(3)}{7} \ (\text{mod } 7) \tag{4.80}$$

Therefore we have

$$29t \equiv -5 \quad (\text{mod } 7) \rightarrow t \equiv 2 \quad (\text{mod } 7) \tag{4.81}$$

Thus we obtain one corresponding solution of Congruence (4.76), namely,

$$y \equiv 3 + 2(7^1) \equiv 17 \ (\text{mod } 49) \tag{4.82}$$

Check: $f(17) = 17^3 + 2(17) + 2 = 4949 = 49 * 101$.

As a second example, suppose we wish to solve the congruence

$$x^3 + 3x + 5 \equiv 0 \ (\text{mod } 9) \tag{4.83}$$

According to Theorem 4.7, we must first solve the congruence

$$x^3 + 3x + 5 \equiv 0 \ (\text{mod } 3) \tag{4.84}$$

By trial and error, $x_1 \equiv 1 \pmod 3$ is the unique solution of Congruence (4.84). In this problem, $f(x) = x^3 + 3x + 5$, $f'(x) = 3x^2 + 3$, $p = 3$, and $n = 1$. Now Congruence (4.75) becomes

$$tf'(1) \equiv \frac{-f(1)}{3} \pmod 3 \tag{4.85}$$

Therefore we have

$$6t \equiv -3 \pmod 3 \quad \rightarrow \quad 0t \equiv 0 \pmod 3 \tag{4.86}$$

There are three solutions, namely, $t \equiv 0, 1, 2 \pmod 3$. The corresponding solutions of Congruence (4.83) are

$$y_1 \equiv 1 + 0(3) \equiv 1 \pmod 9$$

$$y_2 \equiv 1 + 1(3) \equiv 4 \pmod 9$$

$$y_3 \equiv 1 + 2(3) \equiv 7 \pmod 9$$

Check: $f(1) = 9$; $\quad f(4) = 81 = 9 * 9$; $\quad f(7) = 369 = 41 * 9$

As a third example, suppose we wish to solve the congruence

$$x^3 - 5x + 1 \equiv 0 \pmod{27} \tag{4.87}$$

By repeated application of Theorem 4.7, if Congruence (4.87) has a solution, then so does

$$x^3 - 5x + 1 \equiv 0 \pmod 3 \tag{4.88}$$

By inspection, the unique solution of Congruence (4.88) is

$$x \equiv 1 \pmod 3 \tag{4.89}$$

We now seek solutions of the congruence

$$x^3 - 5x + 1 \equiv 0 \pmod 9 \tag{4.90}$$

According to Theorem 4.7, any solution of Congruence (4.90) must have the form

$$x \equiv 1 + 3t \pmod 9 \tag{4.91}$$

where t satisfies the congruence

$$tf'(1) \equiv \frac{-f(1)}{3} \pmod 3 \tag{4.92}$$

Here $f(x) = x^3 - 5x + 1$ and $f'(x) = 3x^2 - 5$, so $f(1) = -3$ and $f'(1) = -2$. Therefore, Congruence (4.92) becomes

$$-2t \equiv 1 \pmod 3 \tag{4.93}$$

or

$$t \equiv 1 \pmod 3 \qquad (4.94)$$

Thus

$$x \equiv 4 \pmod 9$$

is the unique solution of Congruence (4.90).

According to Theorem 4.7, any solution of Congruence (4.87) must have the form

$$x \equiv 4 + 9t \pmod{27} \qquad (4.95)$$

where t satisfies the congruence

$$tf'(4) \equiv \frac{-f(4)}{9} \pmod 3 \qquad (4.96)$$

That is,

$$43t \equiv -5 \pmod 3 \qquad (4.97)$$

or

$$t \equiv 1 \pmod 3 \qquad (4.98)$$

Therefore, the unique solution of Congruence (4.87) is

$$x \equiv 13 \pmod{27} \qquad (4.99)$$

Section 4.5 Exercises

Find all solutions, if any, of each of the following congruences.

1. $x^2 + 3x + 1 \equiv 0 \pmod{121}$
2. $x^3 - x + 4 \equiv 0 \pmod{25}$
3. $x^4 - 4x - 2 \equiv 0 \pmod{25}$
4. $x^3 - x^2 + 6 \equiv 0 \pmod{16}$
5. $x^3 + x + 2 \equiv 0 \pmod 8$
6. $x^5 - 5x + 3 \equiv 0 \pmod{25}$
7. $x^3 + x + 1 \equiv 0 \pmod{81}$
8. $2x^3 - x + 1 \equiv 0 \pmod{169}$
9. $2x^3 - 3x^2 + 1 \equiv 0 \pmod{49}$
10. $2x^3 - 3x^2 - 6 \equiv 0 \pmod{49}$

Section 4.5 Computer Exercises

11. Let p be an odd prime. Let a, b, c, and d be integers such that $(a, p) = 1$. Let $f(x) = ax^3 + bx^2 + cx + d$. Write a computer program that will find all solutions, if any, to the congruence $f(x) \equiv 0 \pmod p$ by trial and error, that is, by successive evaluation of $f(0), f(1), f(2), ..., f(p-1)$.

12. With a, b, c, d, p, and $f(x)$ as in Exercise 11, write a computer program that will find all solutions, if any, to the congruence $f(x) \equiv 0 \pmod{p^2}$.

13. Apply your program from Exercise 12 to find all solutions to the congruence $x^3 \equiv 1 \pmod{p}$ for each prime p such that $5 \le p \le 61$. Is there a pattern?

4.6 Modular Arithmetic; Fermat's Theorem

If the natural number $m > 1$, let us return to the set of least positive residues (mod m), namely, $\{0, 1, 2, ..., m-1\}$, which we denote Z_m. (Algebraists would prefer the notation Z/mZ.) Let us define addition (mod m) on Z_m as follows:

Definition 4.7   Addition Modulo m

If a and b belong to Z_m, then $a \oplus b$ is the least positive residue ((mod m)) of $a + b$. That is to say,

$$a \oplus b = \begin{cases} a + b & \text{if } 0 \le a + b < m \\ a + b + m & \text{if } m \le a + b < 2m \end{cases}$$

For example, in Z_7, we have

$$2 \oplus 4 = 6$$
$$3 \oplus 5 = 1 \quad \text{since } 3 + 5 = 8 \equiv 1 \pmod 7$$
$$6 \oplus 1 = 0 \quad \text{since } 6 + 1 = 7 \equiv 0 \pmod 7$$

Also, in Z_{10}, we have

$$2 \oplus 3 = 5$$
$$5 \oplus 7 = 2 \quad \text{since } 5 + 7 = 12 \equiv 2 \pmod{10}$$
$$8 \oplus 9 = 7 \quad \text{since } 8 + 9 = 17 \equiv 7 \pmod{10}$$

If m is small, it is convenient to describe addition (mod m) by means of an addition table. For example, if $m = 3$, we have

$\oplus$	0	1	2
0	0	1	2
1	1	2	0
2	2	0	1

In general, in such a table one finds $a \oplus b$ by looking for the entry that is in the same row as a and in the same column as b. Glancing at the table above, we see that $1 \oplus 2 = 0$ in Z_3.

One can verify that on Z_m, addition (mod m) has the following properties:

A$_1$: (Commutative property) $a \oplus b = b \oplus a$ for all a and b in Z_m.

A$_2$: (Associative property) $(a \oplus b) \oplus c = a \oplus (b \oplus c)$ for all a, b, and c in Z_m.

A$_3$: (Identity element) $a \oplus 0 = 0 \oplus a = a$ for all a in Z_m.

A$_4$: (Inverse element) For all in Z_m, there is b in Z_m such that $a \oplus b = 0$.

For example, in Z_7, the additive inverse of 2 is 5, since $2 \oplus 5 = 0$; likewise, the additive inverse of 6 is 1, since $6 \oplus 1 = 0$.

If you are familiar with abstract algebra, you should realize that properties A$_1$ through A$_4$ imply that Z_m is an Abelian group of order m with respect to addition (mod m). In fact, if a is a nonzero element of Z_m, then a may be represented as a sum of a 1's. (For example, if $m \geq 4$, then $3 = 1 \oplus 1 \oplus 1$.) This implies that the additive group of Z_m is the cyclic group of order m.

Similarly, we define multiplication (mod m) on Z_m as follows:

Definition 4.8   Multiplication Modulo m

If a and b belong to Z_m, then $a \odot b$ is the least positive residue (mod m) of ab.

For example, in Z_7, we have

$$4 \odot 6 = 3 \quad \text{since } 4(6) = 24 \equiv 3 \pmod 7$$
$$3 \odot 3 = 2 \quad \text{since } 3(3) = 9 \equiv 2 \pmod 7$$
$$2 \odot 4 = 1 \quad \text{since } 2(4) = 8 \equiv 1 \pmod 7$$

Also, in Z_{10}, we have

$$2 \oplus 3 = 6$$
$$5 \oplus 7 = 5 \quad \text{since } 5(7) = 35 \equiv 5 \pmod{10}$$
$$8 \oplus 9 = 2 \quad \text{since } 8(9) = 72 \equiv 2 \pmod{10}$$

Again, if m is small, it is convenient to describe addition (mod m) by means of an addition table. For example, if $m = 3$, we have

$\odot$	0	1	2
0	0	0	0
1	0	1	2
2	0	2	1

One can verify that multiplication (mod m) on Z_m has the following properties:

M_1: (Commutative property) $a \odot b = b \odot a$ for all a and b in Z_m.

M_2: (Associative property) $(a \odot b) \odot c = a \odot (b \odot c)$ for all a, b, and c in Z_m.

M_3: (Identity element) $a \odot 1 = 1 \odot a = a$ for all a in Z_m.

M_4: (Distributive property) $a \odot (b \oplus c) = (a \odot b) \oplus (a \odot c)$ for all a, b, and c in Z_m.

Using the terminology of abstract algebra, one would say that under the operations of addition and multiplication (mod m), Z_m is a commutative ring with identity.

Also, if $m = p$, a prime, then Z_p has an additional propxety:

M_5: (Multiplicative inverse) For all a in Z_p such that $a \neq 0$, there exists b in Z_p such that $a \odot b = b \odot a = 1$.

In this case, the set of nonzero elements of Z_p forms an Abelian group under multiplication (mod p). This implies that Z_p is a special kind of ring, namely, a field.

For example, in Z_{11}, the multiplicative inverse of 3 is 4, since $3 \odot 4 = 1$; also, the multiplicative inverse of 5 is 9, since $5 \odot 9 = 1$.

Let us look at the multiplication table for multiplication (mod 7) in Z_7:

$\odot$	0	1	2	3	4	5	6
0	0	0	0	0	0	0	0
1	0	1	2	3	4	5	6
2	0	2	4	6	1	3	5
3	0	3	6	2	5	1	4
4	0	4	1	5	2	6	3
5	0	5	3	1	6	4	2
6	0	6	5	4	3	2	1

Suppose we wish to compute 5^{20} (mod 7). Using the table above, we see that

$$5^2 \equiv 4 \pmod{7}$$
$$5^4 = (5^2)^2 \equiv 4^2 \equiv 2 \pmod{7}$$
$$5^5 = 5 \cdot 5^4 \equiv 5^2 \equiv 3 \pmod{7}$$
$$5^{10} = (5^5)^2 \equiv 3^2 \equiv 2 \pmod{7}$$
$$5^{20} = (5^{10})^2 \equiv 2^2 \equiv 4 \pmod{7}$$

There is a shortcut, however. If we notice that $5^6 \equiv 1 \pmod 7$, we see that $5^{20} \equiv (5^6)^3 5^2 \equiv 1^3 5^2 = 5^2 \equiv 4 \pmod 7$. More generally, if $(a, 7) = 1$, then $a^6 \equiv 1 \pmod 7$. This congruence is a special case of an important theorem that was discovered by the French mathematician Pierre de Fermat (1601–1665).

Theorem 4.8

Fermat's "Little" Theorem

If p is prime and $(a, p) = 1$, then $a^{p-1} \equiv 1 \pmod p$.

Proof: We will use induction on a to prove that if p is prime, then $a^p \equiv a \pmod p$ for all a. If also $(a, p) = 1$, then by Theorem 4.1, part C_7, a can be cancelled out of this congruence, leaving the desired result

$$a^{p-1} \equiv a \pmod p$$

Since $0^p = 0$, it follows that $0^p \equiv 0 \pmod p$. Next, assuming that $a^p \equiv a \pmod p$, we wish to show that $(a + 1)^p \equiv a + 1 \pmod p$. However,

$$(a + 1)^p = a^p + 1 + \sum_{j=1}^{p-1} \binom{p}{j} a^{p-j}$$

by the Binomial Theorem (Theorem 1.16, part B_3). Theorem 3.18 implies that $\binom{p}{j} \equiv 0 \pmod p$ if $1 \leq j \leq p-1$. Therefore, $(a+1)^p \equiv a^p + 1 \pmod p$. Since $a^p \equiv a \pmod p$, by the induction hypothesis, we have $(a+1)^p \equiv a+1 \pmod p$. ∎

In view of the significance of Fermat's Little Theorem, we now present a second proof:

Second Proof: If p is prime and $p \nmid a$, let

$$T(a, p) = \{a,\ 2a,\ 3a,\ \cdots,\ (p-1)a\}$$

and let $S(a, p)$ be the set of least positive residues (mod p) of the elements of $T(a, p)$. If j is an integer such that $1 \leq j \leq p - 1$, then $p \nmid ja$. Therefore $0 \notin S(a, p)$. Furthermore, if $1 \leq x_1 \leq x_2 \leq p - 1$ and $ax_1 \equiv ax_2 \pmod p$, then, cancelling a, we have $x_1 \equiv x_2 \pmod p$. Since $0 \leq x_2 - x_1 \leq p - 2$, this implies $x_1 = x_2$. Therefore the elements of $T(a, p)$, and also of $S(a, p)$, are distinct (mod p). This implies that the elements of $S(a, p)$ are a permutation, that is, a rearrangement of $\{1, 2, 3, \cdots, p-1\}$. It now follows that

$$\prod_{j=1}^{p-1} aj \equiv \prod_{j=1}^{p} j \quad (\text{mod } p)$$

Simplifying, we have

$$a^{p-1}(p-1)! \equiv (p-1)! \quad (\text{mod } p)$$

Since $(p, (p-1)!) = 1$, we may cancel the common factor $(p-1)!$, obtaining

$$a^{p-1} \equiv 1 \quad (\text{mod } p)$$

The following theorem is a consequence of Theorem 4.8.

Theorem 4.9   If $a \geq 2$ and p and q are primes such that $q|(a^p - 1)$, then $q|(a-1)$ or $p|(q-1)$.

Proof: Exercise.

A particular case of Theorem 4.9 is the following:

Theorem 4.10   If p, q are odd primes such that $q|(2^p - 1)$, then $2p|(q-1)$.
(In other words, we must have $q = 1 + 2kp$ for some integer $k \geq 1$.)

Proof: This follows by setting $a = 2$ in Theorem 4.9 and noting that $2|(q-1)$.

∎

For example, 11 is a prime and $2^{11} - 1 = 2047 = 23 * 89$. Note that each of the prime factors, namely 23 and 89, is 1 more than a multiple of 22, since $23 = 1 + 1(22)$; $89 = 1 + 4(22)$.

Pierre de Fermat (1601–1665)

Pierre de Fermat was born on August 20, 1601 in the village of Beaumont-de-Lomanges (near the city of Toulouse) in southern France. His father was a prosperous leather merchant, and his mother came from a family with good legal connections. In his youth, Fermat received a good classical education: he learned Spanish, Italian, Latin, and Greek, in addition to French. In the late 1620s, he spent some time in Bordeaux, where he studied in depth the mathematical works of François Vieta (1540–1603).

The year 1631 was eventful for Fermat. He received a bachelor of civil law degree from the University of Orleans, married his mother's cousin, Louise de Long, and assumed the offices of Conseiller and Commissaire aux Enquetes in the provincial parliament of Toulouse. His duties included examining petitions addressed to the king. Fermat's legal and parliamentary career prospered as he rose through the ranks in a system based on seniority. In 1638 he became Conseiller aux Enquetes; in 1642 he became a member of the criminal court and the Grande Chambre; and in 1648, he was appointed King's Counselor. At home, he and his wife raised a family of two sons and three daughters.

All the while, Fermat remained devoted to mathematics, his great avocation. Before Newton and Leibniz were born, Fermat conceived some of the main ideas in differential and integral calculus, such as finding the tangent line to a curve at a given point, finding the maximum and minimum values of a polynomial function, and finding the area of a region bounded by a curve. Independently from Descartes, he invented analytic geometry. Together with Pascal, he invented probability theory.

In addition, he made many significant contributions to number theory, such as Fermat's "Little" Theorem (Theorem 4.8), which was later proved by Euler; the method of "descent" for solving Diophantine equations; and the discovery that if p is prime and $p \equiv 1 \pmod 4$, then p is a sum of two squares. As we mentioned in Chapter 1, Fermat's Last Theorem, which states that the equation

$$x^n + y^n = x^n$$

has no solution in positive integers x, y, z if $n \geq 3$, has only recently been proved.

Indeed, Fermat may be said to be responsible for the establishment of number theory as a mathematical discipline in its own right. In a letter to Frenicle de Bessy written in February of 1657, Fermat says:

> There is scarcely anyone who states purely arithmetic questions, scarcely anyone who understands them. Is this not because arithmetic has been treated up to this time geometrically rather than arithmetically? This is certainly indicated by many works ancient and modern. Diophantus himself also indicates this. But he has freed himself from geometry a little more than the others have, in that he limits his analysis to rational numbers only; nevertheless the Zetetica of Vieta, in which the methods of Diophantus are extended to continuous magnitude and therefore to geometry, witness the insufficient separation of arithmetic from geometry.
>
> Now arithmetic has a special domain of its own, the theory of numbers. This was touched upon but only to a slight degree by Euclid in his Elements, and by those who follow him it has not been sufficiently extended, unless perchance it lies hid in those books of Diophantus which the ravages of time have destroyed. Arithmeticians have now to develop or restore it.

Fermat communicated his mathematical discoveries via personal correspondence with other mathematicians such as Descartes, Mersenne, Frenicle, and Roberval. Although Fermat did not know English, he communicated by means of an intermediary with the prominent English mathematicians John Wallis and Lord William Brouncker. (John Wallis (1616–1703) was Savilian Professor of Geometry at Oxford University.) In 1657 Fermat challenged Wallis to prove that if D is not a square, then the equation

$$x^2 - Dy^2 = 1$$

has infinitely many integer solutions. (This equation is known as Pell's equation, although John Pell (1611–1685) had little or nothing to do with it.) Later that same year, Brouncker provided the proof that Fermat requested. In a letter of congratulation to Brouncker, Wallis said that he had "preserved untarnished the fame which Englishmen have won in former times with Frenchmen and has shown that England's champions of wisdom are just as strong as those in war."

Fermat was the last amateur to make significant contributions to number theory. He died in Castres on January 12, 1665. His mathematical papers were published posthumously by his eldest son, Clement-Samuel.

Section 4.6 Exercises

1. (a) Construct the table for multiplication (mod 5) in Z_5.

   (b) Find the multiplicative inverse of every nonzero element.

2. (a) Construct the table for multiplication (mod 11) in Z_{11}.

   (b) Find the multiplicative inverse of every nonzero element.

3. Construct the table for multiplication (mod 10) in Z_{10}. Which elements have multiplicative inverses?

4. Prove that if p is prime and $(a, p) = 1$, then the congruence

   $$ax \equiv b \pmod{p}$$

   has the solution

   $$x \equiv a^{p-2}b \pmod{p}$$

5. Prove Theorem 4.9.   (Hint: Use Theorem 2.12.)

6. If p is prime and $a \geq 2$, prove that

   $$\left(a - 1, \frac{a^p - 1}{a - 1}\right) = \begin{cases} p & \text{if } p \mid (a - 1) \\ 1 & \text{if } p \nmid (a - 1) \end{cases}$$

★7. We say that the prime q is a *primitive factor* of $a^n - 1$ if $q \mid (a^n - 1)$ but $q \nmid (a^m - 1)$ for all m such that $0 < m < n$. Prove that if p is prime and $a \geq 2$, then $a^p - 1$ has a primitive factor.

8. Prove that if p is prime, $n \geq 1$, and $a^d \equiv 1 \pmod{p^n}$, then $a^{pd} \equiv 1 \pmod{p^{n+1}}$.

Section 4.6 Computer Exercises

9. Let $a \geq 2$ and let p be an odd prime. Write a computer program to find the least primitive factor of $a^p - 1$.

10. We wish to determine whether the given odd integer n is prime or composite. If $2^{n-1} \not\equiv 1 \pmod{n}$, then by Theorem 4.8, n is composite. For odd $n < 100$, test the conjecture that if $2^{n-1} \equiv 1 \pmod{n}$, then n is prime. That is, write a computer program that, for each odd composite value of n below 1000, verifies whether $2^{n-1} \equiv 1 \pmod{n}$.

(The exceptional values of n, if any, are called *pseudoprimes to the base 2*.)

4.7 Wilson's Theorem and Fermat Numbers

Quadratic congruences will be studied in detail in Chapter 7, but we are ready to obtain some preliminary results.

Theorem 4.11   Let p be an odd prime. Then $x^2 \equiv a^2 \pmod{p}$ if and only if $x \equiv \pm a \pmod{p}$.

Proof: If $x \equiv \pm a \pmod{p}$, then Theorem 4.1, part C_4 implies that $x^2 \equiv a^2 \pmod{p}$. Conversely, if $x^2 \equiv a^2 \pmod{p}$, $p|(x^2 - a^2)$; that is, $p|(x-a)(x+a)$. By Theorem 3.2, $p|(x-a)$ or $p|(x+a)$, so we have $x - a \equiv 0 \pmod{p}$ or $x + a \equiv 0 \pmod{p}$; that is, $x \equiv \pm a \pmod{p}$. ∎

For example, if $x^2 \equiv 4 \pmod{11}$, then $x \equiv \pm 2 \pmod{11}$; also, if $x^2 \equiv 9 \pmod{11}$, then $x \equiv \pm 3 \pmod{11}$. Note that if m is composite and $x^2 \equiv a^2 \pmod{m}$, then it does not follow that $x \equiv \pm a \pmod{m}$. For example, $3^2 \equiv 1^2 \pmod{8}$, yet $3 \not\equiv \pm 1 \pmod{8}$.

The following theorem concerns odd prime divisors of a sum of two relatively prime squares.

Theorem 4.12   If p is an odd prime, $p|(a^2 + b^2)$, and $(a, b) = 1$, then $p \equiv 1 \pmod{4}$.

Proof: By hypothesis, $p|(a^2 + b^2)$. If $p|a$, then $p|a^2$, so $p|b^2$ and $p|b$. However, then $(a, b) \geq p$, contrary to hypothesis. Therefore, $p \nmid a$. Similarly, $p \nmid b$.

Since $p|(a^2 + b^2)$, we have $-a^2 \equiv b^2 \pmod{p}$. Therefore, $(-a^2)^{\frac{1}{2}(p-1)} \equiv (b^2)^{\frac{1}{2}(p-1)} \pmod{p}$; that is, $(-1)^{\frac{1}{2}(p-1)}a^{p-1} \equiv b^{p-1} \pmod{p}$. Since $p \nmid ab$, Fermat's "Little" Theorem (Theorem 4.8) implies that $a^{p-1} \equiv b^{p-1} \equiv 1 \pmod{p}$. Therefore, $(-1)^{\frac{1}{2}(p-1)} \equiv 1 \pmod{p}$. If $p \equiv 3 \pmod 4$, then $(-1)^{\frac{1}{2}(p-1)} = -1$, so we would have $-1 \equiv 1 \pmod{p}$ and hence $p|2$, an impossibility. Therefore, $p \equiv 1 \pmod 4$. ∎

For example, $20^2 + 9^2 = 481 = 13 \cdot 37$. Note that each of the prime factors, 13 and 37, is congruent to 1 (mod 4).

The following theorem, which is named after the English mathematician John Wilson (1741–1793), was first proved by Lagrange.

Theorem 4.13

Wilson's Theorem

If p is prime, then $(p-1)! \equiv -1 \pmod{p}$.

Proof: One can easily verify that Wilson's theorem holds if $p = 2$ or 3. Now suppose that $p \geq 5$. Since $p - 1 \equiv -1 \pmod{p}$ and $(p-1, p) = 1$, it suffices by virtue of Theorem 4.1, part C_7 to show that $(p-2)! \equiv 1 \pmod{p}$. By Theorem 4.5, for each j such that $1 \leq j \leq p - 1$, there exists a unique k such that $1 \leq k \leq p-1$ and $jk \equiv 1 \pmod{p}$. If $k = j$, then $j^2 \equiv 1 \pmod{p}$, so by Theorem 4.10, $j = 1$ or $p - 1$. Therefore, if $2 \leq j \leq p - 2$, there exists unique $k \neq j$ such that $jk \equiv 1 \pmod{p}$ and $2 \leq k \leq p - 2$. If we multiply together all $\frac{1}{2}(p-3)$ such congruences, we obtain $(p-1)! \equiv -1 \pmod{p}$. ∎

For an alternate proof of Theorem 4.13, see [14].

For example, if $p = 11$, we have

$$2 \cdot 6 \equiv 1 \pmod{11}$$
$$3 \cdot 4 \equiv 1 \pmod{11}$$
$$5 \cdot 9 \equiv 1 \pmod{11}$$
$$7 \cdot 8 \equiv 1 \pmod{11}$$

Multiplying, we obtain

$$9! = 2 \cdot 3 \cdot 4 \cdot 5 \cdot 6 \cdot 7 \cdot 8 \cdot 9 \equiv 1 \pmod{11}$$

Next, we apply Theorems 4.12 and 4.13 to solve a particular quadratic congruence.

Theorem 4.14

If p is an odd prime, then the congruence

$$x^2 \equiv -1 \pmod{p}$$

has the solutions $x \equiv \pm[(p-1)/2]! \pmod{p}$ if $p \equiv 1 \pmod 4$, and has no solutions if $p \equiv 3 \pmod 4$.

Proof: If $x^2 \equiv -1 \pmod p$, then $p|(x^2+1)$. Since $(x,1)=1$, Theorem 4.12 implies that $p \equiv 1 \pmod 4$. By Theorem 4.11, it suffices to show that if $a = [(p-1)/2]!$, then $a^2 \equiv -1 \pmod p$. Let $ab = (p-1)!$ so that

$$b = \frac{(p-1)!}{a} = \prod_{j=1}^{\frac{1}{2}(p-1)} (p-j) \equiv \prod_{j=1}^{\frac{1}{2}(p-1)} (-j) \pmod p$$

so

$$b \equiv (-1)^{\frac{1}{2}(p-1)} \prod_{j=1}^{\frac{1}{2}(p-1)} j \equiv (-1)^{\frac{1}{2}(p-1)} \left(\frac{p-1}{2}\right)! \equiv (-1)^{\frac{1}{2}(p-1)} a \pmod p$$

Since $p \equiv 1 \pmod 4$, $(-1)^{\frac{1}{2}(p-1)} = 1$, so $b \equiv a \pmod p$. Therefore, $a^2 \equiv ab \equiv (p-1)! \equiv -1 \pmod p$ by Wilson's Theorem (Theorem 4.13). ∎

For example, let us solve the quadratic congruence

$$x^2 \equiv -1 \pmod{13}$$

Here $p = 13$, so $[(p-1)/2]! = 6! = 720$. Now $720 \equiv 5 \pmod{13}$, so the solutions are

$$x \equiv \pm 5 \pmod{13}$$

Note that while Theorem 4.14 gives an explicit formula for the solution of the quadratic congruence, the formula is impractical for large p because of the large amount of multiplication needed to compute $[(p-1)/2]! \pmod p$. In Chapter 7 we will present an efficient method for finding solutions to quadratic congruences when these solutions exist.

The result of Section 3.2, Exercise 2 implies that if $2^m + 1$ is prime, then $m = 2^n$ for some $n \geq 0$. The numbers $2^{2^n} + 1$ were first discussed by Fermat and are now named after him.

Definition 4.9   Fermat Numbers

If $n \geq 0$, let $f_n = 2^{2^n} + 1$. Then f_n is called the nth Fermat number. If f_n is prime, then we call it a *Fermat prime*.

Let us compute the first few Fermat numbers.

$$f_0 = 2^{2^0} + 1 = 2^1 + 1 = 2 + 1 = 3$$
$$f_1 = 2^{2^1} + 1 = 2^2 + 1 = 4 + 1 = 5$$
$$f_2 = 2^{2^2} + 1 = 2^4 + 1 = 16 + 1 = 17$$
$$f_3 = 2^{2^3} + 1 = 2^8 + 1 = 256 + 1 = 257$$
$$f_4 = 2^{2^4} + 1 = 2^{16} + 1 = 65536 + 1 = 65537$$

Each of these Fermat numbers is prime. In a letter to Pascal dated August 29, 1654, Fermat stated that every Fermat number is prime. This is the only statement ever made by Fermat that is known to be false. In 1732, Euler showed that f_5 is composite. Indeed, $f_5 = 2^{2^5} + 1 = 2^{32} + 1 = 4294967297 = 641 \cdot 6700417$. It is known that f_n is composite for all n such that $5 \leq n \leq 32$ and for many other values of n. It is not known whether f_n is prime for any $n \geq 33$.

There is an interesting connection between Fermat primes and a classical problem in plane geometry. The problem is to find all odd integers n such that a regular n-sided polygon can be inscribed in a circle using just a straight edge and compass. (A *regular polygon* has sides that all have the same length as well as angles that all have the same measure.) Amazingly, if n is odd and $n \geq 3$, then a regular n-sided polygon can be inscribed in a circle if and only if n is a product of one or more distinct Fermat primes. The proof of this assertion involves some deep results of abstract algebra and is therefore omitted. (see [4], Article 365).

Next, we show that distinct Fermat numbers are relatively prime.

Theorem 4.15   If $m \neq n$, then $(f_m, f_n) = 1$.

Proof: Without loss of generality, assume that $m > n \geq 0$, so $m = n + k$ for some $k \geq 1$. Suppose that p is prime and $p | (f_m, f_n)$. Since f_m and f_n are odd, p is odd. Since $p | f_n$, we have

$$2^{2^n} \equiv -1 \pmod{p}$$

Therefore,

$$(2^{2^n})^{2^k} \equiv (-1)^{2^k} \pmod{p}$$

That is,

$$2^{2^{n+k}} \equiv 1 \pmod{p}$$

or

$$2^{2^m} \equiv 1 \pmod{p}$$

But $p|f_m$ implies $2^{2^m} \equiv -1 \pmod{p}$. Therefore $-1 \equiv 1 \pmod{p}$, which implies that $p = 2$, an impossibility. We conclude that (f_m, f_n) has no prime divisor, so $(f_m, f_n) = 1$. ∎

We can now prove a theorem that complements Theorem 3.11.

Theorem 4.16   There exist infinitely many primes of the form $4k + 1$.

Proof: If $n \geq 1$, let $f_n = 2^{2^n} + 1 = (2^{2^{n-1}})^2 + 1$. Let p_n be the least prime factor of f_n. Since f_n is odd, p_n must be odd. Theorem 4.12 implies that $p_n \equiv 1 \pmod{4}$. Theorem 4.15 implies that $p_n \nmid f_m$ if $m \neq n$. Therefore, $p_n \neq p_m$ if $m \neq n$. Thus the least prime factors p_n of the Fermat numbers f_n with $n \geq 1$ yield an infinite sequence of primes of the form $4k + 1$. ∎

For an alternate proof of Theorem 4.16, which makes use of Fibonacci numbers, see [13].

Section 4.7   Exercises

1. Prove that if n is composite and $n > 4$, then $(n - 1)! \equiv 0 \pmod{n}$. (This is the converse of Wilson's theorem.)

2. Obtain an alternate proof of Theorem 4.16 as follows: Suppose that q_1, q_2, q_3, ... , q_n are the only primes of the form $4k + 1$. Let Q be the product of these primes, and look at $4Q^2 + 1$.

3. Prove that if p is prime, $p \equiv 3 \pmod{4}$, and $p|(a^2 + b^2)$, then $p^2|(a^2 + b^2)$.

4. Let p be prime. Prove that $p \equiv 1 \pmod{4}$ if and only if there exist two consecutive squares whose sum is divisible by p.

5. Prove that if $2^j + 3 = 7^k$, then $j = 2$ and $k = 1$.

6. Let f_n be the nth Fermat number. Prove that if $m > n$, then $f_m \equiv 2 \pmod{f_n}$.

★7. Let p be an odd prime. Say that p has the inverse property if for every j such that $1 < j < \frac{1}{2}p$ there exists k such that $\frac{1}{2}p < k < p - 1$ and $jk \equiv 1 \pmod{p}$. Prove that 5, 7, and 13 are the only primes that have the inverse property.

Section 4.7   Computer Exercises

Consider the congruence

$$x^2 \equiv -1 \pmod{p}$$

where p is a prime such that $p \equiv 1 \pmod{4}$.

8. Write a computer program to solve the congruence above using the formula of Theorem 4.15, namely,

$$x \equiv \pm[(p-1)/2]! \pmod{p}$$

9. Write a computer program to solve the congruence above as follows: Using trial and error, find an integer $q \geq 2$ such that $q^{\frac{1}{2}(p-1)} \equiv -1 \pmod{p}$. Then the solutions are $x \equiv \pm q^{\frac{1}{4}(p-1)} \pmod{p}$.

10. Solve the congruence above with $p = 241$ using the computer programs from Exercises 8 and 9. Which is faster?

4.8 Pythagorean Equation

If a right triangle has legs of lengths x and y and hypotenuse of length z, then x, y, and z satisfy the Pythagorean equation

$$x^2 + y^2 = z^2. \tag{4.100}$$

This statement is one of the most important theorems of plane geometry, and it was introduced to the western world by Pythagoras (569–470 B.C.). In number theory, we are interested in positive integer solutions of this equation, which we call *Pythagorean triples*. The Pythagorean triple with minimal z is $x = 4$, $y = 3$, and $z = 5$.

Suppose that $\{x, y, z\}$ is a Pythagorean triple such that

$$(x, y, z) = d > 1$$

Then there exist integers r, s, and t such that

$$x = dr$$

$$y = ds$$

$$z = dt,$$

and

$$(r, s, t) = 1$$

Now

$$(dr)^2 + (ds)^2 = (dt)^2$$

so

$$r^2 + s^2 = t^2$$

This implies that all solutions of the Pythagorean equation (100) may be obtained from just those solutions which we call *primitive*, that is, those such that $(x, y, z) = 1$.

For example, the minimal solution is primitive, since $(4, 3, 5) = 1$, but the solution $x = 8$, $y = 6$, and $z = 10$ is not primitive, since $(8, 6, 10) = 2 > 1$. If $D = (x, y)$ and the Pythagorean equation (4.100) holds, then also $D|z$, so $D|(x, y, z)$; that is, $D|d$. We therefore note that a solution of the Pythagorean equation (4.100) is primitive if and only if $(x, y) = 1$. Similarly, we have $(x, z) = (y, z) = 1$.

Suppose that $\{x, y, z\}$ is a primitive Pythagorean triple. Therefore, x and y are not both even. If x and y are both odd, then $x^2 \equiv y^2 \equiv 1 \pmod{4}$, so $z^2 \equiv 2 \pmod 4$, an impossibility. Therefore, one of x and y, say x, is even, while y is odd. This implies that z^2 is odd, and hence z is odd. If we write

$$x^2 = z^2 - y^2$$

and let

$$x = 2w$$

we get

$$4w^2 = z^2 - y^2 = (z + y)(z - y)$$

Since y and z are both odd,

$$\frac{1}{2}(z + y) \ and \ \frac{1}{2}(z - y) \text{are integers}$$

Therefore,

$$w^2 = (\frac{z + y}{2})(\frac{z - y}{2})$$

Let

$$t = (\frac{z + y}{2}, \frac{z - y}{2})$$

Therefore,

$$t | \frac{z + y}{2} \ \text{and} \ t | \frac{z - y}{2}$$

$$t | z \ \text{and} \ t | y$$

Thus,

$$t | (y, z)$$

However,

$$(y, z) = (x, y, z) = 1$$

so

$$t = 1$$

Now Theorem 3.19 implies that

$$\frac{1}{2}(z + y) = a^2$$

and

$$\frac{1}{2}(z - y) = b^2$$

with

$$(a, b) = 1$$

Adding these equations, we get

$$z = a^2 + b^2$$

subtracting, we get

$$y = a^2 - b^2$$

Since

$$y < z, \text{ we must have } b > 0.$$

Since

$$w^2 = a^2 b^2, \text{ we have } w = ab, so x = 2ab.$$

Therefore, since

$$y > 0, \text{ we must have } a > b.$$

Since

$$z \text{ is odd, we must have } a \not\equiv b (mod\, 2).$$

To recapitulate, if x, y, and z are natural numbers such that

$$(x, y) = 1$$

and

$$x^2 + y^2 = z^2$$

then

$$x = 2ab$$

$$y = a^2 - b^{2'}$$

$$z = a^2 + b^2$$

where a and b are natural numbers such that

$$a > b$$

$$(a, b) = 1$$

and

$$a \not\equiv b \; (\text{mod} 2)$$

Clearly, the converse also holds. We therefore conclude the following:

Theorem 4.17   Let x and y be natural numbers such that $(x, y) = 1$. Then $x^2 + y^2 = z^2$ if and only if $x = 2ab$, $y = a^2 - b^2$, and $z = a^2 + b^2$, where the natural numbers a, b satisfy $a > b > 0$, $(a, b) = 1$, $a \not\equiv b \; (\text{mod } 2)$.

We can easily generate infinitely many primitive solutions of the Pythagorean equation (100) by letting $a = 2n$ and $b = 1$ or $a = 2n + 1$ and $b = 2$ and letting n range through all natural numbers. We list all primitive solutions of the Pythagorean equation (100) such that $a \leq 7$ in Table 4.1 below.

Table 4.1
Primitive
Pythagorean triples
such that $a \leq 7$

a	b	x	y	z
2	1	4	3	5
3	2	12	5	13
4	1	8	15	17
4	3	24	7	25
5	2	20	21	29
5	4	40	9	41
6	1	12	35	37
6	5	60	11	61
7	2	28	45	53
7	4	56	33	65
7	6	84	13	85

Next, we consider the related equation

$$x^4 + y^4 = z^2 \qquad (4.101)$$

Let $d = (x, y)$, so $d|x$, $d|y$, $d^4|z^2$, and $d^2|z$. This implies that $x = dr$ and $y = ds$ for integers r and s such that $r^4 + s^4 = (z/d^2)^2$. Therefore, it suffices to consider only solutions of Equation (4.101) that are primitive, that is, solutions such that $(x, y) = 1$.

If Equation (4.101) has any solution, then it must have a solution such that z is minimal. Also, Theorem 4.12 implies that $x^2 = 2ab$, $y^2 = a^2 - b^2$, and $z = a^2 + b^2$, where a and b are integers such that $a > b > 0$, $(a, b) = 1$, and $a \not\equiv b \pmod{2}$.

Now $b^2 + y^2 = a^2$ and $(b, y) = (a, b) = 1$, so Theorem 4.17 implies that a is odd; hence b is even. Again, by Theorem 4.17 we have $b = 2uv$, $y = u^2 - v^2$, and $a = u^2 + v^2$ for integers u and v such that $u > v > 0$, $(u, v) = 1$, and $u \not\equiv v \pmod{2}$. However, $(\frac{1}{2}x)^2 = \frac{1}{2}ab$ and $(a, \frac{1}{2}b) = (a, b) = 1$, so by the result of Exercise 21 in Chapter 3, we have $a = s^2$ and $\frac{1}{2}b = t^2$ for natural numbers s and t such that $(s, t) = 1$. Now $t^2 = uv$ and $(u, v) = 1$, so $u = m^2$ and $v = n^2$ for integers m and n such that $(m, n) = 1$. We therefore have $m^4 + n^4 = u^2 + v^2 = a = s^2$; that is, $m^4 + n^4 = s^2$. However, $s \leq a < z$, which contradicts the minimality of z. We conclude the following:

Theorem 4.18 The equation $x^4 + y^4 = z^2$ has no solution in natural numbers.

We might note that the method used to prove Theorem 4.18, namely, using a given solution to generate a smaller solution, is known as *Fermat's method of descent*. As an immediate consequence of Theorem 4.18, we have the following:

Theorem 4.19 The equation $x^4 + y^4 = z^4$ has no solution in natural numbers.

Section 4.8 Computer Exercises

1. Write a computer program to find all primitive Pythagorean triples $\{x, y, z\}$ such that $z < 10000$.

2. Modify your computer program so that it finds all primitive Pythagorean triples $\{x, y, z\}$ such that $z < 10000$ and $|x - y| = 1$.

Section 4.8 Exercises

3. Find all Pythagorean triples such that one member is 17.

4. Find all Pythagorean triples, not necessarily primitive, such that one member is 20.

5. If $\{x, y, z\}$ is a Pythagorean triple, prove that (a) $3|xy$ and (b) $5|xyz$.

6. Extend Table 4.1 so that it includes all $a \leq 12$.

7. Prove that if $n \geq 2$, then the equation $x^{2^n} + y^{2^n} = z^{2^n}$ has no solution in natural numbers.

8. Prove that if $x^2 + y^2 = 2z^2$, then there exist a and b such that $x = (a+b)^2 - 2b^2$, $y = |(a-b)^2 - 2b^2|$, and $z = a^2 + b^2$.

9. Prove that the equation $x^2 + y^2 + z^2 = w^2$ has infinitely many solutions in natural numbers x, y, z, w such that $(x, y, z, w) = 1$.

10. Prove that the equation $x^4 + 4y^4 = z^2$ has no solution in natural numbers.

★ 11. Prove that the equation $3^a + 4^b = 5^c$ has the unique solution $a = b = c = 2$.

12. Prove that the equation $x^4 + y^2 = z^2$ has infinitely many primitive solutions in natural numbers. Find three solutions with x even and three solutions with x odd.

★ 13. The *inradius* of a triangle is the radius of the inscribed circle. Prove that if x, y and z are a Pythagorean triple, then the inradius of the corresponding triangle is an integer.

14. Prove that for every odd prime p there are exactly two right triangles with integer sides and inradius p.

15. Let a, b, c be the lengths of the sides of a triangle such that angle B is twice angle A. Assume furthermore that $(a, b, c) = 1$. (i) Prove that $a(a + c) = b^2$. (ii) Show that all solutions of the latter equation are given by $a = x^2$, $b = xy$, $c = y^2 - x^2$, where x, y are integers such that $0 < x < y < 2x$ and $(x, y) = 1$.

16. Prove that for every positive integer r, there is a primitive Pythagorean triangle with inradius r.

Review Exercises

1. Consider the relation R defined on Z as follows: If $a, b \in Z$, then $(a, b) \in R$ if and only if $7|(a^2 - b^2)$.

    i. Prove that R is an equivalence relation on Z.

    ii. How many equivalence classes are there?

2. Use congruences to prove that $47|(2^{23} - 1)$.

3. Use trial and error to find all solutions of the congruence:
$$x^2 - 2 \equiv 0 \pmod{17}$$

4. Solve the congruence: $27x \equiv 5 \pmod{49}$.

5. Solve the system:

$$\begin{cases} x \equiv 3 \pmod{7} \\ x \equiv 5 \pmod{11} \\ x \equiv 6 \pmod{13} \end{cases}$$

6. Solve the congruence: $x^3 + x + 4 \equiv 0 \pmod{27}$.

7. Construct the table for multiplication (mod 12) in Z_{12}. Which elements have multiplicative inverses?

8. Use congruences to show that $641 | f_5$.

9. Show that a regular polygon with $2^{32} - 1$ sides can be inscribed in a circle using only straight edge and compass.

10. If a primitive Pythagorean triangle has a leg whose length is 21, find the lengths of the two other sides.

Chapter 5

Arithmetic Functions

5.1 Introduction

An *arithmetic function* (also known as a *number theoretic function*) is a function whose domain is the set of natural numbers. In this chapter, we study four such functions: $\tau(n)$, the number-of-divisors function; $\sigma(n)$, the sum-of-divisors function; $\mu(n)$, the *Moebius* function; and $\phi(n)$, *Euler's* function. We conclude with *Euler's Theorem*, which generalizes Fermat's Theorem. In order to develop the properties of these functions, we introduce a concept known as the *Dirichlet product* or *Dirichlet convolution* of two functions.

5.2 Sigma Function, Tau Function, Dirichlet Product

We begin by defining the concept of *summation function*.

Definition 5.1   Summation Function

Let $f(n)$ and $g(n)$ be functions defined on N. We say that $g(n)$ is the *summation function* of $f(n)$ if $g(n)$ is the summation of all $f(d)$ as d runs through the divisors of n. If so, we write

$$g(n) = \sum_{d|n} f(d)$$

For example, let $n = 6$. Since the divisors of 6 are 1, 2, 3, and 6, we have

$$g(6) = \sum_{d|6} f(d) = f(1) + f(2) + f(3) + f(6)$$

Our first specific example of a summation function is the sum-of-divisors function, denoted $\sigma(n)$.

Definition 5.2   Sum-of-Divisors Function

Let $\sigma(n)$ denote the *sum of the divisors* of n. We use the notation:

$$\sigma(n) = \sum_{d|n} d$$

For example,

$$\sigma(6) = \sum_{d|6} d = 1 + 2 + 3 + 6 = 12$$

Also

$$\sigma(4) = \sum_{d|4} d = 1 + 2 + 4 = 7$$

Note that if we define $M(n) = n$ (the identity function with respect to ordinary multiplication), then we have

$$\sigma(n) = \sum_{d|n} d = \sum_{d|n} M(d)$$

that is, $\sigma(n)$ is the summation function of $M(n)$.

Note that if p is prime and $k \geq 1$, then the divisors of p^k are 1, p, p^2, ... , p^k, so that

$$\sigma(p^k) = 1 + p + p^2 + \cdots + p^k = \frac{p^{k+1} - 1}{p - 1}$$

Another summation function is the number-of-divisors function $\tau(n)$.

Definition 5.3   Number-of-Divisors Function

Let $\tau(n)$ denote the *number of divisors* of n. We use the notation:

$$\tau(n) = \sum_{d|n} 1$$

This notation means that to compute $\tau(n)$, add a 1 for every divisor of n. For example,

$$\tau(6) = \sum_{d|6} 1 = 1+1+1+1 = 4$$

Also

$$\tau(4) = \sum_{d|4} 1 = 1+1+1 = 3$$

Note that if we define $u(n) = 1$ (unit function), then we have

$$\tau(n) = \sum_{d|n} 1 = \sum_{d|n} u(d)$$

that is, $\tau(n)$ is the summation function of $u(n)$. Note that if p is prime and $k \geq 1$, then

$$\tau(p^k) = k + 1$$

Suppose that we wish to evaluate $\tau(n)$ and $\sigma(n)$ for composite integers n that have two or more prime factors. For example, let $n = 12$. The divisors of 12 are 1, 2, 3, 4, 6, and 12. Thus by direct evaluation, we have $\tau(12) = 6$ and $\sigma(12) = 1+2+3+4+6+12 = 28$. Clearly, direct evaluation of $\tau(n)$ and $\sigma(n)$ would be cumbersome if n has a large number of divisors. Note that $\tau(4)\tau(3) = 3*2 = 6 = \tau(12)$; also $\sigma(4)\sigma(3) = 7*4 = 28 = \sigma(12)$. Generalizing, it appears that $\tau(mn) = \tau(m)\tau(n)$ and $\sigma(mn) = \sigma(n)\sigma(n)$. We shall see that this is always the case, provided that $(m,n) = 1$. An arithmetic function, f, such that $f(mn) = f(m)f(n)$ whenever $(m,n) = 1$ is called *multiplicative*. We shall see that if $f(n)$ is multiplicative, and if $g(n)$ is the summation function of $f(n)$, then $g(n)$ is also multiplicative. This will provide us with a more efficient alternative to the direct evaluation of $g(n)$.

Definition 5.4   Multiplicative Function

If f is a number-theoretic function such that $f(mn) = f(m)f(n)$ whenever $(m, n) = 1$, then we say f is a *multiplicative function*.

We shall see that the functions $\tau(n)$ and $\sigma(n)$, as well as the functions $\mu(n)$ and $\phi(n)$, which will be defined later on, are all multiplicative functions. The evaluation of a multiplicative function is greatly simplified by applying the following theorem.

Theorem 5.1   If f is a multiplicative function and if $n = \prod_{i=1}^{r} p_i^{e_i}$, then

$$f(n) = \prod_{i=1}^{r} f(p_i^{e_i})$$

Proof: Exercise (induction on r).

Remarks

Theorem 5.1 says that a multiplicative function is completely determined by what it does to powers of primes. For example,

$$\sigma(20) = \sigma((2^2)(5^1)) = \sigma(2^2)\sigma(5^1) = 7 * 6 = 42.$$

∎

Some functions have the even stronger property of being *totally multiplicative*.

Definition 5.5   **Totally Multiplicative Function**

If f is a number-theoretic function such that $f(mn) = f(m)f(n)$ for all m and n, then we say f is a *totally multiplicative function*.

Examples of totally multiplicative functions include

The unit function: **u(n)**, defined by $u(n) = 1$ for all n.

The ordinary identity function: **M(n)**, defined by $M(n) = n$ for all n.

The zero function: **z(n)**, defined by $z(n) = 0$ for all n.

The function: **I(n)**, defined by

$$I(n) = \begin{cases} 1 \text{ if } n = 1 \\ 0 \text{ if } n > 1 \end{cases}$$

Clearly, a function that is totally multiplicative is also multiplicative. The converse does not hold, since each of $\tau(n)$ and $\sigma(n)$ is multiplicative, but not totally multiplicative. The following theorem establishes another property of multiplicative functions:

Theorem 5.2   If f is multiplicative and $f(n) \neq z(n)$, then $f(1) = 1$.

Proof: By hypothesis, there exists n such that $f(n) \neq 0$. Since $(n, 1) = 1$ and f is multiplicative, we have $f(n) = f(n \cdot 1) = f(n)f(1)$. Therefore, $f(n)(1 - f(1)) = 0$. Since $f(n) \neq 0$, we have $1 - f(1) = 0$, that is, $f(1) = 1$. ∎

In order to prove that $\tau(n)$ and $\sigma(n)$ are multiplicative functions, we next introduce the concept of the *Dirichlet product*, which is also known as the *Dirichlet convolution* of two functions.

Definition 5.6   **Dirichlet Product**

Let f and g be arithmetic functions. We define their *Dirichlet product* as

$$(f * g)(n) = \sum_{d|n} f(d)g(n/d)$$

That is, $(f * g)(n)$ is the sum of all products $f(d)g(n/d)$ as d ranges through all divisors of n.

For example,

$$(\tau * \sigma)(4) = \sum_{d|4} \tau(d)\sigma(4/d)$$

$$= \tau(1)\sigma(4) + \tau(2)\sigma(2) + \tau(4)\sigma(1)$$

$$= 1(7) + 2(3) + 3(1)$$

$$= 7 + 6 + 3$$

$$= 16$$

Alternatively, we write

$$(f * g)(n) = \sum_{ab=n} f(a)g(b)$$

That is, $(f * g)(n)$ is the sum of all products $f(a)g(b)$ as a and b range over all positive integers whose product is n.

We next develop some properties of the Dirichlet product.

Properties of the Dirichlet Product

DP_1　$f * g = g * f$

DP_2　$(f * g) * h = f * (g * h)$

DP_3　$f * I = f$

DP_4　$f * u = \sum_{d|n} f(d)$

Proof of DP_1:

$$(f * g)(n) = \sum_{d|n} f(d)g(n/d)$$

Let $c = n/d$, so $d = n/c$. As d ranges through the divisors of n, so does c. Therefore

$$(f * g)(n) = \sum_{c|n} f(n/c)g(c) = \sum_{c|n} g(c)f(n/c) = (g * f)(n)$$

Proof of DP_2:

$$((f * g) * h)(n) = \sum_{dc=n} (f * g)(d)h(c) = \sum_{dc=n} \left(\sum_{ab=d} f(a)g(b) \right)h(c) = \sum_{abc=n} f(a)g(b)h(c)$$

Also,

$$(f * (g * h))(n) = \sum_{ad=n} f(a)(g * h)(d) = \sum_{ad=n} f(a)\left(\sum_{bc=d} g(b)h(c) \right) = \sum_{abc=n} f(a)g(b)h(c)$$

The conclusion now follows.

Proof of DP_3:

$$(f * I)(n) = \sum_{d|n} f(d)I(n/d)$$

If $d < n$, then $n/d > 1$, so $I(n/d) = 0$. Therefore

$$(f * I)(n) = f(n)I(1) = f(n)1 = f(n)$$

Proof of DP$_4$:

$$(f * u)(n) = \sum_{d|n} f(d)u(n/d) = \sum_{d|n} f(d)1 = \sum_{d|n} f(d)$$

Remarks

If we consider the Dirichlet product as a binary operation on the set of all arithmetic functions, then DP$_1$ says that the Dirichlet product is commutative; DP$_2$ says that the Dirichlet product is associative; DP$_3$ says that the function $I(n)$ is the identity element with respect to the Dirichlet product. Furthermore, DP$_4$ says that the summation function of $f(n)$ can be represented as the Dirichlet product of $f(n)$ and $u(n)$. In proving that certain functions are multiplicative, the following theorem is quite useful.

■

Theorem 5.4   If f and g are both multiplicative functions, then so is $f * g$.

Proof: Let $h = f * g$. Then

$$h(mn) = (f * g)(mn) = \sum_{d|mn} f(d)g(mn/d)$$

We must show that if $(m, n) = 1$, then $h(mn) = h(m)h(n)$. If $d|mn$ and $(m, n) = 1$, then $d = ab$, where $a|m$ and $b|n$. Therefore

$$h(mn) = \sum_{a|m, b|n} f(ab)g(mn/ab)$$

$$= \sum_{a|m}(\sum_{b|n} f(ab)g(mn/ab))$$

Since f and g are multiplicative by hypothesis, we have

$$h(mn) = \sum_{a|m}(\sum_{b|n} f(a)f(b)g(m/a)g(n/b))$$

When summing over divisors of n, terms involving divisors of m are constant and so may be factored out of the inner sum. This yields

$$h(mn) = \sum_{a|m} f(a)g(m/a)(\sum_{b|n} f(b)g(n/b)) = \sum_{a|m} f(a)g(m/a)h(n)$$

When summing over divisors of m, $h(n)$ is constant. Therefore

$$h(mn) = h(n) \sum_{a|m} f(a)g(m/a) = h(n)h(m) = h(m)h(n)$$

∎

For example, if f and g are multiplicative, and if $h = f * g$, then

$$h(6) = (f * g)(6)$$
$$= f(1)g(6) + f(2)g(3) + f(3)g(2) + f(6)g(1)$$
$$= 1g(6) + f(2)g(3) + f(3)g(2) + f(6)$$
$$= g(2)g(3) + f(2)g(3) + f(3)g(2) + f(2)f(3)$$
$$= (g(2) + f(2))g(3) + (g(2) + f(2))f(3)$$
$$= (g(2) + f(2))(g(3) + f(3))$$
$$= (1g(2) + f(2)1)(1g(3) + f(3)1)$$
$$= (f(1)g(2) + f(2)g(1))(f(1)g(3) + f(3)g(1))$$
$$= h(2)h(3)$$

As an immediate consequence, we obtain the following:

Theorem 5.5 If f is multiplicative and $g(n) = \sum_{d|n} f(d)$, then g is multiplicative.

Proof: Since $u(n)$ is totally multiplicative, hence multiplicative, the conclusion follows from the hypothesis, Theorem 5.3, part DP_4, and Theorem 5.4. ∎

Now we are ready to prove that the functions $\tau(n)$ and $\sigma(n)$ are multiplicative.

Theorem 5.6 $\tau(n)$ is multiplicative.

Proof:

$$\tau(n) = \sum_{d|n} 1 = \sum_{d|n} u(d)u(n/d) = (u * u)(n)$$

Since $u(n)$ is multiplicative, the conclusion follows from Theorem 5.4. ∎

We can now use the canonical factorization of n to compute $\tau(n)$.

Theorem 5.7 If $n = \prod_{i=1}^{r} p_i^{e_i}$ where the p_i are distinct primes, then

$$\tau(n) = \prod_{i=1}^{r} (e_i + 1)$$

Proof: This follows from the hypothesis and from Theorems 5.6 and 5.1, since $\tau(p^e) = e + 1$ if p is prime. ∎

For example, $\tau(360) = \tau((2^3)(3^2)(5^1)) = 4 \cdot 3 \cdot 2 = 24$.
Also, $\tau(112) = \tau((2^4)(7^1)) = 5 \cdot 2 = 10$.

Theorem 5.8 $\sigma(n)$ is multiplicative.

Proof:

$$\sigma(n) = \sum_{d|n} d = \sum_{d|n} M(d)$$

Since $M(n)$ is totally multiplicative, hence multiplicative, the conclusion follows from Theorem 5.5. ∎

We can now use the canonical factorization of n to compute $\sigma(n)$.

Theorem 5.9 If $n = \prod_{i=1}^{r} p_i^{e_i}$, where the p_i are distinct primes, then

$$\sigma(n) = \prod_{i=1}^{r} \frac{(p_i^{e_i+1} - 1)}{p_i - 1}$$

Proof: This follows the hypothesis and from Theorems 5.8 and 5.1, since $\sigma(p^e) = \frac{p^{e+1}-1}{p-1}$. ∎

For example,

$$\sigma(360) = \sigma((2^3)(3^2)(5^1))$$

$$= (\frac{2^4 - 1}{2 - 1})(\frac{3^3 - 1}{3 - 1})(\frac{5^2 - 1}{5 - 1})$$

$$= 15 \cdot 13 \cdot 6 = 1170$$

Also,

$$\sigma(784) = \sigma((2^4)(5^2))$$

$$= (\frac{2^5 - 1}{2 - 1})(\frac{5^3 - 1}{5 - 1})$$

$$= 31 \cdot 57 = 1767$$

The following theorem follows easily from Theorem 5.9.

Theorem 5.10 m is prime if and only if $\sigma(m) = m + 1$.

Proof: Exercise.

The σ function is involved in one of the oldest unsolved problems in number theory. In Book IX, Proposition 36 of his *Elements*, which was written about 300 B.C., Euclid said that a number is *perfect* if it is equal to the sum of its proper divisors. We can redefine perfect numbers in terms of the σ function.

Definition 5.7   **Perfect Number**

n is *perfect* if $\sigma(n) = 2n$.

As we mentioned earlier, the smallest perfect number is

$$6 = 1 + 2 + 3$$

The next smallest is

$$28 = 1 + 2 + 4 + 7 + 14$$

The following theorem characterizes *even* perfect numbers:

Theorem 5.11   **(Euclid-Euler)**

Let n be even. Then n is perfect if and only if

$$n = 2^{p-1}(2^p - 1)$$

where p and $2^p - 1$ are prime.

Proof: If

$$n = 2^{p-1}(2^p - 1)$$

where the odd factor of the right member is prime, then Theorems 5.8 and 5.9 imply that

$$\sigma(n) = (\sigma(2^{p-1}))(\sigma(2^p - 1))$$
$$= (2^p - 1)(2^p) = 2n$$

Conversely, suppose that $n = 2^{k-1}m$, where $k \geq 2$, m is odd, and $\sigma(n) = 2n$. We have

$$\sigma(2^{k-1}m) = 2(2^{k-1}m)$$

that is,

$$\sigma(2^{k-1}m) = 2^k m$$

Now Theorem 5.8 implies

$$\sigma(2^{k-1})\sigma(m) = 2^k m$$

so that Theorem 5.9 implies

$$(2^k - 1)\sigma(m) = 2^k m$$

Since $(2^k - 1, 2^k) = 1$, Euclid's Lemma (Theorem 2.13) implies that $(2^k - 1)|m$, that is,

$$m = (2^k - 1)d$$

for some $d \geq 1$. Since

$$(2^k - 1)\sigma(m) = (2^k - 1)m + m$$

dividing by $2^k - 1$, we get

$$\sigma(m) = m + d$$

If $d > 1$, then $\sigma(m) \geq m + d + 1$. Therefore $d=1$, so

$$m = 2^k - 1$$

Since

$$\sigma(m) = m + 1$$

it follows, by Theorem 5.10, that m is prime. By the result of Section 3.2, Exercise 1, k is also prime. ∎

For example, since $2^5 - 1 = 31$ is prime, it follows that $2^{5-1}(2^5 - 1) = 16 \cdot 31 = 496$ is perfect. Likewise, since $2^7 - 1 = 127$ is prime, it follows that $2^{7-1}(2^7 - 1) = 64 \cdot 127 = 8128$ is perfect.

Remarks

We have seen in Section 3.2, Exercise 1 that if $2^k - 1$ is prime, then k is prime. The converse of this statement is false, however. For example, 11 is prime, yet $2^{11} - 1 = 2047 = 23 \cdot 89$ is composite. In fact, we will learn in Chapter 7 that if p is a prime such that $p \equiv 3 \pmod 4$, $p \geq 11$, and $q = 2p + 1$ is prime, then q is a nontrivial factor of $2^p - 1$. The first few primes that satisfy these conditions are 11, 23, 83, 131, and 179.

■

Let us take a closer look at those special primes that occur as factors of even perfect numbers.

Definition 5.8　　Mersenne Prime

If p and q are primes such that $q = 2^p - 1$, we say that q is a *Mersenne prime*.

Marin Mersenne (1588–1648)

In the 1620s, he organized a group who met regularly in Paris to discuss scientific matters. He also corresponded with mathematicians who did not live in Paris, such as Descartes and Fermat. Mersenne compiled a list of primes of the form $2^p - 1$, which are called *Mersenne primes*.

At present, 42 Mersenne primes are known to exist. They correspond to the following values of p: 2, 3, 5, 7, 13, 17, 19, 31, 61, 89, 107, 127, 521, 607, 1279, 2203, 2281, 3217, 4253, 4423, 9689, 9941, 11213, 19937, 21701, 23209, 44497, 86243, 110503, 132049, 216091, 756839, 859433, 1257787, 1398269, 2976221, 3021377, 6972593, 13466917, 20996011, 24036583, 25464951. The Mersenne prime corresponding to the last value of p, namely $2^{25464951} - 1$, is currently the largest verified prime. It has more than 7,000,000 decimal digits.

According to Theorem 5.11, there is a one-to-one correspondence between Mersenne primes and even perfect numbers. It has been conjectured that there exist infinitely many Mersenne primes. Remarkably, no one has yet proved the existence of infinitely many primes, p, such that $2^p - 1$ is composite.

Euclid claimed that all perfect numbers are even. Even today, 23 centuries later, no one knows whether an odd number can be perfect. Many necessary conditions for the existence of an odd perfect number have been established, however. For example, an odd perfect number must have at least eight distinct prime factors (see [5]). We now develop some results leading to Euler's theorem regarding the canonical factorization of an odd perfect number.

Theorem 5.12   If the prime $p \equiv 1 \pmod 4$, then $\sigma(p^e) \equiv e + 1 \pmod 4$.

Proof:

$$\sigma(p^e) = \sum_{j=0}^{e} p^j \equiv \sum_{j=0}^{e} 1^j \equiv \sum_{j=0}^{e} 1 \equiv e + 1 \pmod 4 \quad \blacksquare$$

Theorem 5.13   *If the prime $p \equiv -1$ (mod 4), then*

$$\sigma(p^e) \equiv \begin{cases} 1 & \text{(mod 4) if } 2|e \\ 0 & \text{(mod 4) if } 2 \nmid e \end{cases}$$

Proof:

$$\sigma(p^e) = \frac{p^{e+1} - 1}{p - 1} \equiv \frac{(-1)^{e+1} - 1}{-2} \pmod 4$$

The conclusion follows easily from this. $\blacksquare$

Theorem 5.14   If n is odd and perfect, then

$$n = p^a \prod_{i=1}^{r} q_i^{2b_i}$$

where p and the q_i are odd primes and $p \equiv a \equiv 1 \pmod 4$.

Proof: By hypothesis, $\sigma(n) = 2n \equiv 2 \pmod 4$. If

$$n = \prod_{i=1}^{k} p_i^{e_i}$$

then Theorem 5.9 implies that

$$\sigma(n) = \prod_{i=1}^{k} \sigma(p_1^{e_i})$$

The result of Section 4.3, Exercise 9 implies that there exists unique j such that

$$\sigma(p_j^{e_j}) \equiv 2 \pmod 4$$

while $\sigma(p_i^{e_i})$ is odd for all $i \neq j$. Theorem 5.13 implies that

$$p_j \not\equiv -1 \pmod 4 \quad hence \quad p_j \equiv 1 \pmod 4$$

Theorem 5.12 implies that

$$e_j + 1 \equiv 2 \pmod 4 \quad hence \quad e_j \equiv 1 \pmod 4$$

Now suppose that $i \neq j$. If $p_i \equiv 1 \pmod 4$, then Theorem 5.12 implies that $2|e_i$. If $p_i \equiv -1 \pmod 4$, then Theorem 5.13 implies that $2|e_i$. The conclusion now follows if we let $r = k - 1$, $p_j = p$, $e_j = a$, $p_i = q_i$ and $e_i = 2b_i$ for all $i \neq j$. ∎

Remarks

As of the present, no odd perfect number has yet been discovered. Such a number, if it exists, would have more than 300 decimal digits. Theorem 5.14 has been useful in eliminating certain candidates from consideration.

Section 5.2 Exercises

1. Evaluate $\tau(n)$ and $\sigma(n)$ for each n such that $40 \leq n \leq 50$.

2. Prove that $\prod_{d|n} d = n^{\frac{\tau(n)}{2}}$.

3. Prove that if n is a natural number such that $\tau(n) = q$, where q is a prime, then $n = p^{q-1}$ for some prime p.

4. Find the least integer n such that $\tau(n) = 25$. (Do not use trial and error.)

5. If $\omega(n)$ denotes the number of distinct prime factors of n, prove that

$$\tau(n) \geq 2^{\omega(n)}$$

6. Prove that $\tau(n)$ is odd if and only if n is a square.

7. If $\tau(n) = 4$, what can be said about the canonical factorization of n?

8. Find all solutions, if any, to the equation $\sigma(n) = 2^k$ for each of the following values of k:

   (a) 1

   (b) 2

   (c) 3

   (d) 4

   (e) 5

   (f) 6

   (g) 7

   (h) 8

9. Prove that if $\sigma(n)$ is prime, then $n = p^k$, where p is prime and $k \geq 1$.

10. Prove that if $\sigma(p^k) = n$, where p is prime, then $p|(n-1)$.

11. Prove that if n is odd, then $\tau(n) \equiv \sigma(n) \pmod 2$.

12. Prove that if p is prime and $n \geq 2$, then $\sigma(p^{n^2-1})$ is composite.

13. Find all even perfect numbers, if any, of the form $m^3 + 1$. Hint: factor $m^3 + 1$.

14. Prove that n is composite if and only if $\sigma(n) \geq n + \sqrt{n} + 1$.

15. Prove that if $n \equiv 7 \pmod 8$, then $\sigma(n) \equiv 0 \pmod 8$.

16. Prove that if $n \equiv 23 \pmod{24}$, then $\sigma(n) \equiv 0 \pmod{24}$.

17. Prove that $I(n)$ is totally multiplicative.

18. Prove that the ordinary (not Dirichlet) product of two multiplicative functions is also multiplicative.

19. Prove Theorem 5.1.

20. Prove that if $mn > 1$, then $\sigma(mn) > m\sigma(n)$.

21. A natural number, n, is said to be *abundant* if $\sigma(n) > 2n$. Prove that every multiple of an abundant number is also abundant.

22. Prove that if n is perfect, then $\prod_{p|n} \frac{p}{p-1} > 2$.

23. Prove that $(M * M)(n) = n\tau(n)$.

24. Prove that if m is a perfect number, then the equation: $\sigma(n) = 2(n+m)$ has infinitely many solutions.

25. Prove that if p and q are distinct primes such that $\sigma(p^2) = \sigma(q^4)$, then $p = 5$ and $q = 2$.

26. Prove that there are no primes p and q such that $\sigma(p^2) = \sigma(q^6)$.

27. Prove that if p is prime and $\sigma(p^m) = 2^n$, then $m = 1$ and n is prime.

28. Find the least exponent k such that $k \geq 12$ and the equation $\sigma(n) = 2^k$ has no solution.

29. Prove Theorem 5.10.

30. Prove that the function $f(n) = \sum_{d|n} \frac{1}{d}$ is multiplicative.

Section 5.2 Computer Exercises

31. Suppose that p is a prime such that $2^p - 1$ is composite. Write a computer program to find the least prime factor q of $2^p - 1$, using the facts that

   i. $q \equiv 1 \pmod{2p}$

   ii. $2^p \equiv 1 \pmod{q}$

   iii. $q^2 < 2^p - 1$.

32. Let the odd primes be numbered $p_1 = 3$, $p_2 = 5$, $p_3 = 7$, etc. Write a computer program that, given a integer i computes the least value of r such that

$$\prod_{j=i}^{i+r-1} \frac{p_i}{p_i - 1} > 2$$

5.3 Dirichlet Inverse, Moebius Function, Euler's Function, Euler's Theorem

We now return to the study of the Dirichlet product. Let f, g, and h be arithmetic functions such that $h = f * g$. We saw in Theorem 5.4 that if f and g are both multiplicative, then so is h. We now prove the more difficult result that if g and h are both multiplicative, then so is f.

Theorem 5.15

Let $h = f * g$. If g and h are both multiplicative, and neither g nor h is the zero function, then f is also multiplicative.

Proof: We will show that if g is multiplicative but f isn't, then neither is h. Let mn be the smallest number such that $(m, n) = 1$, yet

$$f(mn) \neq f(m)f(n)$$

Since, by hypothesis, g and h are multiplicative and neither is the zero function, we have $g(1) = h(1) = 1$ by Theorem 5.2. Since

$$h(1) = f(1)g(1)$$

this implies that $f(1) = 1$. Therefore $mn > 1$. Now

$$h(mn) = \sum_{d|mn} f(d)g(mn/d) = f(mn)g(1) + \sum_{d|mn, d<mn} f(d)g(mn/d)$$

As in the proof of Theorem 5.4, we have $d = ab$, where $a|m$ and $b|n$. Therefore

$$h(mn) = f(mn) + \sum_{a|m, b|n, ab<mn} f(ab)g(mn/ab)$$

$$= f(mn) + \sum_{a|m, b|n, ab<mn} f(a)f(b)g(m/a)g(n/b)$$

$$= f(mn) - f(m)f(n) + \sum_{a|m}\sum_{b|n} f(a)g(m/a)f(b)g(n/b)$$

As in the proof of Theorem 5.4, the double sum simplifies to a product of two sums, so we have

$$h(mn) = f(mn) - f(m)f(n) + (\sum_{a|m} f(a)g(m/a))(\sum_{b|n} f(b)g(m/b))$$

$$= f(mn) - f(m)f(n) + h(m)h(n)$$

Since $f(mn) \neq f(m)f(n)$, it follows that $h(mn) \neq h(m)h(n)$, so that h is not multiplicative, contrary to hypothesis. ∎

Recall that the function

$$I(n) = \begin{cases} 1 \text{ if } n = 1 \\ 0 \text{ if } n > 1 \end{cases}$$

is the identity function with respect to the Dirichlet product. We next introduce the concept of the *Dirichlet inverse* of an arithmetic function. We determine sufficient conditions for the existence of a Dirichlet inverse of a function, f, and show how to compute the inverse.

Definition 5.9   Dirichlet Inverse

Given an arithmetic function f, if there exists a function g such that $f * g = I$, then we say that g is the *Dirichlet inverse* of f, and we write

$$g = f^{-1}$$

The following theorem guarantees the existence of the Dirichlet inverse for most functions of interest, especially multiplicative functions, and is constructive.

Theorem 5.16 If f is an arithmetic function such that $f(1) \neq 0$, then f^{-1} exists.

Proof: Let $g(1) = \frac{1}{f(1)}$. If $n > 1$, let

$$g(n) = -\frac{\sum_{d|n, d<n} f(n/d)g(d)}{f(1)}.$$

Then $(f * g)(1) = f(1)g(1) = 1$, whereas if $n > 1$, then

$$(f * g)(n) = (g * f)(n) = \sum_{d|n} g(d)f(n/d)$$

$$= g(n)f(1) + \sum_{d|n, d<n} g(d)f(n/d)$$

$$= g(n)f(1) - g(n)f(1) = 0.$$

Therefore, $(f * g)(n) = I(n)$. ■

Next, we see that every nontrivial multiplicative function has a nontrivial multiplicative Dirichlet inverse.

Theorem 5.17 If f is multiplicative and $f(n) \neq z(n)$, then f^{-1} exists, f^{-1} is multiplicative, and $f^{-1} \neq z(n)$.

Proof: Since $f(n)$ is multiplicative and $f(n) \neq z(n)$ by hypothesis, Theorem 5.2 implies that $f(1) = 1$. Therefore, Theorem 5.17 implies that f^{-1} exists. Now $f * f^{-1} = I$. Since f and I are multiplicative, Theorem 5.16 implies that f^{-1} is multiplicative. Finally, if $f^{-1}(n) = z(n)$, then

$$I = f * f^{-1} = f * z = z,$$

an impossibility. ■

Let us now construct the Dirichlet inverse of the unit function $u(n) = 1$. We obtain a function that is known as the *Moebius function* and is denoted $\mu(n)$.

Definition 5.10 **Moebius Function**

Let $\mu(n) = u^{-1}(n)$. Since μ is multiplicative and nonzero, we have $\mu(1) = 1$. If p is prime, then

$$\mu(p) = -\mu(1)u(p) = -1$$

Also

$$\mu(p^2) = -\{\mu(1)u(p^2) + \mu(p)u(p)\} = -(1-1) = 0$$

If $k \geq 3$, then

$$\mu(p^k) = -\sum_{d|p^k, d<p^k} \mu(d)u(p^k/d)$$

$$= -\sum_{j=0}^{k-1} \mu(p^j)$$

$$= -\{\mu(1) + \mu(p) + \sum_{j=2}^{k-1} \mu(p^j)\}$$

$$= -(1-1+0) = 0$$

(The sum in the next to last equation is 0 by induction on k.)

We have seen that

$$\mu(p^k) = \begin{cases} 1 & \text{if } k = 0 \\ -1 & \text{if } k = 1 \\ 0 & \text{if } k > 1 \end{cases}$$

Since μ is multiplicative, we can use Theorem 5.1 to compute $\mu(n)$ for any n.

Theorem 5.18   If $n = \prod_{i=1}^{r} p_i^{e_i}$, then

$$\mu(n) = \begin{cases} (-1)^r & \text{if } e_i = 1 \text{ for all } i \\ 0 & \text{if } e_i > 1 \text{ for some } i. \end{cases}$$

Proof: Theorem 5.16 implies that μ is multiplicative. Theorem 5.1 implies that

$$\mu(n) = \prod_{i=1}^{r} \mu(p_i^{e_i})$$

If all $e_i = 1$, then

$$\mu(n) = \prod_{i=1}^{r} (-1) = (-1)^r$$

If any $e_i > 1$, then $\mu(p_i^{e_i}) = 0$, so $\mu(n) = 0$. ∎

For example,

$$\mu(42) = \mu(2 \cdot 3 \cdot 7) = (-1)^3 = -1$$

$$\mu(44) = \mu(2^2 \cdot 11) = 0$$

$$\mu(46) = \mu(2 \cdot 23) = (-1)^2 = 1$$

$$\mu(47) = -1$$

If $g(n)$ is the summation function of $f(n)$, then the *Moebius inversion formula*, which follows, allows us to express $f(n)$ in terms of a summation involving $g(n)$.

Theorem 5.19 **Moebius Inversion Formula**

$$g(n) = \sum_{d|n} f(d) \qquad \text{if and only if} \qquad f(n) = \sum_{d|n} g(d)\mu(n/d)$$

Proof: By Theorem 5.3, part DP_4, and Definition 5.6, it suffices to show that $g = f * u$ if and only if $f = g * \mu$. if $g = f * u$, then

$$g * \mu = (f * u) * \mu = f * (u * \mu) = f * I = f$$

by Theorem 5.3, parts DP_2 and DP_3. The proof of the converse is similar. ∎

For example, since

$$\sigma(n) = \sum_{d|n} d$$

the Moebius Inversion Formula implies that

$$n = \sum_{d|n} \sigma(d)\mu(n/d)$$

Similarly, since

$$u(n) = \sum_{d|n} I(d)$$

the Moebius Inversion Formula implies that

$$I(n) = \sum_{d|n} u(d)\mu(n/d)$$

We can simplify this result by noting that $I = u * \mu = \mu * u$, so that

$$I(n) = \sum_{d|n} \mu(d)$$

We next define and develop the properties of *Euler's totient function.* Euler's totient occurs in many places, including Euler's Theorem, which generalizes Fermat's Theorem.

Definition 5.11   **Euler's Totient Function**

Let $\phi(n)$ denote the number of integers, k, such that $1 \leq k \leq n$ and $(k, n) = 1$.

For example, $\phi(9) = 6$, since $(k, 9) = 1$ for $k = 1, 2, 4, 5, 7, 8$. Also, $\phi(10) = 4$, since $(k, 10) = 1$ for $k = 1, 3, 7, 9$. Note that

$$\phi(p) = p - 1$$

and

$$\phi(p^k) = p^k - p^{k-1} = p^{k-1}(p - 1)$$

The following theorem is a useful property of $\phi(n)$.

Theorem 5.20

$$\sum_{d|n} \phi(d) = n$$

Proof: Let $S_n = \{1, 2, 3, \ldots, n\}$. If $d|n$, consider the subset $S(d) = \{k \in S_n : (k, n) = d\}$. Let $f(d)$ be the number of elements of $S(d)$. If k belongs to both $S(d_1)$ and $S(d_2)$, where d_1 and d_2 are distinct divisors of n, then $d_1 = (k, n) = d_2$, an impossibility. Therefore the subsets $S(d)$ are pairwise disjoint. Furthermore, for all k such that $1 \leq k \leq n$, there exists d such that $d|n$ and $(k, n) = d$. Therefore, the union of all subsets $S(d)$ where $d|n$ is the set S_n. This implies that

$$\sum_{d|n} f(d) = n$$

Theorem 2.4, parts G_4 and G_5, implies $(k, n) = d$ if and only if $(\frac{k}{d}, \frac{n}{d}) = 1$. By Definition 5.11, the number of such k is $\phi(n/d)$, that is, $f(d) = \phi(n/d)$. Now

$$M(n) = n = \sum_{d|n} \phi(n/d) = \sum_{d|n} u(d)\phi(n/d) = (u * \phi)(n)$$

Thus

$$n = M(n) = (\phi * u)(n) = \sum_{d|n} \phi(d)u(n/d) = \sum_{d|n} \phi(d). \quad \blacksquare$$

For example, the divisors of 12 are 1, 2, 3, 4, 6, and 12. Thus

$$\sum_{d|12} \phi(d) = \phi(1) + \phi(2) + \phi(3) + \phi(4) + \phi(6) + \phi(12)$$

$$= 1 + 1 + 2 + 2 + 2 + 4$$

$$= 12$$

As a by-product of the proof of Theorem 5.20, we can now show that ϕ is multiplicative.

Theorem 5.21 ϕ is multiplicative.

Proof: We saw in the proof of Theorem 5.20 that

$$M(n) = (\phi * u)(n)$$

Since $M(n)$ and $u(n)$ are multiplicative, it follows from Theorem 5.15 that ϕ is also multiplicative. $\blacksquare$

In the next theorem, we give a formula that allows one to compute $\phi(n)$ from the canonical factorization of n.

Theorem 5.22   If $n = \prod_{i=1}^{r} p_i^{e_i}$, then

$$\phi(n) = \prod_{i=1}^{r} p_i^{e_i - 1}(p_i - 1)$$

Proof: This follows from Theorems 5.21 and 5.1, since

$$\phi(p^e) = p^{e-1}(p - 1)$$

if p is prime. $\blacksquare$

For example,

$$\phi(675) = \phi((3^3)(5^2)) = \phi(3^3)\phi(5^2) = (3^2)(2)(5^1)(4) = 360$$

Also,

$$\phi(735) = \phi(3 \cdot 5 \cdot 7^2) = \phi(3)\phi(5)\phi(7^2) = 2 \cdot 4 \cdot (7 \cdot 6) = 336$$

The two following theorems, whose proofs are left as exercises, provide alternate ways of computing $\phi(n)$.

Theorem 5.23

$$\phi(n) = \sum_{d|n} d\mu(n/d)$$

Proof: Exercise.

Theorem 5.24

$$\phi(n) = n \prod_{p|n} (1 - \frac{1}{p})$$

Proof: Exercise.

Euler's Theorem (Theorem 5.25) is a generalization of Fermat's Little Theorem (Theorem 4.8).

Theorem 5.25 (**Euler's Theorem**)

If $m \geq 2$ and $(a, m) = 1$, then $a^{\phi(m)} \equiv 1 \pmod{m}$.

Proof: This proof generalizes the second proof of Fermat's Little Theorem (Theorem 4.8). If $(a, m) = 1$, let

$$H(m) = \{j : 0 < j < m, (j, m) = 1\}$$

$$T(a, m) = \{aj : 0 < j < m, (j, m) = 1\}$$

Let $S(a,m)$ be the set of least positive residues (mod m) of the elements of $T(a,m)$. If $0 < j < m$ and $(j,m) = 1$, then $(aj,m) = 1$, so $0 \notin S(a,m)$. If x_1, x_2 are integers such that $0 < x_1 \leq x_2 < m$ and $(x_1,m) = (x_2,m) = 1$ and $ax_1 \equiv ax_2 \pmod{m}$, then (cancelling a), we have $x_1 \equiv x_2 \pmod{m}$. Since $0 \leq x_2 - x_1 < m$, we must have $x_1 = x_2$. Therefore the elements of $S(a,m)$ are distinct (mod m). It follows that $S(a,m)$ is a permutation of $H(m)$. This implies

$$\prod \{aj : 0 < j < m \,,\, (j,m) = 1\} \equiv \prod \{j : 0 < j < m \,,\, (j,m) = 1\} \pmod{m}$$

Let P_m denote the product of all elements of $H(m)$. Simplifying, we have

$$a^{\phi(m)} P_m \equiv P_m \pmod{m}$$

Now Theorem 2.9 implies $(P_m, m) = 1$. Cancelling P_m, we have

$$a^{\phi(m)} \equiv 1 \pmod{m} \quad \blacksquare$$

For example, since $(3,10) = 1$ and $\phi(10) = 4$, it follows from Euler's Theorem that $3^4 \equiv 1 \pmod{10}$. Similarly, since $(5,18) = 1$ and $\phi(18) = 6$, it follows from Euler's Theorem that $5^6 \equiv 1 \pmod{18}$.

An alternate proof of Euler's Theorem can be given using Theorems 5.26 and 5.27, whose proofs are left as exercises.

Theorem 5.26   If p is prime, $(a,p) = 1$, and $r \geq 1$, then

$$a^{\phi(p^r)} \equiv 1 \pmod{p^r}$$

Proof: Exercise (induction on r).

Theorem 5.27   If $a^{\phi(m)} \equiv 1 \pmod{m}$, $a^{\phi(n)} \equiv 1 \pmod{n}$, and $(m,n) = 1$, then $a^{\phi(mn)} \equiv 1 \pmod{mn}$.

Proof: Exercise.

Leonhard Euler (1707–1783)

Leonhard Euler was born in Basel, Switzerland on April 15, 1707. His father, a Protestant minister, was fond of mathematics and had attended the lectures of Jakob Bernoulli. In 1720, young Euler entered the University of Basel. At that time, Johann Bernoulli was on the faculty. Bernoulli did not give private lessons to Euler, but told him what books to study on his own, and reviewed his progress on a weekly basis.

In 1723, Euler received a Master's degree in philosophy. He preferred mathematics, however. In 1727, Euler applied for a vacant position at the University of Basel, but was turned down. He then accepted a position at the newly organized St. Petersburg Academy in Russia, where he had been recommended by the Bernoulli family.

In St. Petersburg, Euler was active not only in mathematics, but also in many practical matters: mapping the Russian territory, shipbuilding and navigation, the testing of pumps and scales, etc. He wrote on mechanics, hydromechanics, and lunar and planetary motion. His reputation grew, and he became a member of the scientific academies of the major European nations. In 1738, he lost the sight of his right eye.

Due to a change of regime in Russia, Euler's position became less secure. In 1741, he accepted an offer from Frederick II, King of Prussia, to come to the Berlin Academy. In Berlin, in addition to his ongoing research program, Euler had many administrative duties. He supervised the botanical garden and the observatory, hired personnel, and managed the publication of maps and calendars. He was even responsible for the plumbing in the King's summer palace. Euler did not get along that well with King Frederick. When the presidency of the Berlin Academy became vacant in 1759, King Frederick refused to appoint Euler to this position.

In 1766, Euler accepted an invitation from Catherine, Tsarina of Russia, to return to St. Petersburg, where he spent the remaining years of his life. In 1771, Euler became completely blind. He did not, however, curtail his scientific activities. Euler died on September 18, 1783.

Euler's contributions to number theory include the discovery of what is now known as Euler's function, the discovery (but not the proof) of the law of quadratic reciprocity, the use of continued fractions to solve Pell's equation, and the proof of Fermat's Last Theorem for the case $n = 3$.

Euler was also active in algebra, analysis, differential equations, the calculus of variations, and geometry. In addition, Euler introduced some important features of modern mathematical notation, including the use of the letter e to denote the base of the natural logarithms, the use of the letter i to denote the square root of -1, and the use of $f(x)$ to denote functional notation.

An Application to Algebra

The nth *cyclotomic polynomial* may be defined by

$$\Phi_n(x) = \prod_{d|n}(x^d - 1)^{\mu(n/d)}$$

According to this definition, $\Phi_n(x)$ appears to be a rational function of x. For example,

$$\Phi_6(x) = \frac{(x-1)(x^6-1)}{(x^2-1)(x^3\yen-1)}$$

Upon simplifying, however, we see that

$$\Phi_6(x) = \frac{x^3+1}{x+1} = x^2 - x + 1$$

In fact, $\Phi_n(x)$ is always an irreducible monic polynomial with integer coefficients. (See [6], Theorem 6.5.4.) Note that the degree of $\Phi_n(x)$ is

$$\sum_{d|n} d\mu(n/d) = \phi(n)$$

Cyclotomic polynomials occur in the factorization of $x^n - 1$, because of the fact that

$$x^n - 1 = \prod_{d|n} \Phi_d(x) \quad .$$

For example,

$$x^6 - 1 = \prod_{d|6} \Phi_d(x)$$

$$= \Phi_1(x)\Phi_2(x)\Phi_3(x)\Phi_6(x)$$

$$= (x-1)(x+1)(x^2+x+1)(x^2-x+1)$$

Section 5.3   Exercises

1. Evaluate $\mu(n)$ and $\phi(n)$ for each of the following values of n:
   - (a) 42
   - (b) 79
   - (c) 1000
   - (d) 945
   - (e) 91
   - (f) d144.

2. Prove that $\sum_{d|n} |\mu(d)| = 2^{\omega(n)}$.

3. Prove that $\sum_{d|n} \mu(d)\tau(d) = (-1)^{\omega(n)}$.

4. Prove that $\sum_{d|n} \mu(d)\tau(n/d) = 1$.

5. Prove that $\sum_{d|n} \sigma(d)\mu(n/d) = n$.

6. Find a formula for $\sum_{d|n} \mu(d)\sigma(d)$ in terms of the canonical factorization of n.

7. Let $h(n) = \sigma^{-1}(n)$. Find (a) $h(p)$; (b) $h(p^2)$; (c) $h(p^k)$ where $k \geq 3$.

8. Let $g(n) = M^{-1}(n)$. Find (a) $g(p)$, (b) $g(p^2)$, (c) $g(p^k)$ where $k \geq 3$. Prove that $g(n) = \mu(n)M(n)$.

9. Prove that if f and g are number-theoretic functions such that f^{-1} and g^{-1} exist, then $(f * g)^{-1} = f^{-1} * g^{-1}$.

10. Prove that the number of reduced fractions a/b such that $1 \leq a < b \leq n$ is $\sum_{k=1}^{n} \phi(k)$.

11. Prove that if n is odd, then $2^{\omega(n)}|\phi(n)$.

12. Prove that if p is prime and $p \equiv 1 \pmod 3$, then the equation: $\phi(x) = 2p$ has no solution.

13. Solve the equation: $\phi(x) = n$ for each of the following values of n: (a) 1 (b) 2 (c) 4 (d) 6 (e) 8 (f) 10 (g) 12.

14. Prove that $\phi(n)|n$ if and only if $n = 1, 2^a$, or $2^a 3^b$ where a, b are natural numbers.

★ 15. Prove that if $\phi(x) = n$, then $x \leq 2(3^{\log_2 n})$.

16. If $m \geq 2$, find a formula for the sum of all the integers, k, such that $1 < k < m$ and $(k, m) = 1$.

17. Prove that $I(n) = [1/n]$.

18. Prove that $\tau(n) = \sum_{k=1}^{n} [\frac{k}{n}[\frac{n}{k}]]$.

19. Prove that $\phi(n) = \sum_{k=1}^{n} [\frac{1}{(k,n)}]$.

20. Prove that $\frac{\phi(n)}{n} = \sum_{d|n} \frac{\mu(d)}{d}$.

21. Find a formula for $\sum_{d|n} \mu(d)\mu(n/d)$.

22. Prove that $n|\phi(2^n - 1)$ for all $n \geq 1$.

23. Prove that the equation: $\phi(x) = 2^{32}$ has no odd solutions.

24. Prove Theorem 5.23.

25. Prove Theorem 5.24.

26. Prove that if $m|n$, then $\phi(m)|\phi(n)$.

27. Let m and n be integers such that $x = m$ is the unique solution of the equation: $\phi(x) = n$. Prove that $(2 * 3 * 7 * 43)^2|m$. (It is not known whether such an m exists.)

28. Evaluate $\Phi_n(x)$ for all n such that $1 \leq n \leq 16$.

29. Prove that if p is prime, then $\Phi_{p^2}(x) = \Phi_p(x^p)$.

30. Prove Theorem 5.26. (Use induction on r.)

31. Prove Theorem 5.27.

32. Use Theorems 5.26 and 5.27 to give an alternate proof of Theorem 5.25.

★ 33. Prove that $\sum_{k=1}^{n} \phi(k)[\frac{n}{k}] = \frac{n(n+1)}{2}$.

Section 5.3   Computer Exercises

34. Write a computer program to compute $\phi(n)$ using the formula of Exercise 19.

35. Write a computer program that finds all solutions, if any, of the equation, $\phi(x) = 2^n$ for given n.

36. Write a computer program to find all n such that $1 \leq n \leq 300$ and $\phi(n) = \phi(n+1)$. Is there a pattern?

Review Exercises

1. Evaluate $\tau(n)$, $\sigma(n)$, $\mu(n)$, $\phi(n)$ for each of the following:

   (a) $n = 20$

   (b) $n = 105$

   (c) $n = 64$

   (d) $n = 525$

   (e) $n = 107$.

2. If $\tau(n) = 9$, what does this say about the canonical factorization of n?

3. A positive integer n is called *deficient* if $\sigma(n) < 2n$. Prove that if n is an odd positive integer with only one or two distinct prime factors, then n is deficient.

4. Find all solutions, if any, of the equation: $\sigma(n) = 56$.

5. Find all solutions of the equation, $\phi(n) = 20$.

6. Prove that if $n \not\equiv 2 \pmod 4$, then $\mu(n^3 - n) = 0$.

7. Prove that if n is perfect, then

$$n = - \sum_{d|n,\, d<n} \sigma(d)\mu(n/d)$$

8. Prove that $\phi(n^2) = n\phi(n)$ for every positive integer n.

9. Find all positive integers n such that $\phi(\phi(n)) = 1$.

10. Evaluate $\Phi_{18}(x)$ (the 18th cyclotomic polynomial).

Chapter 6

Primitive Roots and Indices

6.1 Introduction

Suppose that a and m are integers such that $0 < a < m$ and $(a, m) = 1$. Then Euler's Theorem implies that $a^{\phi(m)} \equiv 1 \pmod{m}$. In many cases, there is a smaller positive exponent h such that $a^h \equiv 1 \pmod{m}$. For example, if $a = 4$ and $m = 7$, then $(4, 7) = 1$ and $\phi(7) = 6$, but $4^3 \equiv 1 \pmod{7}$.

If h is the least positive exponent such that $a^h \equiv 1 \pmod{m}$, then we say that a *has order* $h \pmod{m}$, or a *belongs to* $h \pmod{m}$. By Euler's Theorem, h exists, and $h \leq \phi(m)$. In fact, we shall see that $h | \phi(m)$. If g is an integer such that $0 < g < m$, $(g, m) = 1$, and g has order $\phi(m)$ $\pmod{m}$, then we call g a *primitive root* $\pmod{m}$. For example, we shall see that 3 and 5 are primitive roots $\pmod{7}$.

Primitive roots are important because of the following property: If g is a primitive root $\pmod{m}$, and if $(t, m) = 1$, then t can be represented as a non-negative power of g $\pmod{m}$, that is, $t \equiv g^k \pmod{m}$ for some integer k such that $0 \leq k \leq \phi(m) - 1$. For example, if we take $m = 7$ and $g = 5$, then we have

$$1 \equiv 5^0 \ , \ 2 \equiv 5^4 \ , \ 3 \equiv 5^5 \ , \ 4 \equiv 5^2 \ , \ 5 \equiv 5^1 \ , \ 6 \equiv 5^3 \pmod{7}$$

Not all natural numbers have primitive roots. For example, there are no primitive roots (mod m) if $m = 8, 12, 15$. We will determine for a given m how many primitive roots exist, if any, and how to generate all primitive roots from a single primitive root. Finally, we develop the properties of *indices*, which are related to primitive roots, and which are helpful in solving certain polynomial congruences.

6.2 Primitive Roots: Definition and Properties

If a and m are integers such that $m \geq 2$ and $(a, m) = 1$, let S be the set of positive integers, k, such that $a^k \equiv 1 \pmod{m}$. Euler's Theorem implies that $\phi(m) \in S$, so S is non empty. We now investigate the properties of the least element of S, which we define as the *order* of a (mod m).

Definition 6.1   **Order** (mod m)

If $(a, m) = 1$, let h be the least positive integer such that $a^h \equiv 1 \pmod{m}$. We say that a *has order* h (mod m), or that a *belongs to* h (mod m).

For example, let us compute the order of 2 (mod 7). Now

$$2^1 \equiv 2 \not\equiv 1 \pmod{7}$$

$$2^2 \equiv 4 \not\equiv 1 \pmod{7}$$

$$2^3 = 8 \equiv 1 \pmod{7}$$

Therefore 2 has order 3 (mod 7). Also,

$$5^1 \equiv 5 \not\equiv 1 \pmod{8}$$

$$5^2 = 25 \equiv 1 \pmod{8}$$

Therefore 5 has order 2 (mod 8).

Note that if a has order h (mod m), then $a^h \equiv 1 \pmod{m}$ and $a^k \not\equiv 1 \pmod{m}$ for all k such that $0 < k < h$. Furthermore, by Euler's Theorem, we have $h \leq \phi(m)$.

Our first two theorems concerning primitive roots will enable us to say more about h, namely, $h | \phi(m)$.

Theorem 6.1   Suppose that $(a, m) = 1$, a has order h (mod m), and $a^k \equiv 1 \pmod{m}$ for $k > 0$. Then $h | k$.

Proof: By the Division Algorithm, we have $k = qh + r$, where $0 \leq r < h$. Now $a^k = a^{qh+r} = (a^h)^q a^r$. Since $a^k \equiv a^h \equiv 1 \pmod{m}$, by hypothesis, it follows that $a^r \equiv 1 \pmod{m}$. This implies $r = 0$, so $h|k$. ∎

Theorem 6.2   If $(a, m) = 1$ and a has order $h \pmod{m}$, then $h|\phi(m)$.

Proof: This follows from the hypothesis, Euler's Theorem, and Theorem 6.1. ∎

Theorem 6.2 is useful in computing the order of $t \pmod{m}$, where t is an integer such that $(t, m) = 1$. For example, if we wish to find the order of $t \pmod{5}$, where $1 \leq t \leq 4$, then Theorem 6.2 tells us that the only possible candidates are the divisors of 4, namely, 1, 2, and 4. Let us compute the order $\pmod{5}$ of each of the integers 1, 2, 3, 4, making use of the multiplication table $\pmod{5}$.

$*$	0	1	2	3	4
0	0	0	0	0	0
1	0	1	2	3	4
2	0	2	4	1	3
3	0	3	1	4	2
4	0	4	3	2	1

We see that $1^1 \equiv 1 \pmod{5}$, so 1 has order 1 $\pmod{5}$; also

$$2^1 \equiv 2 \not\equiv 1 \pmod{5}$$

$$2^2 \equiv 4 \not\equiv 1 \pmod{5}$$

$$2^4 \equiv (2^2)^2 \equiv 4^2 \equiv 1 \pmod{5}$$

Therefore 2 has order 4 (mod 5). In addition

$$3^1 \equiv 3 \not\equiv 1 \pmod{5}$$

$$3^2 \equiv 4 \not\equiv 1 \pmod{5}$$

$$3^4 \equiv (3^2)^2 \equiv 4^2 \equiv 1 \pmod{5}$$

Therefore 3 has order 4 (mod 5). Also,

$$4^1 \equiv 4 \not\equiv 1 \pmod{5}$$

$$4^2 \equiv 1 \pmod{5}$$

Therefore 4 has order 2 (mod 5).

Next, we introduce the concept of a *primitive root* (mod m).

Definition 6.2   **Primitive Root** (mod m)

If g is an integer such that $(g, m) = 1$ and g has order $\phi(m)$ (mod m), then we say that g is a *primitive root* (mod m).

For example, the above computations show that 2 and 3 are primitive roots (mod 5), since 2 and 3 both have order 4 (mod 5), and $\phi(5) = 4$. Similarly, one can show that 3 and 5 are primitive roots (mod 7).

There are moduli for which there are no primitive roots. The least such modulus is $m = 8$. If t is an integer such that $(t, 8) = 1$, then $t \equiv 1, 3, 5, 7$ (mod 8). Now $1^1 \equiv 3^2 \equiv 5^2 \equiv 7^2 \equiv 1$ (mod 8). Note that $\phi(8) = 4$, but no integer has order 4 (mod 8), so there are no primitive roots (mod 8). The reader can verify that there are no primitive roots (mod 12).

In the work ahead, we will see that

   (i) if p is an odd prime and $k \geq 1$, then primitive roots exist (mod p^k) and (mod $2p^k$).

   (ii) if g is a primitive root (mod m), then there are precisely $\phi(\phi(m))$ primitive roots (mod m), all of which can be obtained from g.

   (iii) if $m \neq 2, 4, p^k, 2p^k$ where p is an odd prime, then m has no primitive roots.

If $(a, m) = 1$ and a has order h (mod m), the following theorem tells us how to compute the order of any power of a.

Theorem 6.3   If $(a, m) = 1$, a has order h (mod m), and $k \geq 1$, then a^k has order $h/(h, k)$ (mod m).

Proof: Let a^k have order j (mod m). We will show that $j = h/(h, k)$. Let $d = (h, k)$, so that $h = bd$ and $k = cd$ for some integers b, c such that $(b, c) = 1$. Note that $b = h/(h, k)$. Now $(a^k)^b = (a^{cd})^b = (a^{bd})^c = (a^h)^c \equiv 1^c \equiv 1$ (mod m), by definition of h. Therefore, by definition of j and by Theorem 6.1, it follows that $j | b$. On the other hand, $a^{kj} = (a^k)^j \equiv 1$ (mod m) be definition of j, so Theorem 6.1 implies $h | jk$, that is, $bd | cdj$. By Theorem 2.1, part D_5, we have $b | cj$. Since $(b, c) = 1$, Euclid's Lemma implies $b | j$. Therefore $j = b = h/(h, k)$. ∎

For example, the reader can verify that 2 has order 8 (mod 17). Since $8 = 2^3$, it follows that 8 has order $8/(8, 3) = 8/1 = 8$ (mod 17). Also, since $16 = 2^4$, it follows that 16 has order $8/(8, 4) = 8/4 = 2$ (mod 17).

In elementary algebra, one learns that if $a > 1$, then $a^j = a^k$ if and only if $j = k$. The following theorem states a corresponding property regarding congruences.

Theorem 6.4
If $(a, m) = 1$, let a have order h (mod m). Then $a^j \equiv a^k$ (mod m) if and only if $j \equiv k$ (mod h).

Proof: First assume that $a^j \equiv a^k$ (mod m), and without loss of generality, that $j \geq k$. Therefore $j - k \geq 0$. Cancelling a factor a^k, we obtain $a^{j-k} \equiv 1$ (mod m). By hypotehesis, a has order h (mod m). Therefore Theorem 6.1 implies $h|(j - k)$, that is $j - k \equiv 0$ (mod h), so $j \equiv k$ (mod h). Conversely, suppose that $j \equiv k$ (mod h). Therefore $j = k + th$ for some $t \geq 0$. Now $a^j = a^{k+th} = a^k(a^h)^t \equiv a^k 1^t \equiv a^k$ (mod m). ∎

For example, note that $2^{26} \equiv 2^2 \equiv 4$ (mod 17). Now 2 has order 8 (mod 17), and $26 \equiv 2$ (mod 8). In the opposite direction, since $11 \equiv 3$ (mod 8), it follows that $2^{11} \equiv 2^3 \equiv 8$ (mod 17).

Note that if $(a, m) = 1$, then Theorem 6.4 provides a convenient way to evaluate the least positive residue of a^j (mod m), as follows. First find the order h of a (mod m). Then, find r such that $j = qh + r$. Finally, compute a^r (mod m).

For example, suppose we wish to compute 5^{100} (mod 7). Now 5 has order 6 (mod 7). Since $100 = 16(6) + 4$, it follows that $5^{100} \equiv 5^4 \equiv 2$ (mod 7).

Theorem 6.4 has the following significant consequence:

Theorem 6.5
If $(a, m) = 1$ and a has order h (mod m), then $1, a, a^2, a^3, \cdots, a^{h-1}$ are all distinct (mod m).

Proof: We will show that if $a^j \equiv a^k$ (mod m), where $0 \leq k \leq j \leq h - 1$, then $j = k$. Now Theorem 6.4 implies $j - k = th$ for some integer $t \geq 0$. But $0 \leq j - k \leq h - 1$, so $0 \leq th \leq h - 1$. This implies $t = 0$, so $j = k$. ∎

For example, since $(2, 7) = 1$ and 2 has order 3 (mod 7), it follows that $1, 2, 2^2$ are all distinct (mod 7). Also, since $(5, 7) = 1$ and 5 has order 6 (mod 7), it follows that $1, 5, 5^2, 5^3, 5^4, 5^5$ are all distinct (mod 7). Note that

$$1 \equiv 5^0 \quad (\text{mod } 7)$$

$$2 \equiv 5^4 \quad (\text{mod } 7)$$

$$3 \equiv 5^5 \quad (\text{mod } 7)$$

$$4 \equiv 5^2 \quad (\text{mod } 7)$$

$$5 \equiv 5^1 \quad (\text{mod } 7)$$

$$6 \equiv 5^3 \quad (\text{mod } 7)$$

More generally, we have the following:

Theorem 6.6   If g is a primitive root (mod m), and if a is an integer such that $(a, m) = 1$, then there exists an integer k such that $0 \leq k \leq \phi(m) - 1$ and $a \equiv g^k$ (mod m).

Proof: Let $S = \{j : 1 \leq j < m \text{ and } (j, m) = 1\}$. Note that S has $\phi(m)$ elements. Let r be the least positive residue of a (mod m), so $r \in S$. Let T be the set of least positive residues (mod m) of the set $\{1, g, g^2, \cdots, g^{\phi(m)-1}\}$. Since g has order $\phi(m)$ by hypothesis, Theorem 6.5 implies that the $\phi(m)$ elements of t are distinct (mod m). Now $T \subset S$, and T has the same number of elements as s, namely, $\phi(m)$. It follows that $T = S$, so $r \equiv g^k$ (mod m) for some k such that $0 \leq k \leq \phi(m) - 1$. Since $a \equiv r$ (mod m), we have $a \equiv g^k$ (mod m).   ∎

For example, if $m = 9$, then $S = \{1, 2, 4, 5, 7, 8\}$. Now 2 is a primitive root (mod 9) and we have

$$1 \equiv 2^0 \quad (\text{mod } 9)$$

$$2 \equiv 2^1 \quad (\text{mod } 9)$$

$$4 \equiv 2^2 \quad (\text{mod } 9)$$

$$5 \equiv 2^5 \quad (\text{mod } 9)$$

$$7 \equiv 2^4 \quad (\text{mod } 9)$$

$$8 \equiv 2^3 \quad (\text{mod } 9)$$

If m has a primitive root, g, how many primitive roots are there altogether and how can we find them? The following theorem answers these questions.

Theorem 6.7   If m has at least one primitive root, g, then m has exactly $\phi(\phi(m))$ primitive roots, namely, all g^k such that $0 < k < \phi(m)$ and $(k, m) = 1$.

Proof: Suppose that $0 < a < m$ and $(a, m) = 1$. Since g is a primitive root (mod m) by hypothesis, it follows from Theorem 6.6 that $a \equiv g^k$ (mod m) for some k such that $0 \leq k \leq \phi(m) - 1$. Now Theorem 6.3 implies that g^k has order $\phi(m)/(k, \phi(m))$ (mod m). Therefore g^k is a primitive root (mod m)

if and only if $\phi(m)/(k,\phi(m)) = \phi(m)$, that is, if and only if $(k,\phi(m)) = 1$. The number of such k is $\phi(\phi(m))$.   ∎

For example, 2 is a primitive root (mod 11). Therefore, the total number of primitive roots (mod 11) is $\phi(\phi(11)) = \phi(10) = 4$. These primitive roots may be obtained from the least positive residues (mod 11) of 2^k, where $0 < k < 10$ and $(k,10) = 1$, namely $k = 1, 3, 7, 9$. We thus obtain the primitive roots:

$$2^1 \equiv 2 \pmod{11}$$

$$2^3 \equiv 8 \pmod{11}$$

$$2^7 \equiv 7 \pmod{11}$$

$$2^9 \equiv 6 \pmod{11}$$

Therefore we see that the primitive roots (mod 11) are 2, 6, 7, 8.

A property of primitive roots that will be useful in Chapter 7 is given by the following theorem:

Theorem 6.8   If p is an odd prime, and if g is a primitive root (mod p), then

$$g^{\frac{p-1}{2}} \equiv -1 \pmod{p}$$

Proof: Exercise.

Section 6.2   Exercises

1. For each integer, t, such that $1 \le t \le 12$ and $(t,12) = 1$, find the order of t (mod 13). Which, if any, of these t are primitive roots (mod 13)?

2. For each integer, t, such that $0 < t < 20$ and $(t,20) = 1$, find the order of t (mod 20). Which, if any, of these t are primitive roots (mod 20)?

3. Verify that 2 is a primitive root (mod 19). Find all primitive roots (mod 19). (Express them as least positive residues.)

4. For each of the following integers, assuming that there is at least one primitive root, find the number of primitive roots:
   (a)   49   (d)   54
   (b)   50   (e)   58
   (c)   53

5. Assume that every prime has primitive roots. For each of the following values of k, find all primes that have exactly k primitive roots.

(a) 1   (e) 8
(b) 2   (f) 10
(c) 4   (g) 12
(d) 6

6. Prove that no prime has exactly 14 primitive roots.

7. Prove that if p is an odd prime and $(t, p) = 1$, then t^2 is *not* a primitive root (mod p).

8. Prove Theorem 6.8.

9. If p is an odd prime that has 2^k primitive roots for some $k \geq 1$, what must be true about p?

10. Let n be the number of primitive roots (mod p), where the prime $p \geq 7$. Prove that if $n \equiv 2 \pmod{4}$, then $p \equiv 3 \pmod{4}$.

11. Prove that if t^k is a primitive root (mod p), where p is an odd prime, then t itself must be a primitive root (mod p).

12. Let g and h be primitive roots (mod p), where p is an odd prime. Can gh also be a primitive root (mod p)? (Explain.)

13. Let p be an odd prime such that 2 is a primitive root (mod p). Find necessary and sufficient congruence conditions on p such that 8 is also a primitive root (mod p).

14. Let g be a primitive root (mod p), where p is an odd prime. Prove that $p - g$ is also a primitive root (mod p) if and only if $p \equiv 1 \pmod{4}$.

15. If x is a real number such that $0 < x < 1$, then x has a decimal expansion,

$$x = \sum_{k=1}^{\infty} \frac{a_k}{10^k} = .a_1 a_2 a_3 \cdots$$

that is essentially unique. If x is rational, then the decimal expansion is eventually periodic, that is, there exist m, t such that $a_{t+k} = a_t$ for all $k \geq m$. The least such t is called the *period* of the decimal expansion. For example,

$$\frac{1}{7} = .142857142857142857 \cdots$$

so $\frac{1}{7}$ has period $t = 6$. Prove that if p is a prime such that $(p, 10) = 1$, then the period of the decimal expansion of $\frac{1}{p}$ is the order of 10 (mod p).

★16. Let p be an odd prime such that $p \nmid mn$, and each of m, n has order 2^d (mod p), where $2^d | (p - 1)$. Prove that mn has order 2^c (mod p), where $0 \leq c \leq d - 1$.

17. Let p be an odd prime. Prove that $(p - 1)/2$ is a primitive root (mod p) if and only if $2(-1)^{(p-1)/2}$ is a primitive root (mod p).

Section 6.2 Computer Exercises

18. Write a computer program that, given an integer, t, such that $1 \leq t \leq 256$, will compute the order of t (mod 257).

19. Write a computer program that, given an odd prime, p, (a) finds the least primitive root (mod p) by trial and error, (b) finds all primitive roots (mod p).

20. A famous conjecture by the distinguished mathematician Emil Artin (1898–1962) states that 2 is a primitive root (mod p) for infinitely many primes, p. Write a computer program to find all primes p such that $2 < p < 1000$ and 2 is a primitive root (mod p).

6.3 Primitive Roots: Existence

Let p be an odd prime. A key result given in Theorem 6.13 states that p has primitive roots, in fact, $\phi(p - 1)$ of them. In order to obtain this result, we must first derive some properties of polynomial congruences (mod p). It is well-known that if x represents a real or complex number, then a polynomial in x of degree n can have at most n distinct roots. The following theorem says that this property of polynomials also holds if x is an integer (mod p).

Theorem 6.9

Let p be an odd prime, $n \geq 1$, and $f(x) = \sum_{i=0}^{n} a_i x^i$, where the a_i are integers and $(a_n, p) = 1$. Then the congruence

$$f(x) \equiv 0 \quad (\text{mod } p)$$

has at most n distinct solutions (mod p).

Proof: (Induction on n) If $n = 1$, then $f(x) = a_1 x + a_0$ and $a_1, p) = 1$ by hypothesis. Therefore Theorem 4.5 implies that Congruence (6.1) has exactly one solution. Now assume by induction that Theorem 6.9 holds for all polynomials of degree m, where $1 \leq m < n$. Suppose that (6.1) has $n+1$ distinct solutions, namely, $x \equiv r_1, r_2, r_3, \cdots, r_n, r_{n+1}$ (mod p). Let

$$g(x) = f(x) - a_n \prod_{i=1}^{n} (x - r_i)$$

Therefore

$$g(r_j) = f(r_j) - a_n \prod_{i=1}^{n} (r_j - r_i)$$

sp $g(r_j) \equiv 0 \pmod{p}$ for all j such that $1 \leq j \leq n$. That is to say, $g(x)$ has n distinct roots (mod p). Let $g(x)$ have degree k. By definition, $g(x)$ has degree at most $n - 1$, that is, $0 \leq k \leq n - 1$. By induction hypothesis, it is impossible that $1 \leq k \leq n - 1$. Therefore $k = 0$, so $g(x)$ is a constant function. Since $g(r_j) \equiv 0 \pmod{p}$ for $1 \leq j \leq n$, we must have $g(x) \equiv 0 \pmod{p}$ for all x. This implies that

$$f(x) \equiv a_n \prod_{i=1}^{n} (x - r_i) \pmod{p}$$

However, then

$$f(r_{n+1}) \equiv a_n \prod_{i=1}^{n} (r_{n+1} - r_i) \pmod{p}$$

Since $f(r_{n+1}) \equiv 0 \pmod{p}$ by hypothesis, we have

$$a_n \prod_{i=1}^{n} (r_{n+1} - r_i) \equiv 0 \pmod{p}$$

Since $(a_n, p) = 1$ by hypothesis, it follows that

$$\prod_{i=1}^{n} (r_{n+1} - r_i) \equiv 0 \pmod{p}$$

Since p is prime, we must have $r_{n+1} - r_i \equiv 0 \pmod{p}$ for some index i such that $1 \leq i \leq n$. This contradicts the assumption that the r_i are all distinct (mod p).   ∎

For example, the congruence $x^3 + x \equiv 0 \pmod{17}$ has the 3 solutions: $x \equiv 0, 4, 13 \pmod{17}$.

Remarks

If $f(x)$ is a polynomial of degree n, and m as an integer that has no primitive roots, then the congruence $f(x) \equiv 0 \pmod{n}$ may have as many as $2n$ distinct solutions. For example, the congruence

$$x^2 - 1 \equiv 0 \pmod{8}$$

has 4 solutions, namely, $x \equiv 1, 3, 5, 7 \pmod{8}$.

■

Next, we apply Theorem 6.9 to determine the exact number of roots of two specific polynomials.

Theorem 6.10   If p is prime, then the congruence

$$x^{p-1} - 1 \equiv 0 \pmod{p} \tag{6.1}$$

has exactly $p - 1$ solutions.

Proof: Fermat's Little Theorem (Theorem 4.8) implies that (6/2) has at least $p - 1$ solutions, namely, $x \equiv 1, 2, 3, \cdots, p - 1 \pmod{p}$. Theorem 6.9 implies that (6.2) has at most $p - 1$ solutions. Therefore (102) has exactly $p - 1$ solutions.   ■

The following theorem states a stronger result.

Theorem 6.11   If p is prime and $d | (p - 1)$, then the congruence

$$x^d - 1 \equiv 0 \pmod{p} \tag{6.2}$$

has exactly d solutions.

Proof: If $d = p - 1$, then the conclusion follows from Theorem 6.10. Otherwise, by hypothesis, we have $p - 1 = hd$, with $1 < h < p$. Let

$$f(x) = x^d - 1 \; ; \; g(x) = \frac{x^{p-1} - 1}{x^d - 1} = x^{(h-1)d} + x^{(h-2)d} + \cdots + x^d + 1$$

We will consider the congruences

$$f(x) \equiv 0 \pmod{p} \tag{6.3}$$

and

$$g(x) \equiv 0 \pmod{p} \tag{6.4}$$

Suppose that (6.4) has j solutions while (6.5) has k solutions. We wish to show that $j = d$. Note that the degree of $g(x)$ is $(h - 1)d = p - 1 - d$. Theorem 6.9 implies that $j \leq d$ and $k \leq p - 1 - d$. If t is a solution to (6.4), then $t^d \equiv 1 \pmod{p}$, so $g(t) \equiv h \not\equiv 0 \pmod{p}$. Therefore (6.4) and (6.5) have no common solution. Since $x^{p-1} - 1 = f(x)g(x)$ and p is prime, it follows that $x^{p-1} - 1 \equiv 0 \pmod{p}$ if and only if $f(x) \equiv 0 \pmod{p}$ or $g(x) \equiv 0 \pmod{p}$. Therefore the number of solutions to (6.3) is the sum of the number of solutions to each of the congruences (6.4) and (6.5), that is, $p - 1 = j + k$. Theorem 6.9 implies $k \leq p - 1 - d$, so that $j \geq d$. Therefore $j = d$.   ■

For example, consider the congruence, $x^3 - 1 \equiv 0 \pmod 7$. Since 7 is prime, $7 - 1 = 6$, and $3|6$, it follows that there are 3 solutions, namely, $x \equiv 1, 2, 4 \pmod 7$.

The fact that every prime has primitive roots follows from the following, more general theorem, whose proof is an elaborate counting argument.

Theorem 6.12 If p is prime and if $d|(p-1)$, then the number of integers, k, such that $1 \le k \le p - 1$ and k has order $d \pmod p$ is $\phi(d)$.

Proof: If p is prime and $d|(p-1)$, let $f(d)$ denote the number of intgegers, k, such that $1 \le k \le p - 1$ and k has order $d \pmod p$. If no such k exists, then $f(d) = 0 < \phi(d)$. If such a k does exist, then k is a solution of the congruence

$$x^d - 1 \equiv 0 \pmod p \tag{6.5}$$

In fact, for each j such that $0 \le j \le d - 1, k^j$ is a solution of (6.6). Theorem 6.11 implies that (6.6) has exactly d solutions. Therefore every solution to (6.6) must have the form k^j, with $0 \le j \le d - 1$. Since $(k, p) = 1$ by hypothesis, Theorem 6.5 implies k^j has order $d \pmod p$ if and only if $(j, d) = 1$. The number of such j is $\phi(d)$. Therefore, in this case, $f(d) = \phi(d)$. We have shown that $0 \le f(d) \le \phi(d)$ for each d such that $d|(p-1)$. Now Theorem 6.2 implies $\sum_{d|(p-1)} f(d) = p - 1$. Recall that Theorem 5.20 implies $\sum_{d|(p-1)} \phi(d) = p - 1$. Therefore $\sum_{d|(p-1)} f(d) = \sum_{d|(p-1)} \phi(d) = p - 1$. Since $0 \le f(d) \le \phi(d)$ for each divisor, d, of $p - 1$, it follows that $f(d) = \phi(d)$ for each such d. ∎

For example, let $p = 7$, so that $p - 1 = 6$. If $d|6$, then $d \in \{1, 2, 3, 6\}$. Among the integers from 1 to 6, we have

$\phi(1) = 1$ with order 1 (mod 7), namely 1
$\phi(2) = 1$ with order 2 (mod 7), namely 6
$\phi(3) = 2$ with order 3 (mod 7), namely 2 and 4
$\phi(6) = 2$ with order 6 (mod 7), namely 3 and 5
Applying Theorem 6.12 to the case $d = p - 1$, we obtain

Theorem 6.13 If p is prime, then p has $\phi(p-1)$ primitive roots.

Proof: This follows from Definition 6.2 and Theorem 6.12, taking $d = p-1$. ∎

For example, the prime 7 has $\phi(6) = 2$ primitive roots, namely, 3 and 5. Also, the prime 11 has $\phi(10) = 4$ primitive roots, namely, 2, 6, 7, 8.

Recall from Theorem 6.7 that if m has primitive roots, and if g is one primitive root (mod m), then we can obtain all primitive roots (mod m) by raising g to appropriate exponents. For example, 3 is a primitive root (mod 31), so all primitive roots (mod 31) are of the form 3^j (mod 31), where $1 \leq j \leq 30$ and $(j, 30) = 1$. Since $\phi(30) = 8$, there are 8 such values of j, namely, $j = 1, 7, 11, 13, 17, 19, 23, 29$. The corresponding primitive roots (mod 31) are $3^1 \equiv 3$ (mod 31), $3^7 \equiv 17$ (mod 31), $3^{11} \equiv 13$ (mod 31), $3^{13} \equiv 24$ (mod 31), $3^{17} \equiv 22$ (mod 31), $3^{19} \equiv 12$ (mod 31), $3^{23} \equiv 11$ (mod 31), $3^{29} \equiv 21$ (mod 31).

Table 2 in Appendix B lists the *least* primitive root (mod p), denoted g, for all primes below 1000. If $p > 1000$, there is no convenient formula for computing g. We can, however, determine g by the following "brute force" algorithm. Set $k = 2$. Compute the order, h, of k (mod p). If $h = p - 1$, then $k = g$, and we are done. If $h < p - 1$, then increment k by 1 and then repeat the process. This algorithm can be refined by eliminating from consideration numbers that are powers, such as 4, 8, 9, etc., as well as integers that are "quadratic residues" (mod p). (The latter are defined in Chapter 7.)

For example, let us compute g, the least primitive root (mod 41). Since $2^{20} \equiv 3^8 \equiv 5^{20} \equiv 1$ (mod 41), it follows that $g \geq 6$. Since $6^d \not\equiv 1$ (mod 41) if $d|40$ and $d < 40$, it follows that $g = 6$.

Next, we consider primitive roots (mod p^n) where p is an odd prime and $n \geq 2$. We shall see that such primitive roots always exist, and we shall learn how to find them.

Usually, if g is a primitive root (mod p), then g is also a primitive root (mod p^2). For example, 2 is a primitive root (mod 3) and also (mod 9). Moreover, 3 is a primitive root (mod 7) and also (mod 49). Exceptions do exist, however. For example, 14 is a primitive root (mod 29), but not (mod 29^2). Also, 18 is a primitive root (mod 37), but not (mod 37^2). The following theorem specifies how to find a primitive root (mod p^2), given a primitive root (mod p).

Theorem 6.14   Let g be a primitive root (mod p), where p is an odd prime, and $0 < g < p$. If $g^{p-1} \not\equiv 1$ (mod p^2), then g is a primitive root (mod p^2). If $g^{p-1} \equiv 1$ (mod p^2), then g^{p-2} is a primitive root (mod p^2).

Proof: By hypothesis, $(g, p) = 1$. Therefore Theorem 4.5 implies that there exists h such that $gh \equiv 1$ (mod p). Without loss of generality, we may choose h so that $0 < h < p$. By Fermat's Little Theorem, $h \equiv g^{p-1}h \equiv g^{p-2}gh \equiv g^{p-2}$ (mod p). Since $(p - 2, p - 1) = 1$, Theorem 6.3 implies that h has order $p - 1$ (mod p), that is, h is a primitive root (mod p).

Let g have order j (mod p^2). Then Theorem 6.2 implies $j|\phi(p^2)$, that is, $j|p(p-1)$. Also $g^j \equiv 1$ (mod p^2), so Theorem 4.1, part C_9 implies $g^j \equiv 1$ (mod p). Since g is a primitive root (mod p) by hypothesis, Theorem 6.1 implies $(p-1)|j$. Therefore $j/(p-1)$ is an integer that divides p, so $j/(p-1) = 1$ or p, that is, $j = p-1$ or $j = p(p-1)$. Similarly, h has order $p-1$ or $p(p-1)$ (mod p^2). If $j = p(p-1)$, then g is a primitive root (mod p^2). It remains to show that if $j = p-1$, then h is a primitive root (mod p^2), that is, h has order $p(p-1)$ (mod p^2).

Now $1 < g < p$ and $1 < h < p$, so $1 < gh < p^2$. Also, since $gh \equiv 1$ (mod p), we have $gh = 1 + kp$, with $0 < k < p$. Furthermore,

$$(gh)^{p-1} = (1+kp)^{p-1} = 1 + (p-1)kp + \sum_{i=2}^{p-1} \binom{p-1}{i} (kp)^i$$

so $(gh)^{p-1} \equiv 1 - kp$ (mod p^2). If h has order $p-1$ (mod p^2), then $(gh)^{p-1} \equiv g^{p-1}h^{p-1} \equiv 1$ (mod p^2). But this leads to $p|k$, an impossibility. Therefore h has order $p(p-1)$ (mod p^2); that is, h is a primitive root (mod p^2). ∎

For example, suppose we wish to find a primitive root (mod 25). Now 2 is a primitive root (mod 5) and $2^{5-1} = 2^4 = 16 \not\equiv 1$ (mod 25), so 2 is also a primitive root (mod 25). As a second example, suppose we wish to find a primitive root (mod 29^2), given that 14 is a primitive root (mod 29). Now $14^{29-1} = 14^{28} \equiv 1$ (mod 29^2), so 14 is *not* a primitive root (mod 29^2). To obtain a primitive root (mod 29^2), we solve the congruence, $14h \equiv 1$ (mod 29). This yields $h \equiv 27$ (mod 29). Therefore 27 is a primitive root (mod 29^2).

Given g, a primitive root (mod p), there are other ways to obtain a primitive root (mod p^2). We state the following relevant theorem:

Theorem 6.15   Let g be a primitive root (mod p), where p is an odd prime. Then each of the following is a primitive root (mod p^2):

(i) $g^p - p$

(ii) $g^p - gp$

(iii) at least one of $g, g+p$.

Proof: Omitted, but see [16], [12], [15].

Next, we turn our attention to obtaining primitive roots (mod p^n), where p is an odd prime and $n \geq 3$. If we possess a primitive root (mod p^2), then no further computation is needed, as the next theorem makes clear.

Theorem 6.16   If p is an odd prime, and if g is a primitive root $(\mod p^2)$, then g is also a primitive root $(\mod p^n)$ for all $n \geq 2$.

Proof: (Induction on n) By hypothesis, g is a primitive root $(\mod p^2)$. We must show that if g is a primitive root $(\mod p^n)$, then g is also a primitive root $(\mod p^{n+1})$. Let g have order j $(\mod p^{n+1})$. Then Theorem 6.2 implies $j | \phi(p^{n+1})$. Now $g^j \equiv 1$ $(\mod p^{n+1})$, which implies $g^j \equiv 1$ $(\mod p^n)$. Since g is a primitive root $(\mod p^n)$ by induction hypothesis, Theorem 6.1 implies $\phi(p^n) | j$. Since $\phi(p^{n+1}) = p\phi(p^n)$, we must have $j = \phi(p^n)$ or $j = \phi(p^{n+1})$. Since $(g, p) = 1$, Euler's Theorem (Theorem 5.25) implies $g^{\phi(p^{n-1})} \equiv 1$ $(\mod p^{n-1})$, that is, $g^{\phi(p^{n-1})} = 1 + kp^{n-1}$. Since g is a primitive root $(\mod p^n)$ by hypothesis, it follows that $g^{\phi(p^{n-1})} \not\equiv 1$ $(\mod p^n)$, so $p \nmid k$. Now

$$g^{\phi(p^n)} = (g^{\phi(p^{n-1})})^p = (1 + kp^{n-1})^p = 1 + kp^n + \sum_{i=2}^{p} \binom{p}{i} (kp^{n-1})^i$$

so

$$g^{\phi(p^n)} \equiv 1 + kp^n \quad (\mod p^{n+1})$$

Therefore $g^{\phi(p^n)} \not\equiv 1$ $(\mod p^{n+1})$. This implies $j = \phi(p^{n+1})$, so g is a primitive root $(\mod p^{n+1})$. $\blacksquare$

For example, since 2 is a primitive root $(\mod 3^2)$, it follows that 2 is a primitive root $(\mod 3^n)$ for all n.

The next theorem says that integers that are twice a power of an odd prime also have primitive roots.

Theorem 6.17   Let g be a primitive root $(\mod p^n)$ where p is an odd prime, and $n \geq 1$. If g is odd, then g is also a primitive root $(\mod 2p^n)$. If g is even, then $g + p^n$ is a primitive root $(\mod 2p^n)$.

Proof: Exercise.

For example, since 5 is a primitive root $(\mod 9)$ and 5 is odd, it follows that 5 is also a primitive root $(\mod 18)$. Also, since 2 is a primitive root $(\mod 9)$ but 2 is even, it follows that $2 + 9 = 11$ is a primitive root $(\mod 18)$.

We have seen so far that if p is an odd prime and $n \geq 1$, then p^n and $2p^n$ have primitive roots. It is easily verified that 1 is a primitive root $(\mod 2)$ and 3 is a primitive root $(\mod 4)$. It remains to show, via the following four theorems, that no other natural numbers have primitive roots.

Theorem 6.18 | If $m \geq n \geq 3$ and $(m, n) = 1$, then mn has no primitive roots.

Proof: Let $d = (\phi(m), \phi(n))$, and $t = [\phi(m), \phi(m)]$. Since $m \geq 3$ by hypothesis, it follows that either $m = 2^k$ for some $k \geq 2$, or $p|m$ for some odd prime, p. In either case, we have $2|\phi(m)$. Similarly, we have $2|\phi(n)$. Therefore $2|d$, so $2 \leq d$. Theorem 2.9, part L_7 implies $t = \phi(m)\phi(n)/d$, so we have $t \leq \frac{1}{2}\phi(m)\phi(n)$. Since $(m, n) = 1$ by hypothesis, Theorem 5.21 implies $\phi(mn) = \phi(m)\phi(n)$, so $t \leq \frac{1}{2}\phi(mn)$. Now suppose that a is an integer such that $(a, mn) = 1$, so $(a, m) = (a, n) = 1$. Euler's Theorem (Theorem 5.25) implies $a^{\phi(m)} \equiv 1 \pmod{m}$, hence $a^t \equiv 1 \pmod{m}$. Similarly, $a^t \equiv 1 \pmod{n}$. Therefore $a^t \equiv 1 \pmod{[m, n]}$. But $(m, n) = 1$ implies $[m, n] = mn$. Therefore we have $a^t \equiv 1 \pmod{mn}$. Since $t \leq \frac{1}{2}mn$, it follows that a is *not* a primitive root $\pmod{mn}$. ∎

As a consequence of Theorem 6.18, we have the following:

Theorem 6.19 | Let p and q be odd primes. If n is an integer such that $4p|n$ or $pq|n$, then n has no primitive roots.

Proof: If $4p|n$, then $n = 2^j m$ with $j \geq 2, m$ odd. If $pq|n$, then $n = p^j m$ with $j \geq 1, m$ odd, $(p, m) = 1, m \geq 3$. In either case, the conclusion follows from Theorem 6.18. ∎

The only integers not yet considered are $n = 2^k$, where $k \geq 3$. This case is addressed by the next theorem.

Theorem 6.20 | If t is odd and $n \geq 3$, then $t^{2^{n-2}} \equiv 1 \pmod{2^n}$.

Proof: (Induction on n)   If t is odd, then $t \equiv 1, 3, 5, 7 \pmod{8}$. In each case, we have $t^2 \equiv 1 \pmod{8}$. Therefore the conclusion holds for $n = 3$. Now assume by induction hypothesis that $t^{2^{n-2}} \equiv 1 \pmod{2^n}$, that is, $2^n|(t^{2^{n-2}} - 1)$. Since t is odd by hypothesis, it follows that $2|(t^{2^{n-2}} + 1)$. Therefore $(2^n)2|(t^{2^{n-2}}-1)(t^{2^n}+1)$, that is, $2^{n+1}|(t^{2^{n-1}}-1)$, so that $t^{2^{n-1}} \equiv 1 \pmod{2^{n+1}}$. ∎

As an immediate consequence of Theorem 6.20, we obtain

Theorem 6.21 | If $n = 2^k$ where $k \geq 3$, then n has no primitive roots.

Proof: If t is odd and $k \geq 3$, let t have order $h \pmod{2^k}$. By Theorems 6.20 and 6.1, we have $h|2^{k-2}$, hence $h \leq 2^{k-2}$. But $\phi(2^k) = 2^{k-1}$. The conclusion now follows. ∎

Combining our results, we have

| Theorem 6.22 | Let m be a natural number. Then m has a primitive root if and only if $m = 2, 4, p^n$, or $2p^n$ where p is an odd prime and $n \geq 1$. |

Proof: This follows from Theorems 6.13, 6.14, 6.16, 6.18, 6.19, and 6.21.
∎

Section 6.3   Exercises

1. Find all roots of each of the following congruences:

    (a) $x^3 \equiv 1 \pmod{13}$

    (b) $x^4 \equiv 1 \pmod{17}$

    (c) $x^6 \equiv 1 \pmod{19}$

2. For each integer d such that $d|18$, find all integers t such that $0 < t < 19$ and t has order $d \pmod{19}$.

3. Given that 3 is a primitive root $\pmod{43}$, find all primitive roots $\pmod{43}$.

4. Find all primitive roots $\pmod{m}$ for each of the following values of m: (a) 27 (b) 49 (c) 50.

5. Prove Theorem 6.17.

6. Find a formula, in terms of p and n, for the number of primitive roots $\pmod{p^n}$, where p is an odd prime and $n \geq 2$.

7. Prove that if g is a primitive root $\pmod{p^n}$, where p is an odd prime and $n \geq 2$, then g is a primitive root $\pmod{p}$.

8. Prove that if g is a primitive root $\pmod{m}$, where $m \geq 3$, then $g^{\frac{1}{2}\phi(m)} \equiv -1 \pmod{m}$.

9. Let m be an integer such that $m \geq 5$ and m has primitive roots. Let G be the product of all primitive roots $g \pmod{m}$ such that $0 < g < m$. Prove that $G \equiv 1 \pmod{m}$.

10. Let the congruence $x^2 \equiv 1 \pmod{m}$ have r distinct solutions $\pmod{m}$. Prove that if m has no primitive roots, then $4|r$.

★11. Prove the following generalization of Wilson's Theorem: If $m \geq 2$, let P be the product of all integers, k, such that $0 < k < m$ and $(k, m) = 1$. If m has a primitive root, then $P \equiv -1 \pmod{m}$. Otherwise, $P \equiv 1 \pmod{m}$.

★12. If the prime $p \equiv 3 \pmod 4$, let S_p denote the sum of all integers, g, such that $0 < g < p$ and g is a primitive root $\pmod{p}$. Prove that $S_p \equiv \mu(\frac{p-1}{2}) \pmod{p}$.

13. Use Theorems 6.6 and 6.8 to prove that if p is an odd prime, and if g is a primitive root $\pmod{p}$, then $(p-1)! \equiv g^{\sum_{k=0}^{p-1} k} \pmod{p}$. Simplify to obtain an alternate proof of Wilson's Theorem.

Section 6.3 Computer Exercises

14. Write a computer program to find the least primitive root (mod p) by trial and error, where p is a given odd prime.

15. Write a computer program that, given an odd prime, p , finds all primitive roots (mod p), if any, that are not primitive roots (mod p^2).

6.4 Indices

We conclude this chapter by considering *indices*, which are related to primitive roots. We shall see that indices bear some semblance to logarithms, and are useful in solving certain polynomial congruences.

Definition 6.3   Index

Let g be a primitive root (mod p), where p is an odd prime. Let t be an integer such that $(t, p) = 1$. The *index of t relative to g* is that exponent, k , such that $0 \leq k \leq p - 2$ and $t \equiv g^k$ (mod p). We write $ind_g t = k$, or simply $ind(t) = k$.

For example, recall that 5 is a primitive root (mod 7), and that

$$1 \equiv 5^0 \pmod{7}$$

$$2 \equiv 5^4 \pmod{7}$$

$$3 \equiv 5^5 \pmod{7}$$

$$4 \equiv 5^2 \pmod{7}$$

$$5 \equiv 5^1 \pmod{7}$$

$$6 \equiv 5^3 \pmod{7}$$

Therefore we have

$$ind_5 1 = 0$$

$$ind_5 2 = 4$$

$$ind_5 3 = 5$$

$$ind_5 4 = 2$$

$$ind_5 5 = 1$$

$$ind_5 6 = 3$$

The following theorem states some properties of indices, which recall similar properties of logarithms.

Theorem 6.23

Properties of Indices

Let g be a primitive root (mod p), where p is an odd prime. Then

I_1: $a \equiv b$ (mod p) if and only if $ind_g a = ind_g b$

I_2: $ind_g(g^k) \equiv k$ (mod $p-1$)

I_3: $ind_g 1 = 0$ and $ind_g g = 1$

I_4: $dind_g(ab) \equiv ind_g a + ind_g b$ (mod $p-1$)

I_5: $ind_g(a^r) \equiv r(ind_g a)$ (mod $p-1$)

Proof of I_1: Let $ind_g a = j$, $ind_g b = k$. Then $a \equiv g^j$ (mod p), $b \equiv g^k$ (mod p), and $0 \leq j, k \leq p-2$. If $a \equiv b$ (mod p), then $g^j \equiv g^k$ (mod p). Now Theorem 6.4 implies $j \equiv l$ (mod $p-1$). Since $|j - k| \leq p - 2$, we must have $j = k$, that is, $ind_g a = ind_g b$. Conversely, if $ind_g a = ind_g b$, then $j = k$, so $g^j = g^k$, hence $g^j \equiv g^k$ (mod p), that is, $a \equiv b$ (mod p). ∎

Proof of I_2: Let $r = ind_g(g^k)$. Then, by Definition 6.3, we have $g^r \equiv g^k$ (mod p). Since g is a primitive root (mod p) by hypothesis, Theorem 6.4 implies $r \equiv k$ (mod $p-1$). ∎

Proof of I_3: This follows from the facts that $g^0 = 1$, $g^1 = g$. ∎

Proof of I_4: Since $a \equiv g^j$ (mod p) and $b \equiv g^k$ (mod p), we have $ab \equiv g^j g^k \equiv g^{j+k}$ (mod p). Now part I_2 implies $ind_g(g^{j+k}) \equiv j+k$ (mod $p-1$), that is, $ind_g(ab) \equiv ind_g a + ind_g b$ (mod $p-1$). ∎

Proof of I_5: Using I_4 and induction on r, one can show that $ind_g(\prod_{i=1}^r a_i) \equiv \sum_{i=1}^r ind_g(a_i)$ (mod $p-1$). The conclusion now follows by setting each $a_i = a$ and then simplifying. ∎

If $p = 7$ and $g = 5$, then the numerical data concerning indices shown on the preceding page can be presented in a table of indices as follows:

a	1	2	3	4	5	6
$ind_5(a)$	0	4	5	2	1	3

Let us consider the solution to polynomial congruences of the type $ax^n \equiv b \pmod{p}$ where p is an odd prime and $(a, p) = 1$. By applying the properties of indices, such a polynomial congruence in x is transformed into a linear congruence in $ind(x) \pmod{p-1}$.

Example 6.1   Solve $3x^5 \equiv 5 \pmod 7$.

Solution: Taking indices with respect to $g = 5$, we have

$$ind(3x^5) \equiv ind(5) \qquad\qquad (\text{mod } 6)$$

$$ind(3) + 5ind(x) \equiv ind(5) \qquad\qquad (\text{mod } 6)$$

$$5 + 5ind(x) \equiv 1 \qquad\qquad (\text{mod } 6)$$

$$5ind(x) \equiv -4 \equiv 2 \qquad\qquad (\text{mod } 6)$$

$$ind(x) \equiv 4 \qquad\qquad (\text{mod } 6)$$

$$x \equiv 2 \qquad\qquad (\text{mod } 7)$$

Example 6.2   Solve $3x^4 \equiv 4 \pmod 7$.

Solution: Taking indices with respect to $g = 5$, we have

$$ind(3x^4) \equiv ind(4) \qquad\qquad (\text{mod } 6)$$

$$ind(3) + 4ind(x) \equiv ind(4) \qquad\qquad (\text{mod } 6)$$

$$5 + 4ind(x) \equiv 2 \qquad\qquad (\text{mod } 6)$$

$$4ind(x) \equiv -3 \equiv 3 \qquad\qquad (\text{mod } 6)$$

Since $(4, 6) = 2$ and $2 \nmid 3$, by Theorem 4.4, the linear congruence in $ind(x)$ has no solution. Therefore the original polynomial congruence has no solution.

Section 6.4 Exercises

1. Using 2 as a primitive root (mod 11), construct a table of indices to be used in finding all solutions, if any, of the three following congruences in Exercises 2–4.

2. $3x^3 \equiv 5 \pmod{11}$

3. $6x^4 \equiv 7 \pmod{11}$

4. $5x^2 \equiv 6 \pmod{11}$

5. Using 3 as a primitive root (mod 17), construct a table of indices to be used in finding all solutions, if any, of the four following congruences in Exercises 6–9.

6. $2x^3 \equiv 7 \pmod{17}$

7. $7x^4 \equiv 6 \pmod{17}$

8. $3x^5 \equiv 11 \pmod{17}$

9. $x^4 \equiv 2 \pmod{17}$

10. Prove that if p is prime and if $(a, p(p-1)) = 1$, then the congruence $ax^n \equiv b$ (mod p) always has a unique solution (mod p).

Section 6.4 Computer Exercises

11. If p is a prime such that 2 is a primitive root (mod p), write a computer program to construct a corresponding table of indices.

Review Exercises

1. Verify that 3 is a primitive root (mod 17). Find all primitive roots (mod 17). Express these primitive roots as least positive residues (mod 17).

2. For each of the following values of m, find the number of primitive roots (mod m):

   (a) 25

   (b) 26

   (c) 27

   (d) 28

   (e) 29

3. Which primes, if any, have exactly 6 primitive roots? (Explain.)

4. Find the 5 least integers that exceed 1 but have no primitive roots.

5. Find all primitive roots

   (a) (mod 18)

   (b) (mod 25)

   (c) (mod 27)

6. Given that 18 is a primitive root (mod 37), find a primitive root (mod 37^2).

7. Prove that if p is an odd prime, and g is a primitive root (mod p), then g^{-1} is also a primitive root (mod p).

8. Using 2 as a primitive root (mod 13), construct a table of indices (mod 13).

   Using your solution to the exercise immediately above, find all solutions, if any, of each of the congruences in Exercises 9 and 10.

9. $3x^3 \equiv 10 \pmod{13}$

10. $5x^2 \equiv 4 \pmod{13}$

Chapter 7

Quadratic Congruences

7.1 Introduction

In Chapter 4, among other items, we studied linear congruences, that is, congruences of the type $ax \equiv b \pmod{m}$ where a, b, m are given and x is unknown. This chapter is devoted to *quadratic* congruences. We begin by considering the case where the modulus is an odd prime, p. That is, we consider congruences of the type

$$Ay^2 + By + C \equiv 0 \pmod{p} \tag{7.1}$$

where A, B, C, p are given, p is an odd prime, $(A, p) = 1$, and y is unknown.

Our problem may be simplified by some algebraic manipulation, namely, completing the square. Let the *discriminant* of $Ay^2 + By + C$ be $D = B^2 - 4AC$. If we multiply (7.1) by $4A$ and then add $B^2 - 4AC$, we obtain

$$4A^2y^2 + 4ABy + B^2 \equiv B^2 - 4AC \pmod{p} \tag{7.2}$$

that is,

$$(2Ay + B)^2 \equiv D \pmod{p} \tag{7.3}$$

169

Now let $x = 2Ay + B$, and let a be the least positive residue of D (mod p). This yields the pure quadratic congruence

$$x^2 \equiv a \pmod{p} \tag{7.4}$$

where $0 \le a < p$. If $a = 0$, then the Congruence (7.4) has the unique solution: $x \equiv 0 \pmod{p}$. We consider this case to be degenerate and confine our attention henceforth to the case $0 < a < p$.

For example, suppose we wish to solve the congruence

$$3y^2 + 2y + 3 \equiv 0 \pmod{11} \tag{7.5}$$

Here $A = 3$, $B = 2$, $C = 3$, so $D = B^2 - 4AC = -32$. Therefore $a = -32 + 3(11) = 1$. Letting $x = 6y + 2$, we obtain

$$x^2 \equiv 1 \pmod{11} \tag{7.6}$$

Now Theorem 4.10 implies $x \equiv \pm 1 \pmod{11}$, that is, $x \equiv 1, 10 \pmod{11}$. Therefore $6y + 2 \equiv 1, 10 \pmod{11}$, from which it follows that $y \equiv 5, 9 \pmod{11}$.

Similarly, let us try to solve the congruence

$$2y^2 + y + 3 \equiv 0 \pmod{5} \tag{7.7}$$

Here $A = 2, B = 1, C = 3$, so $D = B^2 - 4AC = -23$ and $a = -23 + 5(5) = 2$. Letting $x = 4y + 1$, we obtain

$$x^2 \equiv 2 \pmod{5} \tag{7.8}$$

Since $1^2 \equiv 4^2 \equiv 1 \pmod{5}$, $2^2 \equiv 3^2 \equiv 4 \pmod{5}$, it follows that the Congruence (7.8) has no solution and hence neither does the Congruence (7.7).

7.2 Quadratic Residues and the Legendre Symbol

Given an odd prime, p, and an integer, a, such that $0 < a < p$, we will be concerned with (i) determining whether the congruence

$$x^2 \equiv a \pmod{p} \tag{7.9}$$

has any solutions and (ii) finding the solutions when they exist. According to Theorem 6.9, (7.9) has at most two distinct solutions. If $x \equiv b \pmod{p}$ is a solution, then so is $x \equiv -b \pmod{p}$. Therefore (7.9) either has two distinct solutions:

$$x \equiv \pm b \pmod{p} \tag{7.10}$$

or has no solutions.

We now introduce the concept of *quadratic residues* and *quadratic non-residues*, which facilitate the study of quadratic congruences.

Definition 7.1   Quadratic Residue (mod p)

If the Congruence (7.9) has solutions, then we say that a is a *quadratic residue* (mod p).

For example, 3 is a quadratic residue (mod 11), since $5^2 \equiv 3$ (mod 11). Also, 2 is a quadratic residue (mod 17), since $6^2 \equiv 2$ (mod 17).

Definition 7.2   Quadratic Nonresidue (mod p)

If the Congruence (9) has no solutions, then we say that a is a *quadratic nonresidue* (mod p).

For example, we have seen that 2 is a quadratic nonresidue (mod 5). The reader can verify that 3 is a quadratic nonresidue (mod 7).

The following theorem tells us that there are equally many quadratic residues and quadratic nonresidues (mod p).

Theorem 7.1 If p is an odd prime, then there are $\frac{1}{2}(p-1)$ quadratic residues and equally many quadratic nonresidues (mod p).

Proof: Let n be the number of quadratic residues (mod p). It suffices to show that $n = \frac{1}{2}(p-1)$, since then the number of quadratic nonresidues (mod p) is $p - 1 - n = p - 1 - \frac{1}{2}(p-1) = \frac{1}{2}(p-1)$. If $(x, p) = 1$, then either $x \equiv i$ (mod p) for some i such that $1 \le i \le \frac{1}{2}(p-1)$, or $x \equiv j$ (mod p) for some j such that $\frac{1}{2}(p+1) \le j \le p-1$. In the latter case, we have $j = p - i$ where $1 \le i \le \frac{1}{2}(p-1)$. Therefore $x \equiv i, p - i$ (mod p) where $1 \le i \le \frac{1}{2}(p-1)$, that is $x \equiv \pm i$ (mod p) where $1 \le i \le \frac{1}{2}(p-1)$. This implies $x^2 \equiv i^2$ (mod p) where $1 \le i \le \frac{1}{2}(p-1)$. Therefore, if a is a quadratic residue (mod p), we must have $a \equiv i^2$ (mod p), where $1 \le i \le \frac{1}{2}(p-1)$. This implies that $n \le \frac{1}{2}(p-1)$.

It remains to show that $n \ge \frac{1}{2}(p-1)$. To do this, we must verify that the least positive residues (mod p) of the integers $1^2, 2^2, 3^2, \cdots, \frac{1}{2}(p-1)^2$ are all distinct (mod p). Suppose that $1 \le j \le i \le \frac{1}{2}(p-1)$ and $i^2 \equiv j^2$ (mod p). We wish to show that $i = j$. Theorem 4.10 implies $i \equiv \pm j$ (mod p). Now by hypothesis, we have $2 \le i + j \le p - 1$. Therefore $i + j \not\equiv 0$ (mod p), that is $i \not\equiv -j$ (mod p). This implies $i \equiv j$ (mod p), that is, $i - j \equiv 0$ (mod p). Since $0 \le i - j \le \frac{p-3}{2}$, we must have $i - j = 0$, that is, $i = j$. ∎

For example, if $p = 7$, then the quadratic residues (mod 7) are the least positive residues (mod 7) of $1^2, 2^2, 3^2$, namely 1, 4, 2. The quadratic nonresidues (mod 7) are 3, 5, 6.

Also, if $p = 11$, then the quadratic residues (mod 11) are the least positive residues (mod 11) of $1^2, 2^2, 3^2, 4^2, 5^2$, namely 1, 4, 9, 5, 3. The quadratic nonresidues (mod 11) are 2, 6, 7, 8, 10.

If p is an odd prime and $(a, p) = 1$, then the *Legendre symbol*, which is defined below, provides a convenient way to indicate whether a is a quadratic residue or a quadratic nonresidue (mod p).

Definition 7.3   **Legendre Symbol**

If p is an odd prime and $(a, p) = 1$, let

$$\left(\frac{a}{p}\right) = \begin{cases} 1 & \text{if } a \text{ is a quadratic residue} & (\text{mod } p) \\ -1 & \text{if } a \text{ is a quadratic nonresidue} & (\text{mod } p) \end{cases}$$

For example, $\left(\frac{3}{11}\right) = 1$ because 3 is a quadratic residue (mod 11). Also, $\left(\frac{2}{5}\right) = -1$, since 2 is a quadratic nonresidue (mod 5). Note that evaluating the Legendre symbol $\left(\frac{a}{p}\right)$ is equivalent to determining whether (7.9) has solutions.

Before proceeding further, we present the following lemma, which will be needed further on.

▶ **Lemma 7.1** *If p is an odd prime, x, y are integers such that $x \equiv y$ (mod p), and $|x| = |y| = 1$, then $x = y$.*

Proof: Since $|x| = |y|$ by hypothesis, it follows that $x = \pm y$. Therefore, to prove that $x = y$, it suffices to prove that $x \neq -y$. If $x = -y$, then $|x - y| = |2x| = 2|x| = 2(1) = 2$. Since $x \equiv y$ (mod p) by hypothesis, it follows that $p|(x - y)$, hence $p|(|x - y|)$. Therefore $p|2$, an impossibility. ■

Euler's Criterion, which follows, provides a method of evaluating $\left(\frac{a}{p}\right)$, but at the expense of approximately $\log_2 p$ multiplications (mod p).

Theorem 7.2   **Euler's Criterion**

If p is an odd prime and $(a, p) = 1$, then $\left(\frac{a}{p}\right) \equiv a^{\frac{p-1}{2}}$ (mod p).

Proof: By hypothesis and Theorem 4.8, we have $a^{p-1} \equiv 1$ (mod p). Therefore Theorem 4.10 implies $a^{\frac{p-1}{2}} \equiv \pm 1$ (mod p). Since $|\left(\frac{a}{p}\right)| = 1$, by Lemma

7.1, it suffices to show that $(\frac{a}{p}) = 1$ if and only if $a^{\frac{p-1}{2}} \equiv 1 \pmod{p}$. Let g be a primitive root $\pmod{p}$.

First, suppose that $(\frac{a}{p}) = 1$. Then $a \equiv b^2 \pmod{p}$ for some b. Theorem 6.5 implies $b \equiv g^k \pmod{p}$ for some k. Thus we have $a \equiv (g^k)^2 \equiv g^{2k} \pmod{p}$. Therefore $a^{\frac{1}{2}(p-1)} \equiv (g^{2k})^{\frac{1}{2}(p-1)} \equiv (g^{p-1})^k \equiv 1^k \equiv 1 \pmod{p}$ by Theorem 4.8.

Conversely, suppose that $a^{\frac{p-1}{2}} \equiv 1 \pmod{p}$. Theorem 6.5 implies $a \equiv g^j \pmod{p}$ for some j. Therefore $(g^j)^{\frac{p-1}{2}} \equiv 1 \pmod{p}$, that is $(g^{\frac{1}{2}(p-1)})^j \equiv 1 \pmod{p}$. But Theorem 6.8 implies $g^{\frac{1}{2}(p-1)} \equiv -1 \pmod{p}$, so we have $(-1)^j \equiv 1 \pmod{p}$. Now Lemma 7.1 implies $(-1)^j = 1$, so $j = 2r$ for some r. Therefore $a \equiv g^{2r} \equiv (g^r)^2 \pmod{p}$, so that $(\frac{a}{p}) = 1$.   ∎

Theorem 7.3   **Properties of the Legendre Symbol**

Let p be an odd prime. Let a, b be integers such that $p \nmid ab$. Then

(i) If $a \equiv b \pmod{p}$, then $(\frac{a}{p}) = (\frac{b}{p})$.

(ii) $(\frac{ab}{p}) = (\frac{a}{p})(\frac{b}{p})$

(iii) $(\frac{a^2}{p}) = 1$

(iv) $(\frac{-1}{p}) = (-1)^{\frac{p-1}{2}}$

Proof of (i):   It suffices to show that if $a \equiv b \pmod{p}$, then $(\frac{a}{p}) = 1$ if and only if $(\frac{b}{p}) = 1$. If $(\frac{a}{p}) = 1$, then $t^2 \equiv a \pmod{p}$ for some t. Since $a \equiv b \pmod{p}$ by hypothesis, we have $t^2 \equiv b \pmod{p}$, so $(\frac{b}{p}) = 1$. Similarly, if $(\frac{b}{p}) = 1$, then $(\frac{a}{p}) = 1$.   ∎

Proof of (ii):   Invoking Euler's Criterion, we have

$$(\frac{ab}{p}) \equiv (ab)^{\frac{p-1}{2}} \equiv a^{\frac{p-1}{2}} b^{\frac{p-1}{2}} \equiv (\frac{a}{p})(\frac{b}{p}) \pmod{p}$$

Since p is odd by hypothesis, and $|(\frac{ab}{p})| = |(\frac{a}{p})(\frac{b}{p})| = 1$, the conclusion now follows from Lemma 7.1.   ∎

Proof of (iii):   Invoking (ii) with $a = b$, we get $(\frac{a^2}{p}) = (\frac{a}{p})^2 = (\pm 1)^2 = 1$.   ∎

Proof of (iv):   This follows from Euler's Criterion and Lemma 7.1.   ∎

By virtue of Theorem 7.3, the problem of evaluating $(\frac{a}{p})$ can be reduced to the problem of evaluating $(\frac{q}{p})$, where q is a prime divisor of a. For example,

$$\left(\frac{130}{151}\right) = \left(\frac{2*5*13}{151}\right) = \left(\frac{2}{151}\right)\left(\frac{5}{151}\right)\left(\frac{13}{151}\right)$$

In the next section, we will learn how to evaluate $(\frac{q}{p})$ when p and q are distinct primes and p is odd.

Section 7.2   Exercises

In each of Exercises 1 through 5, use a change of variable in order to transform the given quadratic congruence into a pure quadratic congruence: $x^2 \equiv a \pmod{p}$.

1. $2y^2 + 3y + 4 \equiv 0 \pmod{11}$

2. $5y^2 - 3y - 7 \equiv 0 \pmod{13}$

3. $6y^2 - y - 13 \equiv 0 \pmod{17}$

4. $4y^2 + 5y - 6 \equiv 0 \pmod{19}$

5. $8y^2 + 7y + 6 \equiv 0 \pmod{23}$

6. Find all quadratic residues and nonresidues (mod p) for each of the following primes: (a) 13, (b) 17, (c) 19.

7. Prove that if p is an odd prime, and if g is a primitive root (mod p), then $(\frac{g}{p}) = -1$.

8. Prove that every quadratic nonresidue (mod p) is a primitive root (mod p) if and only if p is a Fermat prime.

9. Find necessary and sufficient conditions on the odd prime p such that all but one of the quadratic nonresidues is also a primitive root (mod p).

10. Use Euler's Criterion to evaluate each of the following Legendre symbols:
    (a) $(\frac{7}{11})$ (b) $(\frac{5}{17})$ (c) $(\frac{3}{13})$ (d) $(\frac{11}{19})$ (e) $(\frac{6}{23})$.

11. Prove that if the prime $p \geq 7$, then there are two consecutive quadratic residues (mod p).

Section 7.2   Computer Exercises

12. Write a computer program to convert a given quadratic congruence $Ay^2 + By + C \equiv 0 \pmod{p}$ into an equivalent pure quadratic congruence: $x^2 \equiv a \pmod{p}$.

13. Write a computer program to evaluate $(\frac{a}{p})$ using Euler's Criterion if p is an odd prime and $(a, p) = 1$.

7.3 Gauss' Lemma and the Law of Quadratic Reciprocity

We have seen that if p is an odd prime, $a > 1$, and $(a, p) = 1$, then the problem of evaluating $(\frac{a}{p})$ reduces to the problem of evaluating all Legendre symbols $(\frac{q}{p})$ where q is a prime divisor of a. Our next item, namely Gauss' Lemma, is the first of three items needed to prove the Law of Quadratic Reciprocity.

Theorem 7.4

Gauss' Lemma

If p is an odd prime and $(a, p) = 1$, consider the least positive residues (mod p) of the integers $a, 2a, 3a, \dots, (\frac{p-1}{2})a$. Let n be the number of these least positive residues that exceed $\frac{p}{2}$. Then $(\frac{a}{p}) = (-1)^n$.

Proof: Consider the first $\frac{p-1}{2}$ multiples of a, namely, $a, 2a, 3a, \dots, (\frac{p-1}{2})a$. First, we show that these integers are all distinct (mod p). Suppose that $ja \equiv ka \pmod{p}$, where $1 \leq k \leq j \leq \frac{p-1}{2}$. Since $(a, p) = 1$ by hypothesis, it follows that $j \equiv k \pmod{p}$, so $j - k \equiv 0 \pmod{p}$. But by hypothesis, $0 \leq j - k \leq \frac{p-3}{2}$. Thus we must have $j - k = 0$, that is, $j = k$. Now we look at the set of least positive residues (mod p) of $\{a, 2a, 3a, \dots, \frac{(p-1)a}{2}\}$. Let $r_1, r_2, r_3, \dots, r_n$ be those least positive residues that exceed $\frac{p}{2}$, while $s_1, s_2, s_3, \dots, s_m$ are those least positive residues that do not exceed $\frac{p}{2}$. Therefore, $n + m = \frac{p-1}{2}$. By the preceding argument, the r_i and the s_j are all distinct (mod p). For each index, i, we have $\frac{p}{2} < r_i < p$, so $0 < p - r_i < \frac{p}{2}$. We claim that the integers $p - r_i$ are distinct from the s_j. Indeed, if $p - r_i = s_j$ for some indices i, j, then $r_i + s_j \equiv 0 \pmod{p}$. However, by definition of the r_i and the s_j, we have $r_i \equiv ua \pmod{p}$, $s_j \equiv va \pmod{p}$ for some u, v such that $1 \leq u, v \leq \frac{p-1}{2}$. Therefore $ua + va \equiv r_i + s_j \equiv 0 \pmod{p}$, or $(u + v)a \equiv 0 \pmod{p}$. Since $(a, p) = 1$ by hypothesis, this implies $u + v \equiv 0 \pmod{p}$. But this is impossible, since $2 \leq u + v \leq p - 1$.

To recapitulate, the set of integers, $\{p - r_1, p - r_2, \dots, p - r_n, s_1, s_2, \dots, s_m\}$, are all distinct, are $\frac{p-1}{2}$ in number, and are all between 1 and $\frac{p-1}{2}$. Therefore, this set is a permutation of $\{1, 2, 3, \dots, \frac{p-1}{2}\}$. It follows that

$$(p - r_1)(p - r_2) \cdots (p - r_n)s_1 s_2 \cdots s_m = 1 * 2 * 3 * \cdots * \frac{p-1}{2}$$

that is,

$$\prod_{i=1}^{n}(p - r_i) \prod_{j=1}^{m} s_j = (\frac{p-1}{2})!$$

Therefore

$$\prod_{i=1}^{n}(p - r_i) \prod_{j=1}^{m} s_j \equiv (\frac{p-1}{2})! \pmod{p}$$

Simplifying, we have

$$(-1)^n \prod_{i=1}^{n} r_i \prod_{j=1}^{m} s_j \equiv (\frac{p-1}{2})! \pmod{p}$$

In addition, however, we have

$$\prod_{i=1}^{n} r_i \prod_{j=1}^{m} s_j \equiv \prod_{k=1}^{\frac{p-1}{2}} ka \equiv a^{\frac{p-1}{2}} \prod_{k=1}^{\frac{p-1}{2}} k \equiv a^{\frac{p-1}{2}} (\frac{p-1}{2})! \pmod{p}$$

Therefore

$$(-1)^n a^{\frac{p-1}{2}} (\frac{p-1}{2})! \equiv (\frac{p-1}{2})! \pmod{p}$$

which implies that

$$a^{\frac{p-1}{2}} \equiv (-1)^n \pmod{p}$$

The conclusion now follows from Euler's Criterion (Theorem 7.2) and Lemma 7.1. ∎

For example, let us use Gauss' Lemma to evaluate $(\frac{3}{11})$. Since $\frac{1}{2}(11 - 1) = 5$, we look at the first 5 multiples of 3, namely, 3, 6, 9, 12, 15. The least positive residues (mod 11) are 3, 6, 9, 1, 4. Exactly 2 of these least positive residues exceed $\frac{11}{2}$, namely, 6 and 9. Therefore $\frac{3}{11} = (-1)^2 = 1$.

Next, let us evaluate $(\frac{7}{13})$ via Gauss' Lemma. Since $\frac{1}{2}(13 - 1) = 6$, we look at the first 6 multiples of 7, namely, 7, 14, 21, 28, 35, 42. The least positive residues (mod 13) are, 7, 1, 8, 2, 9, 3. exactly 3 of these least positive residues exceed $\frac{13}{2}$, namely, 7, 8, and 9. Therefore $(\frac{7}{13}) = (-1)^3 = -1$.

If we apply Gauss' Lemma to the case $a = 2$, we obtain an explicit formula for $(\frac{2}{p})$, which is given by Theorem 7.5.

Theorem 7.5   If p is an odd prime, then

$$\left(\frac{2}{p}\right) = \begin{cases} 1 & \text{if } p \equiv \pm 1 \pmod{8} \\ -1 & \text{if } p \equiv \pm 3 \pmod{8} \end{cases}$$

Proof: Applying Gauss' Lemma, we look at the first $\frac{p-1}{2}$ multiples of 2, namely, at $\{2, 4, 6, \ldots, p-1\}$. These positive integers are all less than p, so they are their own least positive residues (mod p). Let S_p denote the set of least positive residues (mod p) that exceed $\frac{p}{2}$. If $p = 8k + 1$, so that $\frac{p}{2} = 4k + \frac{1}{2}$, then $S_p = \{4k + 2, 4k + 4, \ldots, 8k\}$. If $p = 8k - 1$, so that $\frac{p}{2} = 4k - \frac{1}{2}$, then $S_p = \{4k, 4k + 2, \ldots, 8k - 2\}$. In either case, these residues are $2k$ in number, so $\left(\frac{2}{p}\right) = (-1)^{2k} = 1$.

If $p = 8k + 3$, so that $\frac{p}{2} = 4k + \frac{3}{2}$, then $S_p = \{4k + 2, 4k + 4, \ldots, 8k + 2\}$. Finally, if $p = 8k + 5$, so that $\frac{p}{2} = 4k + \frac{5}{2}$, then

$$S_p = \{4k + 4, 4k + 6, \ldots, 8k + 4\}.$$

In either case, these residues are $2k + 1$ in number, so $\left(\frac{2}{p}\right) = (-1)^{2k+1} = -1$. ∎

For example, since $13 \equiv -3 \pmod{8}$, it follows that $\left(\frac{2}{13}\right) = -1$. Also, since $17 \equiv 1 \pmod{8}$, it follows that $\left(\frac{2}{17}\right) = 1$.

A more compact formulation of Theorem 7.5 is the following:

Theorem 7.5A If p is an odd prime, then

$$\left(\frac{2}{p}\right) = (-1)^{\frac{p^2-1}{8}}$$

Recall that a Mersenne prime is a prime of the form $2^p - 1$, where p itself is prime. By virtue of Theorem 7.5, we can prove that $2^p - 1$ is composite for certain primes, p.

Theorem 7.6 If the prime $p > 3$, $p \equiv 3 \pmod{4}$, and $q = 2p + 1$ is prime, then $2^p - 1$ is composite.

Proof: Since $p \equiv 3 \pmod{4}$ and $q = 2p + 1$ by hypothesis, it follows that $q \equiv 7 \pmod{8}$. Therefore Theorem 7.5 implies $\left(\frac{2}{q}\right) = 1$. Now Euler's Criterion implies $2^{\frac{q-1}{2}} \equiv 1 \pmod{q}$, that is, $q | (2^p - 1)$. If $n > 3$, one can prove by induction that $2^n - 1 > 2n + 1$. Since $p > 3$ by hypothesis, we have $2^p - 1 > q$. Therefore $2^p - 1$ is composite, having q as a factor. ∎

Remarks

If $2p + 1$ is prime, then $3 \nmid (2p+1)$, so $p \equiv 2 \pmod{3}$. If also $p \equiv 3 \pmod{4}$, then $p \equiv 11 \pmod{12}$. The five smallest primes for which Theorem 7.6 holds are 11, 23, 83, 131, 179.

Now that we know how to evaluate $\left(\frac{2}{p}\right)$ where p is an odd prime, we can restrict our attention to the evaluation of $\left(\frac{a}{p}\right)$, where p is prime, $a \equiv p \equiv 1 \pmod 2$, and $(a, p) = 1$. The following highly technical theorem is the second of three items needed in the proof of the Law of Quadratic Reciprocity.

Theorem 7.7

If p is prime, $a \equiv p \equiv 1 \pmod 2$, and $(a, p) = 1$, then

$$\left(\frac{a}{p}\right) = (-1)^t \qquad where \qquad t = \sum_{k=1}^{\frac{p-1}{2}} \left[\frac{ka}{p}\right]$$

Proof: We use the same notation as in the proof of Gauss' Lemma (Theorem 7.4). We look at the least positive residues (mod p) of the integers ka, where $1 \le k \le \frac{p-1}{2}$. That is to say, we look at the remainders that result when each of the integers ka is divided by p. The r_i (of which there are n) are the remainders that exceed $\frac{p}{2}$, while the s_j (of which there are m) are the remainders that are less than $\frac{p}{2}$. We know by Gauss' Lemma that $\left(\frac{a}{p}\right) = (-1)^n$. Therefore, it suffices to show that $(-1)^t = (-1)^n$, or equivalently, that $t \equiv n \pmod 2$.

By Theorem 1.17, part I$_7$, we have $ka = p[\frac{ka}{p}] + u_k$, where $u_k = r_i$ for some i or $u_k = s_j$ for some j. Therefore

$$\sum_{k=1}^{\frac{p-1}{2}} ka = \sum_{k=1}^{\frac{p-1}{2}} (p[\frac{ka}{p}] + u_k)$$

$$= \sum_{k=1}^{\frac{p-1}{2}} p[\frac{ka}{p}] + \sum_{k=1}^{\frac{p-1}{2}} u_k$$

$$= p \sum_{k=1}^{\frac{p-1}{2}} [\frac{ka}{p}] + \sum_{k=1}^{\frac{p-1}{2}} u_k$$

This yields

$$a \sum_{k=1}^{\frac{p-1}{2}} k = pt + \sum_{i=1}^{n} r_i + \sum_{j=1}^{m} s_j \qquad (7.11)$$

Now recall from the proof of Theorem 7.4 that the integers $p - r_i$ and s_j are precisely all the integers from 1 to $\frac{p-1}{2}$. Therefore we have

$$\sum_{k=1}^{\frac{p-1}{2}} k = \sum_{i=1}^{n}(p - r_i) + \sum_{j=1}^{m} s_j \tag{7.12}$$

so that

$$\sum_{k=1}^{\frac{p-1}{2}} k = np - \sum_{i=1}^{n} r_i + \sum_{j=1}^{m} s_j \tag{7.13}$$

Subtracting equation (7.13) from equation (7.11), we get

$$(a - 1)\sum_{k=1}^{\frac{p-1}{2}} k = p(t - n) + 2\sum_{i=1}^{n} r_i \tag{7.14}$$

Finally, we reduce equation (7.14) (mod 2), noting that a and p are odd by hypothesis. This yields

$$0 \equiv t - n \pmod 2 \qquad hence \qquad t \equiv n \pmod 2 \quad \blacksquare$$

For example, let us use Theorem 7.7 to evaluate $(\frac{3}{11})$. Here $a = 3$ and $p = 11$ so $\frac{p-1}{2} = 5$. Therefore

$$t = \sum_{k=1}^{5} [\frac{3k}{11}]$$

$$= [\frac{3}{11}] + [\frac{6}{11}] + [\frac{9}{11}] + [\frac{12}{11}] + [\frac{15}{11}]$$

$$= 0 + 0 + 0 + 1 + 1$$

$$= 2$$

We conclude that $(\frac{3}{11}) = (-1)^2 = 1$.

The next theorem, which despite its algebraic appearance is really geometric in nature, is the last of the three items needed to prove the Law of Quadratic Reciprocity.

Theorem 7.8   If p and q are distinct odd primes, then

$$\sum_{i=1}^{\frac{p-1}{2}} [\frac{qi}{p}] + \sum_{j=1}^{\frac{q-1}{2}} [\frac{pj}{q}] = (\frac{p-1}{2})(\frac{q-1}{2})$$

Proof: Consider the set of points with integer coordinates (i, j) where $1 \leq i \leq \frac{p-1}{2}$, $1 \leq j \leq \frac{q-1}{2}$. Let line L have the equation, $py = qx$. (Figure 7.1 illustrates the case where $p = 7$, $q = 11$.) First, we will show that none of the points (i, j) lie on line L. If point (i, j) lies on L, then $pj = qi$. By hypothesis, $(p, q) = 1$. Therefore Euclid's Lemma implies $p | i$. But this is impossible, since $1 \leq i \leq \frac{p-1}{2}$. Therefore, each point (i, j) lies either below or above line L. Furthermore, the total number of points, namely, $(\frac{p-1}{2})(\frac{q-1}{2})$, is the sum of the number of points below L and the number of points above L. Let m be the number of points (i, j) below L. For each such point, we have $pj < qi$, so $j < \frac{qi}{p}$. Therefore, for each fixed value of i, if (i, j) is below L, then $1 \leq j \leq [\frac{qi}{p}]$. This implies that $m = \sum_{i=1}^{\frac{p-1}{2}} [\frac{qi}{p}]$. Similarly, let n be the number of points (i, j) above L. For each such point (i, j), we have $pj > qi$, so $i \leq [\frac{pj}{q}]$. Therefore, for each fixed value of j, if (i, j) is above L, then $1 \leq i \leq [\frac{pj}{q}]$. This implies that $n = \sum_{j=1}^{\frac{q-1}{2}} [\frac{pj}{q}]$. Now $m + n = (\frac{p-1}{2})(\frac{q-1}{2})$, from which the conclusion follows.   ∎

For example, in Figure 7.1, we have $p = 11$, $q = 7$, so $\frac{p-1}{2} = 5$, $\frac{q-1}{2} = 3$. The equation of line L is $11y = 7x$. Furthermore, $(\frac{p-1}{2})(\frac{q-1}{2}) = 5 * 3 = 15$.

Figure 7.1

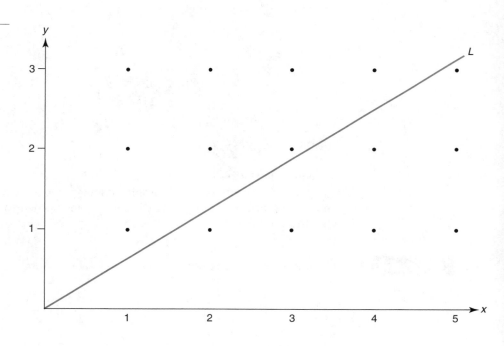

Now

$$m = \sum_{i=1}^{5} \left[\frac{7i}{11}\right]$$

$$= \left[\frac{7}{11}\right] + \left[\frac{14}{11}\right] + \left[\frac{21}{11}\right] + \left[\frac{28}{11}\right] + \left[\frac{35}{11}\right]$$

$$= 0 + 1 + 1 + 2 + 3$$

$$= 7$$

and

$$n = \sum_{j=1}^{3} \left[\frac{11j}{7}\right]$$

$$= \left[\frac{11}{7}\right] + \left[\frac{22}{7}\right] + \left[\frac{33}{7}\right]$$

$$= 1 + 3 + 4$$

$$= 8$$

To recapitulate, if p is an odd prime and $(a, p) = 1$, we have seen that the problem of evaluating $\left(\frac{a}{p}\right)$ can be reduced to the problem of evaluating $\left(\frac{q}{p}\right)$ where q is an odd prime and $0 < q < p$. The Law of Quadratic Reciprocity, which follows, facilitates the evaluation of $\left(\frac{q}{p}\right)$. This theorem is amazing because it gives a relation between the solutions of quadratic congruences to *distinct* odd prime moduli. This statement was made, and partially proved, by Legendre in 1785. The first complete proof was given by Gauss in 1801.

Theorem 7.9 **Law of Quadratic Reciprocity**

If p and q are distinct odd primes, then

$$\left(\frac{p}{q}\right)\left(\frac{q}{p}\right) = (-1)^{\left(\frac{p-1}{2}\right)\left(\frac{q-1}{2}\right)}.$$

Proof: Let

$$m = \sum_{i=1}^{\frac{p-1}{2}} \left[\frac{qi}{p} \right] \quad \text{and} \quad n$$

$$= \sum_{j=1}^{\frac{q-1}{2}} \left[\frac{pj}{q} \right]$$

Theorem 7.7 implies $\left(\frac{q}{p}\right) = (-1)^m$ and $\left(\frac{p}{q}\right) = (-1)^n$, so $\left(\frac{q}{p}\right)\left(\frac{p}{q}\right) = (-1)^{m+n}$. But Theorem 7.8 implies $m+n = \left(\frac{p-1}{2}\right)\left(\frac{q-1}{2}\right)$. The conclusion now follows.

∎

It is more convenient in practice to use the following alternate form of the Law of Quadratic Reciprocity:

Theorem 7.9A   **Law of Quadratic Reciprocity (Alternate Form)**

Let p and q be distinct odd primes. If $p \equiv q \equiv 3 \pmod 4$, then $\left(\frac{q}{p}\right) = -\left(\frac{p}{q}\right)$. Otherwise, that is, if $p \equiv 1 \pmod 4$ or $q \equiv 1 \pmod 4$, then $\left(\frac{q}{p}\right) = \left(\frac{p}{q}\right)$.

_____

Proof: Exercise.

_____

Remarks

The Law of Quadratic Reciprocity establishes a link between the solvability of the congruences

$$x^2 \equiv q \pmod p$$

$$x^2 \equiv p \pmod q$$

where p and q are distinct odd primes. Specifically, if $p \equiv q \equiv 3 \pmod 4$, then exactly one of these two congruences has solutions. On the other hand, if $p \equiv 1 \pmod 4$ or $q \equiv 1 \pmod 4$ (or both), then either both congruences have solutions or neither does.

∎

By repeated use of Theorems 7.9A, 7.5, and 7.3, we can now evaluate $\left(\frac{a}{p}\right)$, where p is an odd prime and $(a, p) = 1$. Without loss of generality, assume that $0 < a < p$. For example, let us evaluate $\left(\frac{30}{43}\right)$.

Solution:

$$\left(\frac{30}{43}\right) = \left(\frac{2 * 3 * 5}{43}\right) = \left(\frac{2}{43}\right)\left(\frac{3}{43}\right)\left(\frac{5}{43}\right)$$

Since $43 \equiv 3 \pmod 8$, we have $\left(\frac{2}{43}\right) = -1$. Also

$$\left(\frac{3}{43}\right) = -\left(\frac{43}{3}\right) = -\left(\frac{1}{3}\right) = -1$$

$$\left(\frac{5}{43}\right) = \left(\frac{43}{5}\right) = \left(\frac{3}{5}\right) = \left(\frac{5}{3}\right) = \left(\frac{2}{3}\right) = -1$$

Therefore $\left(\frac{30}{43}\right) = (-1)^3 = -1$. ∎

Recall that in Section 4.8, we defined the Fermat numbers, $f_n = 2^{2^n} + 1$, where $n \geq 0$. The following theorem, whose proof uses the Law of Quadratic Reciprocity, provides a test for the primality of Fermat numbers.

Theorem 7.10

Pepin's Test for Primality of Fermat Numbers

Let $n \geq 1$. Then f_n is prime if and only if $3^{\frac{1}{2}(f_n - 1)} \equiv -1 \pmod{f_n}$.

Proof: (Sufficiency) Suppose that f_n is prime and $n \geq 1$. Then $2^{2^n} \equiv 1 \pmod 3$, so $f_n \equiv 2 \pmod 3$. Also, $f_n \equiv 1 \pmod 4$. Therefore $\left(\frac{3}{f_n}\right) = \left(\frac{f_n}{3}\right) = \left(\frac{2}{3}\right) = -1$. Now Euler's Criterion implies $3^{\frac{1}{2}(f_n - 1)} \equiv -1 \pmod{f_n}$.

(Necessity) Suppose that $3^{\frac{1}{2}(f_n - 1)} \equiv -1 \pmod{f_n}$. If q is a prime factor of f_n, then $3^{\frac{1}{2}(f_n - 1)} \equiv -1 \pmod q$, so $3^{(f_n - 1)} \equiv 1 \pmod q$. Let 3 have order $m \pmod q$. Then, by Theorem 6.1, $m | (f_n - 1)$, that is, $m | 2^{2^n}$. Therefore $m = 2^k$ for some k such that $0 \leq k \leq 2^n$. Let $k = 2^n - j$ where $j \geq 0$. We have

$$3^{2^{2^n - j}} \equiv 1 \pmod q$$

If $j \geq 1$, then

$$\left(3^{2^{2^n - j}}\right)^{2^{j-1}} \equiv 1^{2^{j-1}} \pmod q$$

that is,

$$3^{2^{2^n - 1}} \equiv 3^{\frac{1}{2}(f_n - 1)} \equiv 1 \pmod q$$

But then $-1 \equiv 1 \pmod q$, which is impossible, since q is odd. Therefore $j = 0$, so 3 has order $2^{2^n} \pmod q$. Now Theorem 6.2 implies $2^{2^n} | (q - 1)$, so $2^{2^n} \leq q - 1$, hence $2^{2^n} + 1 \leq q$, that is $f_n \leq q$. Since $q|f_n$, we must have $q = f_n$, so f_n is prime.   ∎

Remarks

We use the base 3 because 3 is a quadratic nonresidue (mod f_n) for all $n \geq 1$, not just for those f_n that are prime.

■

For example, let us use Pepin's test to verify that $f_4 = 65,537$ is prime. Using Mathematica, we see that $3^{2^{15}} \equiv 65,536 \equiv -1 \pmod{65537}$. We therefore conclude that 65,537 is prime. As a second example, let us test $f_5 = 4,294,967,297$ for primality using Pepin's Test. Using Mathematica, we see that $3^{2^{31}} \equiv 10,324,303 \not\equiv -1 \pmod{f_5}$. Therefore f_5 is composite.

Remarks

Fermat had stated, mistakenly, that f_n is prime for all $n \geq 0$. Euler discovered in 1732 that $f_5 = 2^{32} + 1 = 641 * 6700417$.

■

Carl Friedrich Gauss (1777–1855)

Carl Friedrich Gauss was born in Braunschweig, Germany on April 30, 1777. His father was a casual laborer. A child prodigy, Gauss supposedly corrected his father's errors in addition when he was 3 years old. Young Gauss's talents were made known by his mother to the Duke of Braunschweig, who became Gauss's patron, paying for his education and later supporting his research.

Originally, Gauss had intended to study philology, since his philology teacher in secondary school was more stimulating than his mathematics teacher. In 1796, however, Gauss managed to solve a 2000-year old mathematical problem: how to construct a regular 17-sided polygon using only a straight edge and a compass. This great success inspired him to study higher mathematics.

In 1799, Gauss received a doctorate from the University of Goettingen. In 1801, after returning to Braunschweig, he published a major work, *Disquisitiones Arithmeticae*, in which he introduced congruences and proved the Law of Quadratic Reciprocity. In the same year, he calculated the orbit of the newly discovered asteroid Ceres. Gauss's reputation grew in Europe, and in 1802 he received an offer from the St. Petersburg Academy. Gauss remained in Braunschweig, where the duke increased his stipend.

In 1806, the Duke of Braunschweig died while leading German forces in battle with the French. The death of his patron left Gauss in need of a source of income. In 1807, Gauss became Professor of Astronomy and Director of the Observatory at the University of Goettingen, where he spent the rest of his life. Gauss married twice; both wives suffered from ill health and died at an early age. The two sons from the second marriage emigrated to the United States.

Gauss made major contributions to several branches of mathematics, including number theory, complex analysis, and numerical analysis. He was so admired and respected for his achievements that he came to be known as the "prince of mathematics". Outside mathematics, Gauss did research in geodesy (surveying), magnetism, and dioptrics. His motto was *pauca sed matura* (few but ripe). He is also known for the quotation "Mathematics is the queen of the sciences, and number theory is the queen of mathematics."

Gauss became rich late in life as a result of successful financial speculation. He died on February 23, 1855.

Section 7.3 Exercises

1. Use Gauss' Lemma to evaluate each of the following:

   (a) $\left(\frac{5}{11}\right)$

   (b) $\left(\frac{5}{13}\right)$

   (c) $\left(\frac{6}{17}\right)$

   (d) $\left(\frac{19}{23}\right)$

   (e) $\left(\frac{11}{29}\right)$

2. Use Gauss' Lemma to prove that

$$\left(\frac{3}{p}\right) = \begin{cases} 1 & \text{if } p \equiv \pm 1 \pmod{12} \\ -1 & \text{if } p \equiv \pm 5 \pmod{12} \end{cases}$$

3. Let f_n denote the nth Fermat number, where $n \geq 2$. Prove that if p is prime and $p|f_n$, then $p \equiv 1 \pmod{2^{n+2}}$. (Hint: use Theorem 7.5.)

4. Use Theorems 7.9A and 7.5 to evaluate each of the following:

   (a) $\left(\frac{3}{17}\right)$

   (b) $\left(\frac{20}{29}\right)$

   (c) $\left(\frac{105}{113}\right)$

   (d) $\left(\frac{6}{23}\right)$

   (e) $\left(\frac{11}{37}\right)$

   (f) $\left(\frac{10}{79}\right)$

   (g) $\left(\frac{226}{229}\right)$

(h) $\left(\frac{79}{97}\right)$

(i) $\left(\frac{60}{83}\right)$

(j) $\left(\frac{182}{1263}\right)$

5. Prove that if $2^n - 15 = x^2$, then $n = 4$ or $n = 6$.

6. Let p and q be Fermat primes, with $p < q$. Prove that

$$\left(\frac{p}{q}\right) = \begin{cases} -1 & \text{if } p \leq 5 \\ 1 & \text{if } p \geq 17 \end{cases}$$

7. Prove that if p and q are odd primes and $q = 2p + 1$, then $\left(\frac{p}{q}\right) = \left(\frac{-1}{p}\right)$.

8. Prove that for each $k \geq 1$, there are infinitely many primes, p, such that $p \equiv 1$ (mod 2^k).

9. Find necessary and sufficient congruence conditions on the odd prime p such that each of the following holds:

(a) $\left(\frac{-2}{p}\right) = 1$

(b) $\left(\frac{-3}{p}\right) = 1$

(c) $\left(\frac{5}{p}\right) = 1$

(d) $\left(\frac{15}{p}\right) = 1$

Section 7.3 Computer Exercises

10. Write a computer program to evaluate $\left(\frac{a}{p}\right)$ where p is an odd prime and $0 < a < p$.

11. Given that $f_6 = 2^{64} + 1$ is composite, write a computer program to find the least prime factor of f_6.

7.4 Solution of Quadratic Congruences

Let $0 < a < p$ where p is an odd prime. Once we have verified that $\left(\frac{a}{p}\right) = 1$, that is, the congruence $x^2 \equiv a$ (mod p), has solutions, it remains to find these solutions. The algorithm given below, which is due to Daniel Shanks (1917–1996), provides a method of obtaining the solutions.

Algorithm for Solving Quadratic Congruences (mod p)

Initialization

I_1 : Determine the integers k, m such that $p - 1 = 2^k m$, where $k \geq 1$ and m is odd.

I_2 : Find q, a quadratic nonresidue (mod p).

I_3 : Let $b \equiv q^m$ (mod p).

I_4 : Let $t \equiv a^{\frac{m+1}{2}}$ (mod p).

I_5 : Let $n \equiv a^m$ (mod p).

Main Routine

Step 1: Find the least $j \geq 0$ such that $n^{2^j} \equiv 1$ (mod p).

Step 2: If $j > 0$, go to Step 3; if $j = 0$, go to Step 6.

Step 3: Set $c \equiv b^{2^{k-j-1}}$ (mod p).

Step 4: Replace t by tc and n by nc^2.

Step 5: Go to Step 1.

Step 6: Terminate. The solutions are $x \equiv \pm t$ (mod p).

For example, let us use the Shanks algorithm to solve the congruence, $x^2 \equiv 2$ (mod 97). Since $97 \equiv 1$ (mod 8), Theorem 7.5 implies $(\frac{2}{97}) = 1$.

Solution: $97 - 1 = 96 = 2^5 3$, so $k = 5, m = 3, a = 2$. The least quadratic nonresidue (mod 97) is 5, so let $q = 5$. Now

$$b \equiv q^m \equiv 5^3 \equiv 125 \equiv 28 \quad (\text{mod } 97)$$

$$t \equiv a^{\frac{m+1}{2}} \equiv 2^2 \equiv 4 \quad (\text{mod } 97)$$

$$n \equiv a^m \equiv 2^3 \equiv 8 \quad (\text{mod } 97).$$

Having initialized, we go to the main routine:

$$n^2 \equiv 8^2 \equiv 64 \quad (\text{mod } 97)$$

$$64^2 \equiv 4096 \equiv 22 \quad (\text{mod } 97)$$

$$22^2 \equiv 484 \equiv -1 \quad (\text{mod } 97)$$

$$(-1)^2 \equiv 1 \quad (\text{mod } 97)$$

Therefore $j = 4$. Updating, we have

$$c \equiv b^{2^{k-j-1}} \equiv 28^{2^{5-4-1}} \equiv 28^{2^0} \equiv 28^1 \equiv 28 \qquad \text{(mod 97)}$$

$$c^2 \equiv 28^2 \equiv 784 \equiv 8 \qquad \text{(mod 97)}$$

$$tc \equiv 4(28) \equiv 112 \equiv 15 \to t \qquad \text{(mod 97)}$$

$$nc^2 \equiv 8(8) \equiv 64 \to n \qquad \text{(mod 97)}$$

We have already seen that $64^2 \equiv 22$, $22^2 \equiv -1$, $(-1)^2 \equiv 1$ (mod 97). Therefore $j = 3$. Updating, we have

$$c \equiv b^{2^{k-j-1}} \equiv 28^{2^{5-3-1}} \equiv 28^{2^1} \equiv 28^2 \equiv 784 \equiv 8 \qquad \text{(mod 97)}$$

$$c^2 \equiv 8^2 \equiv 64 \qquad \text{(mod 97)}$$

$$tc \equiv 15(8) \equiv 120 \equiv 23 \to t \qquad \text{(mod 97)}$$

$$nc^2 \equiv 8(8) \equiv 64(64) \equiv 4096 \equiv 22 \to n \qquad \text{(mod 97)}$$

We have seen that $22^2 \equiv -1$, $(-1)^2 \equiv 1$ (mod 97). Therefore $j = 2$. Updating, we have

$$c \equiv b^{2^{k-j-1}} \equiv 28^{2^{5-2-1}} \equiv 28^{2^2} \equiv (28^2)^2 \equiv 8^2 \equiv 64 \qquad \text{(mod 97)}$$

$$c^2 \equiv 64^2 \equiv 22 \qquad \text{(mod 97)}$$

$$tc \equiv 23(64) \equiv 1472 \equiv 17 \to t \qquad \text{(mod 97)}$$

$$nc^2 \equiv 22(22) \equiv 484 \equiv -1 \to n \qquad \text{(mod 97)}$$

Since $(-1)^2 \equiv 1$ (mod 97), we have $j = 1$. Updating for the last time, we have

$$c \equiv b^{2^{k-j-1}} \equiv 28^{2^{5-1-1}} \equiv 28^{2^3} \equiv (28^4)^2 \equiv 64^2 \equiv 22 \qquad \text{(mod 97)}$$

$$c^2 \equiv 22^2 \equiv -1 \qquad \text{(mod 97)}$$

$$tc \equiv 17(22) \equiv 374 \equiv -14 \to t \qquad \text{(mod 97)}$$

$$nc^2 \equiv (-1)(-1) \equiv 1 \to n \qquad \text{(mod 97)}$$

Therefore $j = 0$, so we are done; the solutions are: $x \equiv \pm 14$ (mod 97).

Check: $(\pm 14)^2 = 196 \equiv 2$ (mod 97).

Remarks

We now discuss why Shanks' algorithm for solving quadratic congruences is guaranteed to work. After initialization, we have $t^2 \equiv a^{m+1} \equiv a(a^m) \equiv an$ (mod p). If $n \equiv 1$ (mod p), then $t^2 \equiv a$ (mod p), so the solutions are $x \equiv \pm t$ (mod p). If $n \not\equiv 1$ (mod p), then further steps are necessary. When the values of t and n are altered, the congruence $t^2 \equiv an$ (mod p) is preserved, since $(tc)^2 \equiv a(nc^2)$ (mod p). Initially, we have $n^{2^{k-1}} \equiv (a^m)^{2^{k-1}} \equiv a^{2^{k-1}m} \equiv a^{\frac{p-1}{2}} \equiv 1$ (mod p), since $(\frac{a}{p}) = 1$ by hypothesis. If n has order h (mod p), then Theorem 6.1 implies $h|2^{k-1}$, so $h = 2^j$ for some j such that $0 \le j \le k-1$. We seek an integer, r, such that r also has order 2^j (mod p). Then, by the result of Section 6.2, Exercise 16, nr has order 2^i (mod p) where $0 \le i \le j-1$.

■

Let $r \equiv b^{2^{k-j}}$ (mod p) where $b \equiv q^m$ (mod p), that is, $r \equiv (q^m)^{2^{k-j}} \equiv q^{2^{k-j}m}$ (mod p). Now $r^{2^j} \equiv (q^{2^{k-j}m})^{2^j} \equiv q^{2^k m} \equiv q^{p-1} \equiv 1$ (mod p) by Fermat's Little Theorem. But $r^{2^{j-1}} \equiv q^{\frac{p-1}{2}} \equiv -1$ (mod p), since $(\frac{q}{p}) = -1$ by hypothesis. Therefore r has order 2^j (mod p). Note that $r \equiv c^2$ (mod p) where $c \equiv b^{2^{k-j-1}}$ (mod p). (The exponent on 2 is a non-negative integer, since $j \le k-1$.) Note also that j is reduced by at least 1 in each iteration of the main routine. Therefore at most $k-1$ iterations are needed before the solutions are obtained.

If $p \equiv 3$ (mod 4), then $k = 1$ and $m = \frac{1}{2}(p-1)$, so $\frac{1}{2}(m+1) = \frac{p+1}{4}$. In this case, $j = 0$ immediately, so the solutions of (4) are given by

$$x \equiv \pm a^{\frac{p+1}{4}} \quad (\text{mod } p)$$

For example, in order to solve the congruence, $x^2 \equiv 6$ (mod 43), we first verify that $(\frac{6}{43}) = 1$. Then, since $43 \equiv 3$ (mod 4), the solutions are given by

$$x \equiv \pm 6^{(43+1)/4} \equiv \pm 6^{11} \equiv \pm 7 \quad (\text{mod } 43)$$

Section 7.4   Exercises

Find all solutions, if any, of each of the following congruences, and check your results.

1. $x^2 \equiv 26$ (mod 37)

2. $x^2 \equiv 15$ (mod 47)

3. $x^2 \equiv 2$ (mod 53)

4. $x^2 \equiv 2$ (mod 73)

5. $x^2 \equiv 6$ (mod 97)

6. $x^2 \equiv 21 \pmod{113}$

7. $x^2 \equiv 59 \pmod{137}$

8. $x^2 \equiv 67 \pmod{257}$

9. $x^2 \equiv 62 \pmod{577}$

10. Prove that if p is an odd prime and if q is the least integer such that $0 < q < p$ and $\left(\frac{q}{p}\right) = -1$, then q is prime.

11. Prove that if the prime $p \equiv 1 \pmod 4$ and q is a quadratic nonresidue $\pmod p$, then the solutions of the congruence $x^2 \equiv -1 \pmod p$ are $x \equiv \pm q^{(p-1)/4} \pmod p$.

12. Find necessary and sufficient congruence conditions on the prime p such that 7 is the least quadratic nonresidue $\pmod p$. What is the least such prime?

Section 7.4   Computer Exercises

13. Write a computer program to implement the solution of quadratic congruences $\pmod p$ using the Shanks algorithm.

7.5 Quadratic Congruences with Composite Moduli

Up to this point, we have studied the quadratic congruence $x^2 \equiv a \pmod m$, where $m = p$, a prime. Let us now consider the case where m is composite. To begin with, suppose that m is a power of an odd prime, p, that is, $m = p^n$ for some $n \geq 2$.

We saw in Section 6.4 that if the congruence $f(x) \equiv 0 \pmod{p^n}$ has a solution, then so does the congruence $f(x) \equiv 0 \pmod{p^k}$ for every k such that $1 \leq k \leq n$. In particular, the congruence $f(x) \equiv 0 \pmod p$ must have a solution. Furthermore, if $f(y) \equiv 0 \pmod{p^n}$, then $y \equiv x_i + tp^{n-1} \pmod{p^n}$, where $f(x_i) \equiv 0 \pmod{p^{n-1}}$ and $tf'(x_i) \equiv -\frac{f(x_i)}{p^{n-1}} \pmod p$.

In this context, we now consider the congruence

$$x^2 \equiv a \pmod{p^n} \tag{7.15}$$

where p is an odd prime, $n \geq 2$, $0 < a < p^n$, and $\left(\frac{a}{p}\right) = 1$. We shall see that (7.15) always has exactly two solutions:

$$x \equiv \pm b_n \pmod{p^n}$$

which are obtained by first solving (7.15) when $n = 1$ and then "working one's way up."

Theorem 7.11 Let p be an odd prime, $0 < a < p^n$, $n \geq 1$, $\left(\frac{a}{p}\right) = 1$. Then the congruence

$$x^2 \equiv a \pmod{p^n} \tag{7.16}$$

always has exactly two solutions, namely,

$$x \equiv \pm b_n \pmod{p^n} \tag{7.17}$$

where $(b_n, p) = 1$.

Proof: (Induction on n)   By hypothesis, $\left(\frac{a}{p}\right) = 1$, so the conclusion holds for $n = 1$. That is, the congruence $x^2 \equiv a \pmod p$ has two solutions: $x \equiv \pm b_1 \pmod p$, where $(b_1, p) = 1$. By induction hypothesis, suppose that the congruence

$$x^2 \equiv a \pmod{p^{n-1}}$$

has only the two solutions: $x \equiv \pm b_{n-1} \pmod{p^{n-1}}$, where $(b_{n-1}, p) = 1$. Let $f(x) = x^2 - a$, so $f'(x) = 2x$. According to Theorem 4.7, we obtain solutions of (7.16) for each t such that

$$tf'(b_{n-1}) \equiv -\frac{f(b_{n-1})}{p^{n-1}} \pmod p$$

that is,

$$2b_{n-1}t \equiv -\frac{(b_{n-1}^2 - a)}{p^{n-1}} \pmod p$$

Now $(b_{n-1}, p) = 1$ by induction hypothesis, so Theorem 4.5 implies that the last congruence has a unique solution $\pmod p$,

$$t \equiv t_n \pmod p$$

Therefore, if we let $b_n = b_{n-1} + t_n p^{n-1}$, we conclude that (7.16) has exactly two solutions, namely,

$$x \equiv \pm b_n \pmod{p^n}$$

Since $(b_{n-1}, p) = 1$, it follows that $(b_n, p) = 1$.   ∎

For example, suppose we wish to solve the congruence

$$x^2 \equiv 2 \pmod{7^3}$$

Since $7 \equiv -1 \pmod 8$, Theorem 7.5 implies $\left(\frac{2}{7}\right) = 1$, that is, the congruence

$$x^2 \equiv 2 \pmod 7$$

has two solutions, namely, $x \equiv \pm 3 \pmod 7$, so $b_1 = 3$. Next, we solve the congruence

$$x^2 \equiv 2 \pmod{7^2}$$

According to Theorem 7.11, the solutions to this last congruence are $x \equiv \pm b_2 \pmod{7^2}$ where $b_2 \equiv b_1 + 7t_1 \pmod{7^2}$ and

$$2b_1 t_1 \equiv -\frac{(b_1^2 - 2)}{7} \pmod 7 \rightarrow 6t_1 \equiv -1 \pmod 7$$

This yields $t_1 \equiv 1 \pmod 7$, so $b_2 \equiv 3 + 7(1) \equiv 10 \pmod{7^2}$. According to Theorem 7.11, the solutions to

$$x^2 \equiv 2 \pmod{7^3}$$

are

$$x \equiv \pm b_3 \pmod{7^3}$$

where

$$b_3 \equiv b_2 + 49t_1 \pmod{7^3}$$

and

$$2b_2 t_2 \equiv -\frac{(b_2^2 - 2)}{7^2} \pmod 7 \rightarrow 20t_2 \equiv -2 \pmod 7$$

This yields $t_2 \equiv 2 \pmod 7$, so $b_3 \equiv 10 + 49(2) \equiv 108 \pmod{7^3}$. Therefore, our desired solutions are $x \equiv \pm 108 \pmod{7^3}$.

Check: $108^2 - 2 = 11662 = 34 * 343$

Next, we consider the congruence

$$x^2 \equiv a \pmod m \tag{7.18}$$

where $m = \prod_{i=1}^{r} p_i^{e_i}$, the p_i are distinct odd primes, and each $e_i \geq 1$. In fact, let $m_i = p_i^{e_i}$ for each i, so that $m = \prod_{i=1}^{r} m_i$ and the m_i are pairwise relatively prime. If (7.18) has a solution, then so does each congruence in the system

$$\begin{cases} x^2 \equiv a \pmod{m_1} \\ x^2 \equiv a \pmod{m_2} \\ x^2 \equiv a \pmod{m_3} \\ \qquad \cdots \\ x^2 \equiv a \pmod{m_r} \end{cases} \tag{7.19}$$

Conversely, if each of the r congruences in the system (7.19) has a solution, and if the moduli are pairwise relatively prime, then (7.18) also has a solution. Now let each congruence

$$x^2 \equiv a \pmod{m_i}$$

have a solution, $x \equiv c_i \pmod{m_i}$. (Theorem 7.11 guarantees the existence of such solutions provided that $\left(\frac{a}{p_i}\right) = 1$ for each i.) By the Chinese Remainder Theorem (Theorem 4.6), the system of simultaneous linear congruences

$$\begin{cases} x \equiv c_1 \pmod{m_1} \\ x \equiv c_2 \pmod{m_2} \\ x \equiv c_3 \pmod{m_3} \\ \quad \cdots \\ x \equiv c_r \pmod{m_r} \end{cases} \tag{7.20}$$

has a unique solution (mod m) for each set of parameters $c_1, c_2, \ldots, c_r$. According to Theorem 7.10, each of the r congruences in (7.19) has two solutions. Therefore we obtain 2^r sets of values of the parameters c_i, and hence 2^r solutions of (7.20).

For example, consider the congruence

$$x^2 \equiv 2 \pmod{119} \tag{7.21}$$

Since $119 = 7 * 17$, the congruence (7.21) is equivalent to the system of simultaneous congruences

$$\begin{cases} x^2 \equiv 2 \pmod{7} \\ x^2 \equiv 2 \pmod{17} \end{cases} \tag{7.22}$$

Now $x^2 \equiv 2 \pmod{7}$ if $x \equiv \pm 3 \pmod{7}$, and $x^2 \equiv 2 \pmod{17}$ if $x \equiv \pm 6 \pmod{7}$. We therefore obtain four systems of simultaneous linear congruences, namely,

$$\begin{cases} x \equiv 3 \pmod{7} \\ x \equiv 6 \pmod{17} \end{cases} \tag{7.23}$$

$$\begin{cases} x \equiv -3 \pmod{7} \\ x \equiv -6 \pmod{17} \end{cases} \tag{7.24}$$

$$\begin{cases} x \equiv 3 \pmod{7} \\ x \equiv -6 \pmod{17} \end{cases} \tag{7.25}$$

$$\begin{cases} x \equiv -3 & (\text{mod } 7) \\ x \equiv 6 & (\text{mod } 17) \end{cases} \qquad (7.26)$$

The solution of (7.23) is $x \equiv -11$ (mod 119). Therefore the solution of (7.24) is $x \equiv 11$ (mod 119). The solution of (7.25) is $x \equiv 45$ (mod 119). Therefore the solution of (7.26) is $x \equiv -45$ (mod 119). In conclusion, the solutions of (7.21) are $x \equiv \pm 11, \pm 45$ (mod 119).

As a second example, consider the congruence

$$x^2 \equiv 3 \quad (\text{mod } 55) \qquad (7.27)$$

Since $55 = 5 * 11$, the congruence (7.27) is equivalent to the system of simultaneous linear congruences

$$\begin{cases} x^2 \equiv 3 & (\text{mod } 5) \\ x^2 \equiv 3 & (\text{mod } 11) \end{cases} \qquad (7.28)$$

But since $(\frac{3}{5}) = -1$, the first congruence in (7.28) has no solution. Therefore (7.27) has no solution.

Section 7.5   Exercises

Find all solutions, if any, of each of the following congruences:

1. $x^2 \equiv 3$ (mod 11^2)

2. $x^2 \equiv 5$ (mod 19^2)

3. $x^2 \equiv 3$ (mod 13^3)

4. $x^2 \equiv 2$ (mod 17^3)

5. $x^2 \equiv 7$ (mod 33)

6. $x^2 \equiv 3$ (mod 143)

7. $x^2 \equiv 7$ (mod 57)

8. $x^2 \equiv 10$ (mod 403)

7.6 Jacobi Symbol

Consider the congruence:

$$x^2 \equiv a \quad (\text{mod } m) \qquad (7.29)$$

where m is odd and $(a, m) = 1$. In the preceding section, we learned (1) how to determine whether (7.29) has solutions if $m = p^n$ where p is an odd prime and $n \geq 1$, and (2) how to find solutions when they exist. The Jacobi symbol, which generalizes the Legendre symbol, sheds some additional light on how to determine whether (7.29) has solutions when m has two or more distinct prime factors.

Definition 7.4   Jacobi Symbol

Let $m = \prod_{i=1}^{r} p_i$ where the p_i are odd primes, not necessarily distinct. Let $(a, m) = 1$. Let $(\frac{a}{p_i})$ denote the Legendre symbol for each i such that $1 \leq i \leq r$. Then

$$(\frac{a}{m}) = \prod_{i=1}^{r} (\frac{a}{p_i})$$

is called a *Jacobi symbol*.

For example, $(\frac{2}{15}) = (\frac{2}{3})(\frac{2}{5}) = (-1)^2 = 1$. Also $(\frac{3}{77}) = (\frac{3}{7})(\frac{3}{11}) = (-1)(1) = -1$.

We now discuss the relevance of the value of the Jacobi symbol $(\frac{a}{m})$ to Congruence (7.29) above. If $(\frac{a}{m}) = -1$, then necessarily some factor $(\frac{a}{p_i}) = -1$, so the congruence

$$x^2 \equiv a \pmod{p_i} \tag{7.30}$$

has no solutions, from which it follows that (7.29) has no solutions. Unfortunately, if $(\frac{a}{m}) = 1$, it does not necessarily follow that (7.29) has solutions. This is so because it may be that an even number of factors $(\frac{a}{p_i}) = -1$. For example, although $(\frac{2}{15}) = 1$, nevertheless the congruence $x^2 \equiv 2 \pmod{15}$ has no solutions. In this sense, the Jacobi symbol provides less information than the Legendre symbol.

The Jacobi symbol shares many properties in common with the Legendre symbol. Before establishing the properties of the Legendre symbol, we need the following lemma:

▶ **Lemma 7.2** *Let* $A_r = \prod_{i=1}^{r} a_i$ *where all the* a_i *are odd. Let* $B_r = \frac{1}{2}(A_r - 1)$. *Let* $b_i = \frac{1}{2}(a_i - 1)$ *for each* i. *Then* $B_r \equiv \sum_{i=1}^{r} b_i \pmod{2}$.

Proof: (Induction on r)   If $r = 1$, then $A_1 = a_1$, so $B_1 = \frac{1}{2}(A_1 - 1) = \frac{1}{2}(a_1 - 1) = b_1$, hence $B_1 \equiv b_1 \pmod{2}$. Now $B_{r+1} = \frac{1}{2}(A_{r+1} - 1) = \frac{1}{2}(a_{r+1}A_r - 1) = \frac{1}{2}[(2b_{r+1} + 1)A_r - 1] = b_{r+1}A_r + \frac{1}{2}(A_r - 1) = b_{r+1}A_r + B_r$. By induction hypothesis, we have $B_r \equiv \sum_{i=1}^{r} b_i$. Therefore

$$B_{r+1} \equiv b_{r+1}A_r + \sum_{i=1}^{r} b_i \equiv \sum_{i=1}^{r+1} b_i \pmod{2} \quad \blacksquare$$

The next theorem lists properties of the Jacobi symbol.

Theorem 7.12

Properties of the Jacobi Symbol

Let m be odd, $m > 1$. Then

$\mathbf{J_1}$: If $a \equiv b \pmod{m}$, then $(\frac{a}{m}) = (\frac{b}{m})$.

$\mathbf{J_2}$: $(\frac{ab}{m}) = (\frac{a}{m})(\frac{n}{m})$

$\mathbf{J_3}$: $(\frac{a}{mn}) = (\frac{a}{m})(\frac{a}{n})$

$\mathbf{J_4}$: $(\frac{-1}{m}) = (-1)^{(\frac{m-1}{2})}$

$\mathbf{J_5}$: $(\frac{2}{m}) = (-1)^{(\frac{m^2-1}{8})}$

Proof: Exercise. (Use Lemma 7.2 and the corresponding properties of the Legendre symbol.)

For example, $(\frac{55}{21}) = (\frac{55}{3})(\frac{55}{7}) = (\frac{1}{3})(\frac{-1}{7}) = 1(-1) = -1$.

The Law of Quadratic Reciprocity may be extended to the Jacobi symbol as follows:

Theorem 7.13

If each of m, n is odd and exceeds 1 , then

$$(\frac{m}{n})(\frac{n}{m}) = (-1)^{(\frac{m-1}{2})(\frac{n-1}{2})}$$

Proof: Exercise. (Use Lemma 7.2 and Theorem 7.9.)

For example, $(\frac{55}{21}) = (\frac{13}{21}) = (\frac{21}{13}) = (\frac{8}{13}) = (\frac{2}{13}) = -1$.

The Jacobi symbol sometimes provides a quicker way to evaluate a Legendre symbol. For example, suppose we wish to evaluate $(\frac{1155}{1187})$. This is a Legendre symbol, since 1187 is prime. Now $1155 = 3 * 5 * 7 * 11$, so using Theorem 7.3, part (ii), we have

$$\left(\frac{1155}{1187}\right) = \left(\frac{3}{1187}\right)\left(\frac{5}{1187}\right)\left(\frac{7}{1187}\right)\left(\frac{11}{1187}\right)$$

Now

$$\left(\frac{3}{1187}\right) = -\left(\frac{1187}{3}\right) = -\left(\frac{2}{3}\right) = -(-1) = 1$$

$$\left(\frac{5}{1187}\right) = \left(\frac{1187}{5}\right) = \left(\frac{2}{5}\right) = -1$$

$$\left(\frac{7}{1187}\right) = -\left(\frac{1187}{7}\right) = -\left(\frac{4}{7}\right) = -1$$

$$\left(\frac{11}{1187}\right) = -\left(\frac{1187}{11}\right) = -\left(\frac{10}{11}\right) = -\left(\frac{2}{11}\right)\left(\frac{5}{11}\right) = -(-1)\left(\frac{11}{5}\right)$$

$$= \left(\frac{1}{5}\right) = 1$$

Therefore $\left(\frac{1155}{1187}\right) = 1(-1)(-1)1 = 1$.

If we treat $\left(\frac{1155}{1187}\right)$ as a Jacobi symbol, we obtain

$$\left(\frac{1155}{1187}\right) = -\left(\frac{1187}{1155}\right) = -\left(\frac{32}{1155}\right) = -\left(\frac{2}{1155}\right) = -(-1) = 1$$

Suppose we wish to find a quadratic nonresidue, $t \pmod m$, where m is odd and $m \geq 3$, but either (1) it is not known whether m is prime or composite, or (2) m is known to be composite, but its prime factors are unknown. As we have seen previously, the least such t will be prime. Using Theorems 7.12 and 7.13, we can evaluate the Jacobi symbol $\left(\frac{t}{m}\right)$. If $\left(\frac{t}{m}\right) = -1$, then we know that t is a quadratic nonresidue $\pmod m$. If $\left(\frac{t}{m}\right) = 1$, then no conclusion can be drawn. In this case, one should replace t by another candidate, namely, the next prime.

For example, suppose we wish to find a quadratic nonresidue $\pmod{11111}$, without determining whether 11111 is prime or composite. It is easily verified, using the properties of the Jacobi symbol, that

$$\left(\frac{2}{11111}\right) = \left(\frac{3}{11111}\right) = \left(\frac{5}{11111}\right) = 1; \quad \left(\frac{7}{11111}\right) = -1$$

Section 7.6 Exercises

1. Evaluate each of the following Jacobi symbols:

    (a) $\left(\frac{30}{1001}\right)$

    (b) $\left(\frac{91}{561}\right)$

    (c) $\left(\frac{130}{231}\right)$

(d) $\left(\frac{70}{141}\right)$

(e) $\left(\frac{136}{455}\right)$

2. Evaluate $\left(\frac{5}{77}\right)$. Does the congruence $x^2 \equiv 5 \pmod{77}$ have solutions? (Explain.)

3. Evaluate $\left(\frac{3}{133}\right)$. Does the congruence $x^2 \equiv 3 \pmod{133}$ have solutions? (Explain.)

4. Evaluate each of the following Legendre symbols by treating them as Jacobi symbols:

   (a) $\left(\frac{105}{113}\right)$

   (b) $\left(\frac{385}{401}\right)$

   (c) $\left(\frac{91}{101}\right)$

   (d) $\left(\frac{77}{103}\right)$

   (e) $\left(\frac{65}{79}\right)$

5. Prove Theorem 7.12.

6. Prove Theorem 7.13.

7. Prove that if m is odd, $m > 3$, and $(m, 3) = 1$, then

$$\left(\frac{3}{m}\right) = \begin{cases} 1 & \text{if } m \equiv \pm 1 \pmod{12} \\ -1 & \text{if } m \equiv \pm 5 \pmod{12} \end{cases}$$

8. Without factoring 1111111, find a quadratic nonresidue (mod 1111111).

Section 7.6   Computer Exercises

9. Write a computer program that makes use of the Jacobi symbol to find a quadratic nonresidue (mod m), where the given odd natural number $m \geq 3$.

Review Exercises

1. Find all quadratic residues and quadratic nonresidues (mod 11).

2. Use Euler's Criterion to evaluate each of the following Legendre symbols:

   (a) $\left(\frac{3}{7}\right)$

   (b) $\left(\frac{2}{11}\right)$

   (c) $\left(\frac{5}{13}\right)$

   (d) $\left(\frac{10}{17}\right)$

   (e) $\left(\frac{3}{19}\right)$

3. Use Gauss' Lemma to evaluate all the Legendre symbols in Exercise 2.

4. Use the Law of Quadratic Reciprocity to evaluate each of the following:

(a) $\left(\frac{7}{13}\right)$

(b) $\left(\frac{3}{11}\right)$

(c) $\left(\frac{7}{23}\right)$

(d) $\left(\frac{5}{29}\right)$

(e) $\left(\frac{15}{31}\right)$

5. Prove that if p is an odd prime, then $\left(\frac{-2}{p}\right) = 1$ if and only if $p \equiv 1, 3 \pmod{8}$.

6. Use the Shanks algorithm to solve the congruences:

(a) $x^2 \equiv 3 \pmod{61}$

(b) $x^2 \equiv 2 \pmod{113}$

7. Find all solutions, if any, of each of the following congruences:

(a) $x^2 \equiv 5 \pmod{11^2}$

(b) $x^2 \equiv 2 \pmod{77}$

(c) $x^2 \equiv 11 \pmod{35}$

8. Evaluate each of the following Jacobi symbols:

(a) $\left(\frac{102}{1001}\right)$

(b) $\left(\frac{65}{561}\right)$

(c) $\left(\frac{85}{231}\right)$

9. Evaluate each of the following Legendre symbols by treating them as Jacobi symbols:

(a) $\left(\frac{91}{113}\right)$

(b) $\left(\frac{105}{401}\right)$

(c) $\left(\frac{35}{103}\right)$

10. Evaluate $\left(\frac{5}{39}\right)$. Does the congruence $x^2 \equiv 5 \pmod{39}$ have solutions? (Explain.)

Chapter 8

Sums of Squares

8.1 Introduction

In the early part of this chapter, we will give necessary and sufficient conditions on the natural number n so that n can be represented as a sum of two squares of nonnegative integers. (For example, we have $34 = 5^2 + 3^2$, $53 = 7^2 + 2^2$, $64 = 8^2 + 0^2$.) In the latter part of this chapter, we will present Lagrange's famous four-square theorem: every natural number can be represented as a sum of four squares of nonnegative integers. (For example, $39 = 5^2 + 3^2 + 2^2 + 1^2$.)

8.2 Sums of Two Squares

Let us look at the integers from 1 to 50 and determine which of them can be represented as a sum of two squares of nonnegative integers. That is, for each integer n such that $1 \leq n \leq 50$, we seek integers x, y such that $x \geq y \geq 0$ and $n = x^2 + y^2$. The results are given in the following table:

    None of the integers that do not appear in the table, that is, none of 3, 6, 7, 11, 12, 14, 15, 19, 21, 22, 23, 24, 27, 28, 30, 31, 33, 35, 38, 39, 42, 43, 44, 46, 47, and 48, can be represented as a sum of two squares.

n	x	y	n	x	y	n	x	y
1	1	0	16	4	0	34	5	3
2	1	1	17	4	1	36	6	0
4	2	0	18	3	3	37	6	1
5	2	1	20	4	2	40	6	2
8	2	2	25	4	3	41	5	4
9	3	0	26	5	1	45	6	3
10	3	1	29	5	2	49	7	0
13	3	2	32	4	4	50	7	1

Since $1 = 1^2 + 0^2$, we henceforth consider natural numbers $n \geq 2$. We wish to develop criteria, based on the canonical factorization of n, to determine whether n is a sum of two squares.

Let $n = \prod_{i=1}^{r} p_i^{e_i}$ where the p_i are distinct primes. We claim that (1) $n = x^2 + y^2$ if and only if (2) $e_i \equiv 0 \pmod 2$ for all i (if any) such that $p_i \equiv 3 \pmod 4$.

First, we show that (1) implies (2).

Theorem 8.1   Let $n \geq 2$. If $n = x^2 + y^2$, p is a prime such that $p^e || n$, and $p \equiv 3 \pmod 4$, then $2|e$.

Proof: Let p be a prime such that $p^e || n$, where $n = x^2 + y^2$. It suffices to show that if $p \equiv e \equiv 1 \pmod 2$, then $p \equiv 1 \pmod 4$. Let $(x, y) = d$. Then $x = du$, $y = dv$ for some integers u, v such that $(u, v) = 1$. Now $n = d^2 m$, where $m = u^2 + v^2$. Let $p^j || d$, where $j \geq 0$. Therefore $p^{e-2j} || m$. Since e is odd by hypothesis, it follows that $e - 2j \geq 1$. Therefore $p|m$, that is, $p|(u^2 + v^2)$. Since $(u, v) = 1$, Theorem 4.12 implies $p \equiv 1 \pmod 4$. ∎

For example, $18^2 + 9^2 = 405 = 3^4 5^1$. Also, $21^2 + 14^2 = 637 = 7^2 13^1$.

In order to show that (2) implies (1), some preliminaries are needed. The first of these preliminaries, namely, Lemma 8.1 below, appears in the *Arithmetica* of Diophantus, about 250 A.D.

▶ **Lemma 8.1** *If each of the natural numbers u, v is a sum of two squares, then so is uv.*

Proof: Let $u = a^2 + b^2$, $v = c^2 + d^2$. Then

$$uv = (a^2 + b^2)(c^2 + d^2)$$

$$= a^2c^2 + a^2d^2 + b^2c^2 + b^2d^2$$

$$= a^2c^2 + 2abcd + b^2d^2 + a^2d^2 - 2abcd + b^2c^2$$

$$= (ac + bd)^2 + (ad - bc)^2 \quad \blacksquare$$

Remarks

Similarly, $uv = (ac - bd)^2 + (ad + bc)^2$. These two representations of uv as a sum of two squares are easily seen to be distinct unless $a = b$ or $c = d$. For example, since

$$65 = 5 * 13, \quad 5 = 2^2 + 1^2, \quad 13 = 3^2 + 2^2$$

it follows that

$$65 = (6 + 2)^2 + (4 - 3)^2 = 8^2 + 1^2$$

$$= (6 - 2)^2 + (4 + 3)^2$$

$$= 4^2 + 7^2$$

$\blacksquare$

Next, we will show that if the prime $p \equiv 1 \pmod 4$, then there exist unique integers x, y such that $x > y > 0$ and $p = x^2 + y^2$. We will present two algorithms for obtaining p as a sum of two squares. In the first of these algorithms, we represent a multiple of p as a sum of two squares, then a smaller multiple, continuing until we obtain such a representation for p itself.

Theorem 8.2

If the prime $p \equiv 1 \pmod 4$, then there exist unique integers x, y such that $x > y > 0$ and $p = x^2 + y^2$.

Proof: We will show that Algorithm 1, given below, produces such a representation of p, and that it is unique. $\blacksquare$

Algorithm 1

Initialization: Let q be a quadratic nonresidue (mod p). (According to Theorem 7.1, these are not hard to find.) Let t be an integer such that $0 < t < p$ and $t \equiv q^{\frac{p-1}{4}}$ (mod p). Let $x_1 = Min\{t, p - t\}$, so that $0 < x < \frac{p}{2}$. Then

$$x_1^2 \equiv t^2 \equiv q^{\frac{p-1}{2}} \equiv -1 \pmod{p}$$

by Euler's Criterion. If we let $y_1 = 1$, then we have $x_1^2 + y_1^2 \equiv 0 \pmod{p}$, so $x_1^2 + y_1^2 = k_1 p$ for some $k_1 \geq 1$. Furthermore, $k_1 p \leq (\frac{p-1}{2})^2 + 1$, so $k_1 \leq \frac{p-1}{4}$.

Main Routine

Key Step: If $x_i^2 + y_i^2 = p$ for some $i \geq 1$, then we are done. We have $p = x^2 + y^2$ with $x = x_i$, $y = y_i$. Now suppose

$$x_i^2 + y_i^2 = k_i p \quad \text{with} \quad 1 < k_i < p .$$

If k_i is even, then we must have $x_i \equiv y_i \pmod 2$, so that $\frac{x_i \pm y_i}{2}$ are integers. Let

$$x_{i+1} = \left| \frac{x_i + y_i}{2} \right|$$

$$y_{i+1} = \left| \frac{x_i - y_i}{2} \right|$$

Then

$$x_{i+1}^2 + y_{i+1}^2 = \frac{(x_i + y_i)^2}{4} + \frac{(x_i - y_i)^2}{4}$$

$$= \frac{x_i^2 + y_i^2}{2}$$

$$= \frac{k_i p}{2}$$

that is, $x_i^2 + y_i^2 = k_{i+1} p$ with $1 \leq k_{i+1} \leq \frac{k_i}{2}$. Now return to the Key Step, replacing $i + 1$ with i.

If k_i is odd, choose integers A_i, B_i so that $A_i \equiv x_i \pmod{k_i}$, $B_i \equiv y_i \pmod{k_i}$, $|A_i| < \frac{k_i}{2}$, $|B_i| < \frac{k_i}{2}$. Now

$$A_i^2 + B_i^2 \equiv x_i^2 + y_i^2 \equiv 0 \pmod{k_i} \rightarrow A_i^2 + B_i^2 = k_{i+1} k_i$$

If $k_{i+1} = 0$, then $A_i = B_i = 0$, so $x_i \equiv y_i \equiv 0 \pmod{k_i}$. But then $k_i p = x_i^2 + y_i^2 = k_i^2 m$, so $p = k_i m$. Since p is prime by hypothesis and $k_i > 1$, we must have $m = 1$, so $k_i = p$, contrary to hypothesis. Therefore $1 \leq k_{i+1} \leq \frac{k_i}{2}$. Furthermore

$$A_i x_i + B_i y_i \equiv x_i^2 + y_i^2 \equiv 0 \pmod{k_i}$$

and

$$A_i y_i - B_i x_i \equiv x_i y_i - y_i x_i \equiv 0 \pmod{k_i}$$

so that

$$x_{i+1} = \frac{|A_i x_i + B_i y_i|}{k_i}$$

$$y_{i+1} = \frac{|A_i y_i - B_i x_i|}{k_i}$$

are integers. Thus we have

$$x_{i+1}^2 + y_{i+1}^2 = \frac{(A_i x_i + B_i y_i)^2}{k_i^2} + \frac{(A_i y_i - B_i x_i)^2}{k_i^2}$$

$$= \frac{(A_i^2 + B_i^2)(x_i^2 + y_i^2)}{k_i^2}$$

$$= \frac{(k_{i+1} k_i)(k_i p)}{k_i^2}) = k_{i+1} p$$

with $1 \le k_{i+1} \le \frac{k_i}{2}$. Now return to the Key Step, replacing $i+1$ with i.

Since the k_i are positive integers such that $k_1 > k_2 > k_3 > \cdots$, it follows from Theorem 1.8 that $k_r = 1$ for some $r \ge 1$. Furthermore, since each $k_{i+1} \le \frac{k_i}{2}$, it follows that $r \le 1 + \log_2(\frac{p-1}{4})$. Since r, the total number of iterations needed, is proportional to the logarithm of the input, we can say that Algorithm 1 is efficient. ■

Uniqueness: We conclude by showing that the representation of p as a sum of two squares is unique. If also $p = v^2 + w^2$, with $v > w > 0$, then $x^2 \equiv -y^2 \pmod{p}$ and $v^2 \equiv -w^2 \pmod{p}$, so $x^2 w^2 \equiv y^2 v^2 \pmod{p}$. Now each of v, w, x, y is between 0 and $\sqrt{p}$, so $-p < xw - yv < p$ and $0 < xw + yv < 2p$. If $xw \equiv yv \pmod{p}$, then $xw - yv = 0$, so $w = \frac{yv}{x}$. But then $x^2 + y^2 = p = v^2 + w^2 = v^2 + v^2 \frac{y^2}{x^2} = (\frac{v^2}{x^2})(x^2 + y^2)$. This implies $v = x$, so $w = y$. If $xw \equiv -yv \pmod{p}$, then $xw + yv = p$. However, then

$$p^2 + (xv - yw)^2 = (xw + yv)^2 + (xv - yw)^2$$

$$= (x^2 + y^2)(v^2 + w^2)$$

$$= p^2$$

so

$$xv - yw = 0$$

But then $w = (\frac{x}{y})v > v$, contrary to hypothesis. ■

For example, let us use Algorithm 1 to represent the prime 317 as the sum of two squares. Since $317 \equiv 5 \pmod 8$, it follows that $\left(\frac{2}{1013}\right) = -1$, so we may take $q = 2$. Now

$$t \equiv 2^{\frac{317-1}{4}} = 2^{79} \equiv 203 \pmod{1013}$$

Now $x_1 = Min\{203, 317 - 203\} = 114$, and $y_1 = 1$. This yields

$$x_1^2 + y_1^2 = 114^2 + 1^2 = 12997 = 41(317) \rightarrow k_1 = 41.$$

Next, we choose A_1, B_1 such that $|A_1| \leq 20$, $|B_1| \leq 20$, and $A_1 \equiv 114 \pmod{41}$, $B_1 \equiv 1 \pmod{41}$, that is, $A_1 = -9$, $B_1 = 1$. Then

$$x_2 = \frac{|A_1 x_1 + B_1 y_1|}{k_1} = \frac{|-9(114) + 1(1)|}{41} = 25$$

$$y_2 = \frac{|A_1 y_1 - B_1 x_1|}{k_1} = \frac{|(-9)1 - 1(114)|}{41} = 3$$

This yields

$$x_2^2 + y_2^2 = 25^2 + 3^2 = 634 = 2(317)$$

Finally, let

$$x_3 = \frac{|x_2 + y_2|}{2} = \frac{25 + 3}{2} = 14$$

$$y_3 = \frac{|x_2 - y_2|}{2} = \frac{25 - 3}{2} = 11$$

Now $x_3^2 + y_3^2 = 14^2 + 11^2 = 196 + 121 = 317$, so we are done.

We have demonstrated that condition (1) implies (2). Now we are ready to prove that (2) implies (1) by way of the following theorem:

Theorem 8.3 Let $n \geq 2$. Then $n = x^2 + y^2$ if and only if for every prime p, if any, such that $p \equiv 3 \pmod 4$ and $p^e || n$, it is true that $2 | e$.

Proof: Theorem 8.1 implies necessity, so it suffices to prove sufficiency. By hypothesis, $n = AB^2$, where $A = 1$ or $A = \prod_{i=1}^r p_i^{e_i}$, where no $p_i \equiv 3 \pmod 4$, and $B^2 = \prod_{j=1}^s q_j^{2f_j}$, and where all $q_j \equiv 3 \pmod 4$. Noting that $2 = 1^2 + 1^2$, and using Theorem 8.2 and Lemma 8.1 repeatedly, we see that A is a sum of two squares. Since $B^2 = B^2 + 0^2$, Lemma 8.1 implies n is a sum of two squares. ∎

Remarks

If the prime $p \equiv 1 \pmod 4$, then there is an alternate method for representing p as a sum of two squares, which we call Algorithm 2.

Algorithm 2

Initialization: (Same as in Algorithm 1)

Let q be a quadratic nonresidue (mod p). Let t be an integer such that $0 < t < p$ and $t \equiv q^{\frac{p-1}{4}}$ (mod p). Let $x_1 = Min\{t, p-t\}$, so that $0 < x < \frac{p}{2}$. Then

$$x_1^2 \equiv t^2 \equiv q^{\frac{p-1}{2}} \equiv -1 \pmod{p}$$

by Euler's Criterion. If we let $y_1 = 1$, then we have $x_1^2 + y_1^2 \equiv 0 \pmod{p}$, so $x_1^2 + y_1^2 = k_1 p$ for some $k_1 \geq 1$. Furthermore, $k_1 p \leq (\frac{p-1}{2})^2 + 1$, so $k_1 \leq \frac{p-1}{4}$.

Main Routine: If $p = x_1^2 + 1$, then we are done. Otherwise, we begin to compute (p, x_1). Let r_k be the first remainder such that $r_k < \sqrt{p}$. Then $p = r_k^2 + r_{k+1}^2$.

We omit the proof that Algorithm 2 works, but we refer the interested reader to [17].

For example, let us use Algorithm 2 to represent the prime 317 as a sum of two squares. Again, we have $x_1 = 114$. We begin to compute $(317, 114)$. Now

$$317 = 2(114) + 89 \ ; \ 114 = 1(89) + 25 \ ; \ 89 = 2(25) + 14 \ ; \ 25 = 1(14) + 11.$$

Since $14 < \sqrt{317}$, we have $317 = 14^2 + 11^2$.

Suppose that we are given a natural number $n \geq 2$, and we wish to determine whether n is a sum of two squares. In order to use Theorem 8.3, we first factor n as a product of primes. If either (1) n has no prime factors p such that $p \equiv 3 \pmod 4$, or (2) each such factor carries an even exponent, then n is a sum of two squares. Otherwise, n is not a sum of two squares.

For example, let us determine which of the integers from 116 to 120 is a sum of two squares.

$$116 = 2^2 29^1 = 10^2 + 4^2$$

$$117 = 3^2 13^1 = 9^2 + 6^2$$

$$118 = 2^1 59^1 \neq x^2 + y^2$$

$$119 = 7^1 17^1 \neq x^2 + y^2$$

$$120 = 2^3 3^1 5^1 \neq x^2 + y^2$$

Let $r_2(n)$ denote the number of representations of n as a sum of two squares of *integers*, where representations that differ only in the order of terms are considered distinct. The following theorem expresses $r_2(n)$ in terms of the canonical factorization of n:

Theorem 8.4   Let n satisfy the hypothesis of Theorem 8.3, that is

$$n = 2^a l m \quad , \quad a \geq 0 \quad , \quad l = \prod_{i=1}^{r} p_i^{e_i} \quad , \quad m = \prod_{j=1}^{s} q_j^{2f_j}$$

where each prime $p_i \equiv 1 \pmod 4$ and each prime $q_j \equiv 3 \pmod 4$. Then

$$r_2(n) = 4\tau(l) = 4 \prod_{i=1}^{r}(e_i + 1)$$

Proof: Omitted, but see Theorem 3.22 on p. 166 of [9].

For example, $r_2(25) = r_2(5^2) = 4(3) = 12$. The representations are

$$25 = (\pm 4)^2 + (\pm 3)^2 = (\pm 3)^2 + (\pm 4)^2 = (\pm 5)^2 + 0^2 = 0^2 + (\pm 5)^2$$

Joseph Louis Lagrange (1736–1813)

Joseph Louis Lagrange was born in Italy on January 25, 1736. (His name at birth was Luigi Lagrangia.) His father, a local government official, wanted Lagrange to study law. Lagrange, however, preferred mathematics, for which he showed an early aptitude. He began publishing his research in 1754. The following year, Lagrange became Professor of Mathematics at the Royal Artillery School in Turin. He did research on the calculus of variations, fluid mechanics, differential equations, and planetary motion. In 1765, Lagrange won a prize offered by the Paris Academy of Science for his research on the moons of Jupiter. Despite his growing reputation and his desire to remain in Turin, Lagrange never got a raise in salary while he remained there.

In 1766, through the intercession of the influential French mathematician d'Alembert, Lagrange obtained a post at the Berlin Academy of Sciences. (Euler had just left Berlin to return to St. Petersburg.) Lagrange remained in Berlin for over twenty years, working mostly on mechanics and number theory. In 1768, he gave the solution to Pell's equation with continued fractions and proved that only quadratic irrationals have periodic continued fractions. In 1770, he proved the four-square theorem: every natural number is a sum of four squares. In 1771, he gave the first proof of what is now known as Wilson's Theorem and its converse.

In 1786, King Frederick II of Prussia died; the following year, Lagrange moved to Paris, where he became a member of the Paris Academy of Sciences. The French Revolution, with all its upheavals, soon followed. In 1790, Lagrange was named chair of a commission charged with the standardization of weights and measures. This commission was retained even when the Paris Academy of Sciences was abolished, and Lagrange was kept as chair, even though other members of the commission were purged. In 1793, through the intervention of Lavoisier, Lagrange was exempted from a government order to arrest all foreigners and seize their property. Lagrange survived the French Revolution unscathed, perhaps because of the following credo: *I believe that, in general, one of the first principles of every wise man is to conform strictly to the laws of the country in which he is living, even when they are unreasonable.*

Lagrange taught at the newly opened Ecole Polytechnique from 1794 to 1799. Under Napoleon's rule, he became a senator and, in 1808, a Count of the Empire. He died on April 10, 1813.

Section 8.2 Exercises

1. Represent each of the following primes as a sum of two squares:

    (a) 233

    (b) 49

    (c) 613

    (d) 1009

    (e) 1409

2. For each integer from 131 to 140

    (a) determine whether it can be represented as a sum of two squares, and

    (b) if so, find such a representation.

3. Represent each of the following integers as a sum of two squares of integers in as many ways as possible:

    (a) 65

    (b) 85

    (c) 221

    (d) 1073

    (e) 1105

    (f) 1885

    (g) 5525

4. Suppose that the odd prime $p = x^2 + y^2$, with $x > y > 0$. Find integers u, v (in terms of x and y) such that $2p = u^2 + v^2$, with $u > v > 0$.

5. Prove that if the prime $p = \frac{u^2+1}{2}$, then p is the sum of the squares of two consecutive integers.

6. Prove that if the prime $p = x^2 + y^2$ and $p \equiv \pm 1 \pmod{10}$, then $5 | xy$.

7. Find the least three consecutive natural numbers such that none is a square, but each is the sum of two squares.

8. Let p be prime. Prove that there is a right triangle with integer sides and hypotenuse p if and only if $p \equiv 1 \pmod 4$.

9. Prove that there are arbitrarily large gaps between consecutive integers each of which is a sum of two squares. (Hint: The Chinese Remainder Theorem is helpful.)

Section 8.2   Computer Exercises

10. Write a computer program that, given a prime, p , such that $p \equiv 1 \pmod 4$, finds the integers x, y such that $x > y > 0$ and $p = x^2 + y^2$ using

   (a) Algorithm 1

   (b) Algorithm 2

8.3 Sums of Four Squares

Let us try to represent each of the integers from 1 to 50 as a sum of four squares of non-negative integers. That is, for each integer n such that $1 \le n \le 50$, we seek integers w, x, y, z such that $w \ge x \ge y \ge z \ge 0$ and $n = w^2 + x^2 + y^2 + z^2$. The results are presented in the table on page 211.

It appears that every natural number may be represented as the sum of the squares of four non-negative integers. This is indeed the case, as we shall see shortly. First, we need the following lemma, which is due to Lagrange.

▶ **Lemma 8.2** *If each of two natural numbers is the sum of four squares, then so is their product.*

Proof: Let $u = a^2 + b^2 + c^2 + d^2$, $v = w^2 + x^2 + y^2 + z^2$. Then

$$uv = (a^2 + b^2 + c^2 + d^2)(w^2 + x^2 + y^2 + z^2)$$

$$= (aw + bx + cy + dz)^2 + (ax - bw + cz - dy)^2 +$$

$$(ay - bz - cw + dx)^2 + (az + by - cx - dw)^2 \quad \blacksquare$$

n	w	x	y	z	n	w	x	y	z	n	w	x	y	z
1	1	0	0	0	18	3	3	0	0	35	5	3	1	0
2	1	1	0	0	19	3	3	1	0	36	6	0	0	0
3	1	1	1	0	20	4	2	0	0	37	6	1	0	0
4	2	0	0	0	21	4	2	1	0	38	6	1	1	0
5	2	1	0	0	22	3	3	2	0	39	6	1	1	1
6	2	1	1	0	23	3	3	2	1	40	6	2	0	0
7	2	1	1	1	24	4	2	2	0	41	5	4	0	0
8	2	2	0	0	25	5	0	0	0	42	5	4	1	0
9	3	0	0	0	26	5	1	0	0	43	5	3	3	0
10	3	1	0	0	27	5	1	1	0	44	6	2	2	0
11	3	1	1	0	28	5	1	1	1	45	6	3	0	0
12	2	2	2	0	29	5	2	0	0	46	6	3	1	0
13	3	2	0	0	30	5	2	1	0	47	6	3	1	1
14	3	2	1	0	31	5	2	1	1	48	6	2	2	2
15	3	2	1	1	32	4	4	0	0	49	7	0	0	0
16	4	0	0	0	33	4	4	1	0	50	7	1	0	0
17	4	1	0	0	34	5	3	0	0					

For example, let $u = 3$ and $v = 5$, so that $a = b = c = 1$, $d = 0$ and $w = 2$, $x = 1$, $y = z = 0$. Then

$$uv = 15 = (2 + 1 + 0 + 0)^2 + (1 - 2 + 0 - 0)^2 + (0 + 0 - 2 - 0)^2$$
$$+ (0 + 0 - 1 - 0)^2 = 3^2 + 1^2 + 2^2 + 1^2$$

Once it is shown that every prime is a sum of four squares, it will follow from Lemma 8.2 and Theorem 3.10 that every natural number is a sum of four squares. First, we need some preparatory lemmas.

▶ **Lemma 8.3** *Let the prime $p \equiv 3$ (mod 4). Then there is a prime, q, such that $q \not\equiv 3$ (mod 4) and $\left(\frac{q}{p}\right) = -1$.*

Proof: By hypothesis, either $p \equiv 3$ (mod 8) or $p \equiv 7$ (mod 8). If $p \equiv 3$ (mod 8), then $\left(\frac{2}{p}\right) = -1$, so let $q = 2$. If $p \equiv 7$ (mod 8), let q be the least prime such that $q \equiv 2p - 1$ (mod $4p$). Since $(2p - 1, 4p) = 1$, such a q exists by Dirichlet's Theorem (Theorem 3.13). Therefore $q \equiv 1$ (mod 4) and $q \equiv -1$ (mod p), so $\left(\frac{q}{p}\right) = \left(\frac{-1}{p}\right) = -1$. ■

For example, if $p = 7$, then the least such q is 5; also, if $p = 31$, then the least such q is 13.

Next, we will show constructively that if p is prime, then p has a representation as a sum of four squares. We already know that $2 = 1^2 + 1^2 + 0^2 + 0^2$. Furthermore, if $p \equiv 1 \pmod{4}$, then by Theorem 8.2, we have $p = x^2 + y^2 + 0^2 + 0^2$. Therefore it suffices to consider as prime, p, such that $p \equiv 3 \pmod{4}$.

Theorem 8.5   If the prime $p \equiv 3 \pmod{4}$, then there exist integers a, b, c, d such that $p = a^2 + b^2 + c^2 + d^2$.

Proof: We will show that Algorithm 3, given below, produces such a representation.

Algorithm 3

Initialization: Find the least prime, q, such that $q \not\equiv 3 \pmod{4}$ and $\left(\frac{q}{p}\right) = -1$. (Such a q exists, by Lemma 8.3.) By Euler's Criterion (Theorem 7.2), we have

$$q^{\frac{p-1}{2}} \equiv -1 \pmod{p}$$

$$\rightarrow q^{\frac{p+1}{2}} \equiv -q \pmod{p}$$

$$\rightarrow (q^{\frac{p+1}{4}})^2 + q \equiv 0 \pmod{p}$$

Theorem 8.2 implies $q = r^2 + s^2$, so we have

$$(q^{\frac{p+1}{4}})^2 + r^2 + s^2 + 0^2 \equiv 0 \pmod{p}$$

Now we choose integers a_1, b_1, c_1, d_1 such that $a_1 \equiv q^{\frac{p+1}{4}} \pmod{p}$, $b_1 \equiv r \pmod{p}$, $c_1 \equiv s \pmod{p}$, $d_1 = 0$, and $|a_1| < \frac{p}{2}$, $|b_1| < \frac{p}{2}$, $|c_1| < \frac{p}{2}$. Now

$$a_1^2 + b_1^2 + c_1^2 + d_1^2 \equiv 0 \pmod{p} \rightarrow 0 < k_1 p = a_1^2 + b_1^2 + c_1^2 + d_1^2 < 3\left(\frac{p}{2}\right)^2$$

$$\rightarrow 0 < k_1 < \frac{3p}{4}$$

Main Routine

Key Step: If $a_i^2 + b_i^2 + c_i^2 + d_i^2 = p$ for some $i \geq 1$, then we are done. If $a_i^2 + b_i^2 + c_i^2 + d_i^2 = k_i p$ for some $k_i > 1$, then we will find integers $a_{i+1}, b_{i+1}, c_{i+1}, d_{i+1}$ such that

$$a_{i+1}^2 + b_{i+1}^2 + c_{i+1}^2 + d_{i+1}^2 < k_{i+1} p \quad with \quad 0 < k_{i+1} < k_i$$

If k_i is even, then by hypothesis, we must have $a_i \equiv b_i \pmod 2$, $c_i \equiv d_i \pmod 2$. Let

$$a_{i+1} = |\frac{a_i + b_i}{2}|$$

$$b_{i+1} = |\frac{a_i - b_i}{2}|$$

$$c_{i+1} = |\frac{c_i + d_i}{2}|$$

$$d_{i+1} = |\frac{c_i - d_i}{2}|$$

Then

$$a_{i+1}^2 + b_{i+1}^2 + c_{i+1}^2 + d_{i+1}^2 = \frac{1}{2}(a_1^2 + b_1^2 + c_1^2 + d_1^2)$$

$$= \frac{1}{2}k_i p \rightarrow k_{i+1}$$

$$= \frac{1}{2}k_i$$

If k_i is odd, choose integers A_i, B_i, C_i, D_i such that

$$A_i \equiv a_i \pmod {k_i}, \ |A_i| < \frac{k_i}{2}, \ B_i \equiv b_i \pmod {k_i}, \ |B_i| < \frac{k_i}{2},$$

$$C_i \equiv c_i \pmod {k_i}, \ |C_i| < \frac{k_i}{2}, \ D_i \equiv d_i \pmod {k_i}, \ |D_i| < \frac{k_i}{2}$$

Now

$$A_i^2 + B_i^2 + C_i^2 + D_i^2 \equiv a_i^2 + b_i^2 + c_i^2 + d_i^2 \equiv 0 \pmod {k_i}$$

$$\rightarrow A_i^2 + B_i^2 + C_i^2 + D_i^2 = k_{i+1}k_i$$

If $k_{i+1} = 0$, then $A_i = B_i = C_i = D_i = 0$, so

$$a_i \equiv b_i \equiv c_i \equiv d_i \equiv 0 \pmod {k_i}$$

$$\rightarrow \ k_i^2 | (a_i^2 + b_i^2 + c_i^2 + d_i^2) \rightarrow k_i^2 | k_i p \rightarrow k_i | p$$

which is impossible, since $0 < k_i < p$. Also

$$A_i^2 + B_i^2 + C_i^2 + D_i^2 < 4(\frac{k_i}{2})^2 \rightarrow k_{i+1}k_i < k_i^2 \rightarrow k_{i+1} < k_i$$

Now

$$a_i A_i + b_i B_i + c_i C_i + d_i D_i \equiv a_i^2 + b_i^2 + c_i^2 + d_i^2 \equiv 0 \pmod{k_i}$$

$$a_i B_i - b_i A_i + c_i D_i - d_i C_i \equiv a_i b_i - b_i a_i + c_i d_i - d_i c_i \equiv 0 \pmod{k_i}$$

$$a_i C_i - b_i D_i - c_i A_i + d_i B_i \equiv a_i c_i - b_i d_i - c_i a_i + d_i b_i \equiv 0 \pmod{k_i}$$

$$a_i D_i + b_i C_i - c_i B_i - d_i A_i \equiv a_i d_i + b_i c_i - c_i b_i - d_i a_i \equiv 0 \pmod{k_i}$$

Let

$$a_{i+1} = \frac{|a_i A_i + b_i B_i + c_i C_i + d_i D_i|}{k_i}$$

$$b_{i+1} = \frac{|a_i B_i - b_i A_i + c_i D_i - d_i C_i|}{k_i}$$

$$c_{i+1} = \frac{|a_i C_i - b_i D_i - c_i A_i + d_i B_i|}{k_i}$$

$$d_{i+1} = \frac{|a_i D_i + b_i C_i - c_i B_i - d_i A_i|}{k_i}$$

Now $a_{i+1}, b_{i+1}, c_{i+1}, d_{i+1}$ are all integers. Furthermore $a_{i+1}^2 + b_{i+1}^2 + c_{i+1}^2 + d_{i+1}^2$

$$= \frac{1}{k_i^2}\{(a_i A_i + b_i B_i + c_i C_i + d_i D_i)^2 + (a_i B_i - b_i A_i + c_i D_i - d_i C_i)^2$$

$$+ (a_i C_i - b_i D_i - c_i A_i + d_i B_i)^2 + (a_i D_i + b_i C_i - c_i B_i - d_i A_i)^2\}$$

$$= \frac{1}{k_i^2}(a_i^2 + b_i^2 + c_i^2 + d_i^2)(A_i^2 + B_i^2 + C_i^2 + D_i^2)$$

$$= \frac{1}{k_i^2}(k_i p)(k_{i+1} k_i = k_{i+1} p$$

Now return to Key Step, replacing $i + 1$ with i.

Since the k_i are positive integers such that $k_1 > k_2 > k_3 > \cdots$, it follows from Theorem 1.8 that $k_r = 1$ for some $r \geq 1$. ∎

For example, let us represent the prime 71 as a sum of four squares. We begin by seeking the least prime q such that $q \not\equiv 3 \pmod 4$ and $(\frac{q}{p}) = -1$. We have

$$(\frac{2}{71}) = 1 \quad ; \quad (\frac{5}{71}) = 1 \quad ; \quad (\frac{13}{71}) = -1$$

so $q = 13$. Now $13^{\frac{71+1}{4}} \equiv 13^{18} \equiv 49 \pmod{71}$ and $13 = 3^2 + 2^2$. We take $a_1 = Min\{49, 71 - 49\} = 22$, $a_2 = 3$, $a_3 = 2$. Now

$$22^2 + 3^2 + 2^2 + 0^2 = 497 = 7(71) .$$

We have $a_1 = 22$, $b_1 = 3$, $c_1 = 2$, $d_1 = 0$, $k_1 = 7$. Therefore $A_1 = 1$, $B_1 = 3$, $C_1 = 2$, $D_1 = 0$.

This implies

$$a_2 = \frac{|22(1) + 3(3) + 2(2) + 0(0)|}{7} = 5$$

$$b_2 = \frac{|22(3) - 1(3) + 2(0) - 2(0)|}{7} = 9$$

$$c_2 = \frac{|22(2) - 3(0) - 2(1) - 0(3)|}{7} = 6$$

$$d_2 = \frac{|22(0) + 3(2) - 2(3) - 0(1)|}{7} = 0$$

$$a_2^2 + b_2^2 + c_2^2 + d_2^2 = 5^2 + 9^2 + 6^2 + 0^2 = 142 = 2(71) \rightarrow k_2 = 2$$

$$a_3 = \frac{|5 + 9|}{2} = 7 \, , \, b_3 = \frac{|5 - 9|}{2} = 2 \, , \, c_3 = \frac{|6 + 0|}{2} = 3 \, , \, d_3 = \frac{|6 - 0|}{2} = 3$$

$$a_3^2 + b_3^2 + c_3^2 + d_3^2 = 7^2 + 2^2 + 3^2 + 3^2 = 71$$

so we are done.

We are now ready to present the main result of this section, namely, Lagrange's Four Square Theorem.

Theorem 8.6

Every natural number can be represented as a sum of four squares.

Proof: First, note that $1 = 1^2 + 0^2 + 0^2 + 0^2$. If $n \geq 2$, then n is a product of one or more primes, so the conclusion follows from the repeated application of Theorem 8.5, Theorem 8.2, and Lemma 8.2. ∎

Remarks

Theorem 8.6 is the best possible in the sense that there are infinitely many natural numbers that cannot be represented as a sum of three or fewer squares, as the following theorem shows.

Theorem 8.7

If $n = 4^k m$ with $k \geq 0$ and $m \equiv 7 \pmod 8$, then $n \neq a^2 + b^2 + c^2$.

Proof: Suppose that $4^k m = a^2 + b^2 + c^2$, with $k \geq 0$ and $m \equiv 7 \pmod 8$. If $k \geq 1$, then $a^2 + b^2 + c^2 \equiv 0 \pmod 4$. Therefore either a, b, c are all even, or two are odd and one is even. In the latter case, we would have $a^2 + b^2 + c^2 \equiv 2 \pmod 8$, an impossibility. Therefore $a = 2a_1$, $b = 2b_1$, $c = 2c_1$, which yields $4^{k-1} m = a_1^2 + b_1^2 + c_1^2$. If $k \geq 2$, we may continue in like fashion, ultimately obtaining $m = a_k^2 + b_k^2 + c_k^2$. Without loss of generality, assume that $a_k^2 \geq b_k^2 \geq c_k^2 \pmod 8$. If $w^2 \equiv r \pmod 8$, where $0 \leq r \leq 7$, then $r \in \{0, 1, 4\}$. Therefore the possible values of $a_k^2 + b_k^2 + c_k^2 \pmod 8$ are 0+0+0, 1+0+0, 1+1+0, 1+1+1, 4+0+0, 4+1+0, 4+1+1, 4+4+0, 4+4+1, 4+4+4. Therefore $m \equiv r \pmod 8$, where $r \in \{0, 1, 2, 3, 4, 5, 6\}$, that is, $m \not\equiv 7 \pmod 8$, contrary to hypothesis. ∎

Remarks

The converse of Theorem 8.7 is also true, that is, if $n \neq 4^k m$ with $k \geq 0$ and $m \equiv 7 \pmod 8$, then n may be represented as a sum of three squares. The proof of the converse is beyond the scope of this text and is therefore omitted.

∎

If $1 \leq k \leq 4$, let S_k denote the set of all natural numbers that can be represented as the sum of k squares of non-negative integers. Suppose that whenever $u \in S_k$ and $v \in S_k$, then also $uv \in S_k$. In this case, we say that S_k is closed under multiplication. Since $x^2 y^2 = (xy)^2$, it follows that S_1 is closed under multiplication. Lemmas 8.1 and 8.2 say respectively that S_2 and S_4 are closed under multiplication. The same cannot be said for S_3, however. For example, $3 = 1^2 + 1^2 + 1^2$ and $5 = 2^2 + 1^2 + 0^2$, so $3 \in S_3$, $5 \in S_3$. But $3 * 5 = 15 \equiv 7 \pmod 8$, so $15 \notin S_3$.

A natural number may have several representations as a sum of four squares. For example,

$$31 = 5^2 + 2^2 + 1^2 + 1^2$$
$$= 3^2 + 3^2 + 3^2 + 2^2$$

and

$$55 = 7^2 + 2^2 + 1^2 + 1^2$$
$$= 6^2 + 3^2 + 3^2 + 1^2$$
$$= 5^2 + 5^2 + 2^2 + 1^2$$

Suppose that we allow squares of all integers (positive, negative, and zero), and that we count representations as distinct when they differ only in the order of terms. Let $r_4(n)$ be the number of such representations as a sum of four squares. Then $r_4(2) = 24$, since we can write

$$2 = (\pm 1)^2 + (\pm 1)^2 + 0^2 + 0^2$$

as well as the six permutations of these four representations.

We conclude by stating without proof Jacobi's formula for $r_4(n)$:

Theorem 8.8 $\quad r_4(n) = 8 \sum \{d : d|n \nmid d\}$

Proof: Omitted, but see [9], p. 323.

For example, $r_4(55) = 8\sigma(55) = 8\sigma(5 * 11) = 8 * 6 * 12 = 576$.

Section 8.3 Exercises

1. Prove that if $n = a^2 + b^2 + c^2 + d^2$ with $a \geq b \geq c \geq d \geq 0$, then $\frac{\sqrt{n}}{2} \leq a \leq \sqrt{n}$.

2. Prove that if $n > 0$ and $n \equiv 7 \pmod 8$, then $4n$ can be represented as a sum of four odd squares.

3. Prove that at least one of every two consecutive integers can be represented as a sum of three or fewer squares.

4. Represent each of 60, 92, 124 as a sum of four odd squares.

5. Using Algorithm 3, represent each of the following primes as a sum of four squares:

   (a) 107

   (b) 191

   (c) 263

   (d) 431

   (e) 479

   (f) 1031

   (g) 8831

6. Let $n \equiv 3 \pmod 8$. Prove that if $n = a^2 + b^2 + c^2 + d^2$, then $abcd \equiv 0 \pmod 4$.

7. Let $n \equiv 7 \pmod 8$. Prove that if $n = a^2 + b^2 + c^2 + d^2$, then $abcd \equiv 2 \pmod 4$.

8. If $n \equiv 7 \pmod 8$, and we are interested only in representations of n as a sum of four *positive* squares, then Theorem 8.8 implies the number of such representations is $\frac{1}{2}\sigma(n)$. Find all such representations of the prime 127.

9. Let the prime $p \equiv 23 \pmod{24}$, and $p = a^2 + b^2 + c^2 + d^2$ where $a \geq b \geq c \geq d \geq 1$. Prove that at most two of $\{a, b, c, d\}$ are equal.

Section 8.3   Computer Exercises

10. Write a computer program that, given an odd prime, p, finds a representation of p as a sum of four squares of nonnegative integers.

11. Let the prime $p \equiv 3 \pmod{8}$. We have seen in the proof of Theorem 8.4 that if q is a prime such that $q \not\equiv 3 \pmod{4}$ and $\left(\frac{q}{p}\right) = -1$, then we can obtain a representation of p as a sum of four squares. Write a computer program to test the conjecture that all such representations of p may be obtained for $p < 100$ by picking sufficiently many values of q.

Review Exercises

1. Represent each of the following primes as a sum of two squares using both Algorithms 1 and 2:

    (a) 193

    (b) 229

    (c) 241

    (d) 373

    (e) 449

2. For each integer from 301 to 310

    (a) determine whether it can be represented as a sum of two squares.

    (b) if so, find such a representation.

3. Represent each of the following integers as a sum of two squares in as many ways as possible:

    (a) 145

    (b) 205

    (c) 377

    (d) 493

    (e) 1885

4. For each prime p that follows, find the least prime q such that $q \not\equiv 3 \pmod{4}$ and $\left(\frac{q}{p}\right) = -1$.

    (a) 47

    (b) 53

    (c) 71

(d) 79

(e) 131

5. Find all representations of each of the following integers as a sum of four nonnegative squares:

(a) 28

(b) 47

(c) 107

(d) 127

6. For each integer n given below, do the following:

    i. Compute $r_4(n)$ using Jacobi's formula (Theorem 8.8).

    ii. Find all representations of n as a sum of four squares of integers.

(a) 40

(b) 53

(c) 63

(d) 75

(e) 88

7. Find two two-digit integers, each of which is a sum of three squares, but whose product is not a sum of three squares.

Chapter 9

Continued Fractions

9.1 Introduction

The awkward-looking expression:

$$4 + \cfrac{1}{7 + \cfrac{1}{2 + \cfrac{1}{3}}}$$

is an example of a *simple finite continued fraction*. Despite their uninviting appearance, soon to be remedied by appropriate notation, continued fractions provide (1) the best possible approximations of irrational numbers by rational numbers, (2) the solution of the Diophantine equation known as Pell's equation, (3) a method for factoring large integers.

9.2 Finite Continued Fractions

Let the integers $a_0 \geq 0$, $a_i > 0$ for $1 \leq i \leq n$. We define finite continued fractions in the following inductive manner:

Definition 9.1   **Finite Continued Fraction**

$$[a_0] = a_0 \quad \text{and}$$

$$[a_0, a_1, a_2, \cdots, a_n] = a_0 + \cfrac{1}{[a_1, a_2, \cdots, a_n]} \text{ for } n \geq 1$$

For example,

$$[2] = 2$$

$$[7, 2] = 7 + \frac{1}{[2]} = 7 + \frac{1}{2} = \frac{15}{2}$$

$$[3, 7, 2] = 3 + \frac{1}{[7, 2]} = 3 + \frac{2}{15} = \frac{47}{15}$$

$$[8, 3, 7, 2] = 8 + \frac{1}{[3, 7, 2]} = 8 + \frac{15}{47} = \frac{391}{47}$$

Note that in general, we have

$$[a_0, a_1] = a_0 + \cfrac{1}{a_1}$$

$$[a_0, a_1, a_2] = a_0 + \cfrac{1}{[a_1, a_2]} = a_0 + \cfrac{1}{a_1 + \cfrac{1}{a_2}}$$

$$[a_0, a_1, a_2, \cdots, a_{n-1} a_n] = a_0 + \cfrac{1}{a_1 + \cfrac{1}{a_2 + \cdots + \cfrac{1}{a_{n-1} + \cfrac{1}{a_n}}}}$$

Our first theorem regarding continued fractions shows that a given continued fraction with two or more terms can be represented by another continued fraction with one less term.

Theorem 9.1   Let $n \geq 1$. Let $a_0, a_1, a_2, \cdots, a_n$ be real numbers such that $a_0 \geq 0$, $a_i > 0$ for $1 \leq i \leq n$. Then

$$[a_0, a_1, a_2, \cdots, a_{n-2}, a_{n-1}, a_n] = [a_0, a_1, a_2, \cdots, a_{n-2}, a_{n-1} + \frac{1}{a_n}].$$

Proof: (Induction on n)

$$[a_0, a_1] = a_0 + \frac{1}{a_1} = [a_0 + \frac{1}{a_1}]$$

so the theorem holds for $n = 1$. Then

$$[a_0, a_1, a_2, \cdots, a_n] = a_0 + \cfrac{1}{[a_1, a_2, \cdots, a_n]}$$

$$= a_0 + \cfrac{1}{[a_1, a_2, \cdots, a_{n-2}, a_{n-1} + \frac{1}{a_n}]}$$

by induction hypothesis, hence

$$[a_0, a_1, a_2, \cdots, a_{n-2}, a_{n-1}, a_n] = [a_0, a_1, a_2, \cdots, a_{n-2}, a_{n-1} + \frac{1}{a_n}]$$

by Definition 9.1. ∎

An alternate statement of the conclusion of Theorem 9.1 is

$$[a_0, a_1, a_2, \cdots, a_{n-2}, a_{n-1}, a_n] = [a_0, a_1, a_2, \cdots, a_{n-2}, [a_{n-1}, a_n]]$$

It follows that for any k such that $0 \leq k \leq n - 1$, we have

$$[a_0, a_1, a_2, \cdots, a_{n-2}, a_{n-1}, a_n] = [a_0, a_1, a_2, \cdots, a_k, [a_{k+1}, \cdots, a_{n-2}, a_{n-1}, a_n]]$$

Repeated application of Theorem 9.1 provides one way of evaluating finite continued fractions. For example,

$$[5, 3, 2] = [5, [3, 2]] = [5, 3 + \frac{1}{2}] = [5, \frac{7}{2}] = 5 + \frac{2}{7} = \frac{37}{7}$$

Also

$$[1, 2, 3, 4] = [1, 2, [3, 4]] = [1, 2, 3 + \frac{1}{4}] = [1, 2, \frac{13}{4}] = [1, [2, \frac{13}{4}]]$$

$$= [1, 2 + \frac{4}{13}] = [1, \frac{30}{13}] = 1 + \frac{13}{30} = \frac{43}{30}$$

Later, we will develop a more convenient way of evaluating finite continued fractions. As a corollary to Theorem 9.1, we have the following:

Theorem 9.2 If n and the a_i are as in the hypothesis of Theorem 9.1, then

$$[a_0, a_1, a_2, \cdots, a_{n-2}, a_{n-1}, 1] = [a_0, a_1, a_2, \cdots, a_{n-2}, a_{n-1} + 1]$$

Proof: This follows from Theorem 9.1 with $a_n = 1$. ∎

For example, $[7, 5, 3, 1] = [7, 5, 4]$. Also, $[8, 5, 1] = [8, 6]$.

The a_i are called the *terms* or *partial quotients* of the continued fraction. Clearly, if all the a_i are rational numbers, then $[a_0, a_1, a_2, \cdots, a_{n-2}, a_{n-1}, a_n]$ is rational. Also, $[a_0, a_1, a_2, \cdots, a_{n-2}, a_{n-1}, a_n] > 0$ unless $a_0 = n = 0$. We shall see that every positive rational number has an essentially unique representation as a finite continued fraction with integer terms. In this text, we will consider only *simple* continued fractions, that is, continued fractions whose numerators are all 1's.

The history of continued fractions dates back to the remote past. In the fourth century B.C., Baudhayana, an Indian mathematician, used continued fractions to approximate $\frac{1}{2}\sqrt{\pi}$. Archimedes (287–212 B.C.) used continued fractions to obtain a well-known approximation to π, namely, $\frac{22}{7}$. During the Middle Ages, several Indian mathematicians were able to solve certain Diophantine equations by means of algorithms that are equivalent to the use of continued fractions. These mathematicians include Aryabhata (476–550), Brahmagupta (598–665), and Bhaskara (1114–1185). The first systematic treatment of continued fractions, based on efforts by Lord William Brouncker (1620–1684) and John Wallis (1616–1703), was given by the latter in his *Arithmetica Infinitorum*, published in 1655.

Given a positive rational number a/b, one may find its representation as a finite continued fraction with integer terms by employing Euclid's Algorithm to compute (a, b), and using the *quotients* that arise. For example, let us find the continued fraction representation for $100/13$.

$$100 = 7(13) + 9 \qquad \text{that is,} \qquad \frac{100}{13} = 7 + \frac{9}{13}$$

$$13 = 1(9) + 4 \qquad\qquad\qquad \frac{13}{9} = 1 + \frac{4}{9}$$

$$9 = 2(4) + 1 \qquad\qquad\qquad \frac{9}{4} = 2 + \frac{1}{4}$$

$$4 = 4(1) \qquad\qquad\qquad \frac{4}{1} = 4$$

Therefore $100/13 = [7, 1, 2, 4]$ (or $[7, 1, 2, 3, 1]$).

As a second example, let us find the continued fraction representation for $27/50$.

$$27 = 0(50) + 27 \qquad \text{that is,} \qquad \frac{27}{50} = 0 + \frac{27}{50}$$

$$50 = 1(27) + 23 \qquad\qquad\qquad \frac{50}{27} = 1 + \frac{23}{27}$$

$$27 = 1(23) + 4 \qquad\qquad\qquad \frac{27}{23} = 1 + \frac{4}{23}$$

$$23 = 5(4) + 3 \qquad\qquad\qquad \frac{23}{4} = 5 + \frac{3}{4}$$

$$4 = 1(3) + 1 \qquad\qquad\qquad \frac{4}{3} = 1 + \frac{1}{3}$$

$$3 = 3(1) \qquad\qquad\qquad \frac{3}{1} = 3$$

Therefore $27/50 = [0, 1, 1, 5, 1, 3]$.

This procedure may be formalized as follows:

Step 1: Set $x = a/b$ (the given positive rational).

Step 2: Compute and store $[x]$ (the integer part of x).

Step 3: Compute $x - [x]$ (the fractional part of x).

Step 4: If $x - [x] \neq 0$, then replace x by $1/(x - [x])$ and go to Step 2; If $x - [x] = 0$, then terminate. (The answer is the sequence of integers stored while performing Step 2.)

Remarks

Note that this procedure consists of alternately subtracting the integer part of a real number that is greater than 1, and then inverting the resulting fractional part until eventually zero is obtained.

■

In the work that follows, we prove that the finite continued fraction representation of a positive rational number is essentially unique.

▶ **Lemma 9.1** *If $n \geq 1$ and $a_0 \geq 1$, then $[a_0, a_1, a_2, \cdots, a_n] > 1$.*

Proof: (Induction on n)

$[a_0, a_1] = a_0 + \frac{1}{a_1} > 1$ by hypothesis. Similarly, $[a_{n-1}, a_n] > 1$. Now Theorem 9.1 implies $[a_0, a_1, a_2, \cdots, a_{n-2}, a_{n-1}, a_n] = [a_0, a_1, a_2, \cdots, a_{n-2}, [a_{n-1}, a_n]]$, from which the conclusion follows by induction hypothesis. ■

For example, $[1, 2, 3] = [1, [2, 3]] = [1, \frac{7}{3}] = 1 + \frac{3}{7} > 1$.

The following lemma shows that, with minor exceptions, a finite continued fraction with integer terms is never itself an integer.

▶ **Lemma 9.2** *If $n \geq 0$, let a_0, a_1, a_2, $\cdots$, a_n be integers such that $a_0 \geq 0$ and $a_i > 0$ for all $i > 0$. Then $[a_0, a_1, a_2, \cdots, a_n] = k$, an integer, if and only if (i) $n = 0$ and $a_0 = k$ or (ii) $n = 1$ and $a_0 = k - 1$.*

Proof: (Sufficiency) $k = [k] = [k - 1, 1]$ by Definition 9.1 and Theorem 9.2. (Necessity) If $n = 0$, then by hypothesis and Definition 9.1, we have $k = [a_0] = a_0$. If $n \geq 1$, then by hypothesis and Definition 9.1, we have $a_0 + \frac{1}{[a_1, a_2, \cdots, a_n]} = k$, so $\frac{1}{[a_1, a_2, \cdots, a_n]} = k - a_0$, an integer. If $n \geq 2$, then by hypothesis and Lemma 9.1, we have $[a_1, a_2, \cdots, a_n] > 1$, so $0 < k - a_0 < 1$, an impossibility. Therefore $n = 1$ and $\frac{1}{a_1} = \frac{1}{[a_1]} = k - a_0$, so $a_1 = 1$ and $a_0 = k - 1$. ■

Next, we show that if two finite continued fractions with integer terms have the same value, and neither has 1 as its last term, then they must be identical, that is, have exactly the same terms.

Theorem 9.3 If $[a_0, a_1, a_2, \cdots, a_m] = [b_0, b_1, b_2, \cdots, b_n]$, where all a_i and b_j are integers, and $a_m b_n \neq 1$, then $m = n$ and $a_i = b_i$ for all i such that $0 \leq i \leq m$.

Proof: Without loss of generality, assume that $m \leq n$. If $m = 0$, then by hypothesis, we have $a_0 = [a_0] = [b_0, b_1, b_2, \cdots, b_n]$. Now Lemma 9.2 implies that either $b_0 = a_0$ and $n = 0$, or $b_0 = a_0 - 1$ and $b_1 = n = 1$. However the latter possibility is excluded by hypothesis. If $m \geq 1$, then by hypothesis and Definition 9.1, we have

$$a_0 + \cfrac{1}{[a_1, a_2, \cdots, a_m]} = b_0 + \cfrac{1}{[b_1, b_2, \cdots, b_n]}$$

Now Lemma 9.1 implies that each of $[a_1, a_2, \cdots, a_m], [b_1, b_2, \cdots, b_n]$ exceeds 1. Therefore $a_0 = b_0$ and $[a_1, a_2, \cdots, a_m] = [b_1, b_2, \cdots, b_n]$. By repeated application of this argument, we get $a_1 = b_1$, $a_2 = b_2$, $\cdots$, $a_{m-1} = b_{m-1}$, $a_m = [a_m] = [b_m, b_{m+1}, b_{m+2}, \cdots, b_n]$. Since $b_n > 1$ by hypothesis, Lemma 9.2 implies that $a_m = b_m$ and $m = n$. ∎

We are now ready to combine the results of Theorems 9.2 and 9.3 in Theorem 9.4.

Theorem 9.4 Every positive rational number has exactly two representations as a finite continued fraction with integer terms: one whose last term exceeds 1, and one whose last term is 1.

Proof: Given a positive rational number c/d, let $c = r_{-1}$, and $d = r_0$. Then, by the Euclidean algorithm, we have

$$\frac{c}{d} = \frac{r_{-1}}{r_0} = a_0 + \frac{r_1}{r_0} \quad \text{with} \quad 0 < r_1 < r_0$$

$$\frac{r_0}{r_1} = a_1 + \frac{r_2}{r_1} \quad \text{with} \quad 0 < r_2 < r_1$$

$$\frac{r_1}{r_2} = a_2 + \frac{r_3}{r_2} \quad \text{with} \quad 0 < r_3 < r_2$$

$$\vdots$$

$$\frac{r_{n-2}}{r_{n-1}} = a_{n-1} + \frac{r_n}{r_{n-1}} \quad \text{with} \quad 0 < r_n < r_{n-1}$$

$$\frac{r_{n-1}}{r_n} = a_n > 1$$

This sequence of $n + 1$ equations yields $\frac{c}{d} = [a_0, r_0/r_1]$, which equals

$$[a_0, a_1, r_2/r_1] = \cdots = [a_0, a_1, a_2, \cdots, a_{n-1}, r_{n-1}/r_n] = [a_0, a_1, a_2, \cdots, a_n]$$

By Theorem 9.2, we also have

$$\frac{c}{d} = [a_0, a_1, a_2, \cdots, a_{n-1}, a_n - 1, 1]$$

Theorem 9.3 guarantees that c/d has no additional representations as a finite continued fraction with integer terms. ∎

Earlier, when using the Euclidean algorithm, we obtained the result $100/13 = [7, 1, 2, 4]$. Now let us consider the reverse problem of finding the positive rational number that corresponds to a given finite continued fraction with integer terms. One possibility is to work backwards, that is, to employ Theorem 9.1 repeatedly. For example,

$$[7, 1, 2, 4] = [7, 1, 2 + \frac{1}{4}] = [7, 1, \frac{9}{4}] = [7, 1 + \frac{4}{9}]$$

$$= [7, \frac{13}{9}] = [7 + \frac{9}{13}] = [\frac{100}{13}] = \frac{100}{13}$$

If we apply this method to $[a_0, a_1, a_2, \cdots, a_n]$ where all the a_i are integers, then n additions of fractions are required. A second method, which we are about to present, is more convenient, since it involves only operations on integers and allows us to work forwards, that is, from a_0 to a_n. We must first define what are known as the *convergents* to a continued fraction.

Definition 9.2   **Convergent** Let $A = [a_0, a_1, a_2, \cdots, a_n]$ where the a_i are real and all positive, except possibly a_0, which may be zero. If $0 \leq k \leq n$, let $C_k = [a_0, a_1, a_2, \cdots, a_k]$. C_k is called the kth *convergent* to A.

For example, let $A = [8, 3, 7, 2]$. Then we have

$$C_0 = [8] = 8$$

$$C_1 = [8, 3] = \frac{25}{3}$$

$$C_2 = [8, 3, 7] = \frac{183}{22}$$

$$C_3 = [8, 3, 7, 2] = \frac{391}{47} = A$$

Next, we develop an iterative method to generate the convergents of a given continued fraction.

| Theorem 9.5 |

Let $A = [a_0, a_1, a_2, \cdots, a_n]$ where the a_i are real and all positive, except possibly a_0, which may be zero. Define sequences $\{p_n\}, \{q_n\}$ by

$$p_{-2} = 0,\, p_{-1} = 1,\, p_k = a_k p_{k-1} + p_{k-2} \text{ for } k \geq 0$$

$$q_{-2} = 1,\, q_{-1} = 0,\, q_k = a_k q_{k-1} + q_{k-2} \text{ for } k \geq 0$$

If $0 \leq k \leq n$ and $C_k = [a_0, a_1, a_2, \cdots, a_k]$, that is, C_k is the kth convergent to A, then $C_k = p_k/q_k$.

Proof: (Induction on k) If $k = 0$, then $p_0/q_0 = a_0/1 = a_0 = [a_0] = C_0$. If $k = 1$, then

$$\frac{p_1}{q_1} = \frac{a_1 p_0 + p_{-1}}{a_1 q_0 + q_{-1}} = \frac{a_1 a_0 + 1}{a_1} = a_0 + \frac{1}{a_1} = [a_0, a_1] = C_1$$

If $k \geq 2$, then

$$\frac{p_k}{q_k} = \frac{a_k p_{k-1} + p_{k-2}}{a_k q_{k-1} + q_{k-2}} = \frac{a_k(a_{k-1}p_{k-2} + p_{k-3}) + p_{k-2}}{a_k(a_{k-1}q_{k-2} + q_{k-3}) + q_{k-2}}$$

$$= \frac{(a_k a_{k-1} + 1)p_{k-2} + a_k p_{k-3}}{(a_k a_{k-1} + 1)q_{k-2} + a_k q_{k-3}} = \frac{(a_{k-1} + \frac{1}{a_k})p_{k-2} + p_{k-3}}{(a_{k-1} + \frac{1}{a_k})q_{k-2} + q_{k-3}}$$

$$= [a_0, a_1, a_2, \cdots, a_{k-2}, a_{k-1} + \frac{1}{a_k}]$$

by induction hypothesis. Therefore Theorem 9.1 implies that $p_k/q_k = [a_0, a_1, a_2, \cdots, a_k] = C_k$. ∎

Our first example of a continued fraction was $[8, 3, 7, 2]$. Let us now recompute this quantity using Theorem 9.5. We get

$p_0 = a_0 p_{-1} + p_{-2} = 8(1) + 0 = 8;$ $q_0 = a_0 q_{-1} + q_0 = 8(0) + 1 = 1$

$p_1 = a_1 p_0 + p_{-1} = 3(8) + 1 = 25;$ $q_1 = a_1 q_0 + q_{-1} = 3(1) + 0 = 3$

$p_2 = a_2 p_1 + p_0 = 7(25) + 8 = 183;$ $q_2 = a_2 q_1 + q_0 = 7(3) + 1 = 22$

$p_3 = a_3 p_2 + p_1 = 2(183) + 25 = 391;$ $q_3 = a_3 q_2 + q_1 = 2(22) + 3 = 47$

Therefore $[8, 3, 7, 2] = p_3/q_3 = 391/47$.

We now develop some properties of convergents. Recall that p_n/q_n is the nth convergent to a continued fraction, and that F_n is the nth Fibonacci number.

▶ **Lemma 9.3** $q_n \geq F_{n+1}$ for all $n \geq 0$.

Proof: (Induction on n) $q_0 = 1 = F_1$ and $q_1 = a_1 q_0 + q_{-1} = a_1(1) + 0 = a_1 \leq 1 = F_2$. If $n \geq 2$, then $q_n = a_n q_{n-1} + q_{n-2} \geq q_{n-1} + q_{n-2} \geq F_n + F_{n-1}$ by induction hypothesis. However $F_n + F_{n-1} = F_{n+1}$, so $q_n \geq F_{n+1}$. ∎

▶ **Lemma 9.4** $q_n \geq \alpha^{n-1}$ for all $n \geq 0$.

Proof: This follows from Lemma 9.3 and Theorem 1.15. ∎

▶ **Lemma 9.5** (i) $q_n > 0$ for all $n \neq -1$ and (ii) $q_n > q_{n-1}$ for all $n \geq 2$.

Proof: $q_0 = a_0 q_{-1} + q_{-2} = a_0(0) + 1 = 1$ and $q_1 = a_1 q_0 + q_{-1} = a_1(1) + 0 = a_1 \geq 1$. Since $q_n = a_n q_{n-1} + q_{n-2}$, (i) follows by induction on n. If $n \geq 2$, then $n - 2 \geq 0$, so (i) implies $q_{n-2} \geq 1$. Therefore $q_n > a_{n-1} q_{n-1} \geq q_{n-1}$. Thus (ii) has been proven. ∎

The next several theorems present identities concerning the numerators and denominators of consecutive and alternate convergents.

Theorem 9.6

(a) If $n \geq -1$, then $\quad p_n q_{n-1} - p_{n-1} q_n = (-1)^{n-1}$

(b) If $n \geq 1$, then

$$\frac{p_n}{q_n} - \frac{p_{n-1}}{q_{n-1}} = \frac{(-1)^{n-1}}{q_n q_{n-1}}$$

Proof: If $n \geq 1$, then Lemma 9.3 implies that $q_{n-1} \geq 1$ and $q_n \geq 1$. Therefore (b) follows from (a), so it suffices to prove (a). We use induction on n. Now

$$p_{-1} q_{-2} - p_{-2} q_{-1} = 1(1) - 0(0) = 1 = (-1)^{-2}.$$

If $n \geq 0$, then

$$p_n q_{n-1} - p_{n-1} q_n = (a_n p_{n-1} + p_{n-2}) q_{n-1} - p_{n-1}(a_n q_{n-1} + q_{n-2})$$

$$= -(p_{n-1} q_{n-2} - p_{n-2} q_{n-1}) = -(-1)^{n-2} = (-1)^{n-1}$$

by induction hypothesis. ∎

For example, looking back at the convergents to $[8, 3, 7, 2]$, we have

$$p_0 q_{-1} - p_{-1} q_0 = 8(0) - 1(1) = -1$$

$$p_1 q_0 - p_0 q_1 = 25(1) - 8(3) = 1$$

$$p_2 q_1 - p_1 q_2 = 183(3) - 25(22) = -1$$

$$p_3 q_2 - p_2 q_3 = 391(22) - 183(47) = 1$$

Theorem 9.7   If $n \geq 0$, then $(p_n, q_n) = (p_n, p_{n-1}) = (q_n, q_{n-1}) = 1$.

Proof: The conclusion follows directly from Theorem 9.6, part (a).   ∎

Theorem 9.8

(a) If $n \geq 0$,   then $p_n q_{n-2} - p_{n-2} q_n = (-1)^n a_n$

(b) If $n \geq 2$,   then

$$\frac{p_n}{q_n} - \frac{p_{n-2}}{q_{n-2}} = \frac{(-1)^n a_n}{q_n q_{n-2}}$$

Proof: $q_2 > q_1 \geq F_2 = 1$ by Lemmas 9.3 and 9.5; also $q_0 = 1$. If $n \geq 3$, then $q_n > q_{n-1} > q_{n-2} \geq \alpha^{n-3} \geq \alpha^0 = 1$ by Lemmas 9.4 and 9.5. Therefore (b) follows from (a), so it suffices to prove (a). Now $p_0 q_{-2} - p_{-2} q_0 = a_0(1) - 0(1) = a_0 = (-1)^0 a_0$. If $n \geq 1$, then

$$p_n q_{n-2} - p_{n-2} q_n = (a_n p_{n-1} + p_{n-2}) q_{n-2} - p_{n-2}(a_n q_{n-1} + q_{n-2})$$

$$= a_n(p_{n-1} q_{n-2} - p_{n-2} q_{n-1}) = (-1)^{n-2} a_n = (-1)^n a_n$$

by Theorem 9.6(a).   ∎

For example, again referring to the convergents to $[8, 3, 7, 2]$, we have

$$p_1 q_{-1} - p_{-1} q_1 = 25(0) - 1(3) = -3 = -a_1$$

$$p_2 q_0 - p_0 q_2 = 183(1) - 8(22) = 7 = a_2$$

$$p_3 q_1 - p_1 q_3 = 391(3) - 25(47) = -2 = -a_3$$

As consequences of Theorem 9.8(b), we obtain two results that will be useful when we encounter infinite continued fractions.

Theorem 9.9   Let $C_n = p_n/q_n$ be the nth convergent to a continued fraction. Then

(a) the subsequence $\{C_{2n-1}\}$ is strictly decreasing, and

(b) the subsequence $\{C_{2n}\}$ is strictly increasing.

Proof: By Theorem 9.8 (b), we have

$$C_{2n+1} - C_{2n-1} = \frac{p_{2n+1}}{q_{2n+1}} - \frac{p_{2n-1}}{q_{2n-1}} = \frac{(-1)^{2n+1}a_{2n+1}}{q_{2n+1}q_{2n-1}} < 0$$

$$C_{2n+2} - C_{2n} = \frac{p_{2n+2}}{q_{2n+2}} - \frac{p_{2n}}{q_{2n}} = \frac{(-1)^{2n}a_{2n+2}}{q_{2n+2}q_{2n}} > 0 \ \blacksquare$$

Even more can be said, namely,

Theorem 9.10   Let $C_k = [a_0, a_1, a_2, \cdots, a_k]$ where $0 \le k \le n$. Suppose j, m are integers such that $0 \le j \le n$, $0 \le m \le n$, j is even, and m is odd. Then

(a) $C_j < C_m$

(b) $C_j \le C_n < C_m$ if n is even

(c) $C_j < C_n \le C_m$ if n is odd.

Proof: Let $j = 2h$ and $m = 2t - 1$. Then $C_m - C_j = C_{2t-1} - C_{2h}$. If $t \le h$, then

$$C_{2t-1} - C_{2h} \ge C_{2h-1} - C_{2h} = \frac{(-1)^{2h}}{q_{2h}q_{2h-1}} > 0$$

by Theorems 9.9(a) and 9.6(b). If $t > h$, then similarly

$$C_{2t-1} - C_{2h} \ge C_{2t-1} - C_{2t} = \frac{(-1)^{2h}}{q_{2t}q_{2t-1}} > 0$$

This proves (a). Now (b) and (c) follow from (a) and the hypothesis. $\blacksquare$

For example, returning to the convergents of $[8, 3, 7, 2]$, we have

$$\frac{p_0}{q_0} < \frac{p_2}{q_2} < \frac{q_3}{p_3} < \frac{p_1}{q_1}$$

that is

$$\frac{8}{1} < \frac{183}{22} < \frac{391}{47} < \frac{25}{3}$$

Section 9.2   Exercises

1. Find continued fraction representations with integer terms (and $a_n > 1$) for each of the following:

    (a) $\frac{100}{37}$

    (b) $\frac{1001}{45}$

    (c) $\frac{21}{13}$

    (d) $\frac{13}{35}$

    (e) $\frac{1000}{301}$

2. Find rational numbers having the following continued fraction representations:

    (a) $[1, 2, 3, 4, 5]$

    (b) $[1, 1, 2, 2, 3, 3]$

    (c) $[1, 1, 1, 1, 1, 1]$

    (d) $[1, 3, 5, 7, 9]$

3. Prove that if x is a rational number whose continued fraction representation consists of n $1's$, then $x = F_{n+1}/F_n$, where F_n denotes the nth Fibonacci number.

4. The Pell sequence $\{P_n\}$ is defined inductively for $n \geq 0$ as follows: $P_0 = 0, P_1 = 1, P_n = 2P_{n-1} + P_{n-2}$ for $n \geq 2$. Prove that if x is a rational number whose continued fraction representation consists of n $2's$, then $x = P_{n+1}/P_n$, where P_n denotes the nth Pell number.

5. Prove that if x is a positive real number such that $x = [a_0, a_1, a_2, \cdots, a_{n-1}, x]$, then x is a quadratic irrational.

6. Let a, b be integers such that $0 < a < b$. How does the continued fraction representation of a/b compare to that of b/a?

Section 9.2   Computer Exercises

7. Write a computer program to obtain the continued fraction representation of a positive rational number.

8. Write a computer program to evaluate $[1, 2, 3, \cdots, n]$ for $1 \leq n \leq 50$ and to convert the resulting fraction to a decimal approximation. What seems to be happening?

9.3 Infinite Continued Fractions

The notion of an infinite continued fraction follows naturally from the notion of a finite continued fraction via the limit process.

Definition 9.3   Infinite Continued Fractions

Let $a_0 \geq 0$ and $a_i > 0$ for all $i > 0$. Then we define the *infinite continued fraction*, denoted $[a_0, a_1, a_2, \cdots]$ as follows:

$$[a_0, a_1, a_2, \cdots] = \lim_{n \to \infty} [a_0, a_1, a_2, a_n]$$

Before we present our first example of an infinite continued fraction, let us recall some properties of Fibonacci numbers. If F_n denotes the nth Fibonacci number, $\alpha = (1 + \sqrt{5})/2$, $\beta = (1 - \sqrt{5})/2$, then, according to Theorem 1.14, $F_n = (\alpha^n - \beta^n)(\alpha - \beta)$. From this formula, one can easily prove that

$$\lim_{n \to \infty} \frac{F_{n+1}}{F_n} = \alpha$$

Now, using the result of Section 9.2, Exercise 3, we have

$$[1, 1, 1, \cdots] = \lim_{n \to \infty} [1, 1, 1, \cdots, 1] = \lim_{n \to \infty} \frac{F_{n+1}}{F_n} = \alpha = \frac{1 + \sqrt{5}}{2}$$

One might question whether the limit whose existence is implied by Definition 9.3 always exists. The following theorem answers this question affirmatively and gives an idea of how well an infinite continued fraction is approximated by its nth convergent.

Theorem 9.11   If $a_0 \geq 0$ and $a_i > 0$ for all $i > 0$, then $L = \lim_{n \to \infty} [a_0, a_1, a_2, \cdots, a_n]$ exists. Furthermore,

$$\left| L - \frac{p_n}{q_n} \right| < \frac{1}{q_n q_{n+1}}$$

for all $n \geq 1$.

Proof: Let $C_{2n} = [a_0, a_1, a_2, \cdots, a_{2n}]$ and $C_{2n+1} = [a_0, a_1, a_2, \cdots, a_{2n+1}]$. Theorem 9.10 implies that the sequence $\{C_{2n}\}$ is monotone increasing and bounded from above by C_1. Therefore $\{C_{2n}\}$ converges to a limit, L_1. Likewise, the sequence $\{C_{2n+1}\}$ is monotone decreasing and bounded from below by C_0. Therefore $\{C_{2n+1}\}$ converges to a limit, L_2. For all n, we have

$$\frac{p_{2n}}{q_{2n}} = C_{2n} < L_1 \quad \text{and} \quad \frac{p_{2n+1}}{q_{2n+1}} = C_{2n+1} > L_2$$

so that

$$|L_2 - L_1| < \frac{p_{2n+1}}{q_{2n+1}} - \frac{p_{2n}}{q_{2n}} = \frac{1}{q_{2n}q_{2n+1}} \leq \frac{1}{\alpha^{4n-1}}$$

by Theorem 9.6(b) and Lemma 9.4. Since $\alpha > 1$ and n is arbitrary, we must have $|L_2 - L_1| = 0$, so $L_2 = L_1 = L$. Furthermore, since L falls between p_n/q_n and p_{n+1}/q_{n+1} for all n, we have

$$|L - \frac{p_n}{q_n}| < |\frac{p_n}{q_n} - \frac{p_{n+1}}{q_{n+1}}| = \frac{1}{q_n q_{n+1}} \quad \blacksquare$$

For example, suppose we wish to estimate the golden ratio, α , to 3 decimal places using a convergent to its infinite continued fraction. That is, we seek the least integer n such that $\alpha - (p_n/q_n)| < \frac{1}{2}(10^{-3})$. By Theorem 9.11, we may choose n so that $1/q_n q_{n+1} < \frac{1}{2}(10^{-3})$, that is, $q_n q_{n+1} > 2000$. By the result of Exercise 3 from Section 9.2, we have $q_n = F_n$, so we want $F_n F_{n+1} > 2000$. The least such n is 10. Since $p_n = F_{n+1}$, our estimate for α is $F_{11}/F_{10} = 89/55 = 1.618$.

We saw previously that every positive rational number has an essentially unique representation as a finite continued fraction. We now establish the relation between positive irrational numbers and infinite continued fractions.

Theorem 9.12

If the real number x has a representation as an infinite continued fraction, then x is irrational.

Proof: Let $x = [a_0, a_1, a_2, \cdots]$. As in the proof of Theorem 9.11, we have

$$\frac{p_{2n}}{q_{2n}} < x < \frac{p_{2n+1}}{q_{2n+1}}$$

for all n, so that

$$0 < x - \frac{p_{2n}}{q_{2n}} < \frac{p_{2n+1}}{q_{2n+1}} - \frac{p_{2n}}{q_{2n}} = \frac{1}{q_{2n}q_{2n+1}}$$

If $x = a/b$, then

$$0 < \frac{a}{b} - \frac{p_{2n}}{q_{2n}} < \frac{1}{q_{2n}q_{2n+1}}$$

which implies

$$0 < aq_{2n} - bq_{2n+1} < \frac{b}{q_{2n+1}} \leq \frac{b}{\alpha^{2n}}$$

by Lemma 9.4. However if n is sufficiently large, then $b < \alpha^{2n}$, so that $0 < aq_{2n} - bp_{2n} < 1$, an impossibility. ■

It remains to be shown that every positive irrational number has a unique representation as an infinite continued fraction with integer terms. First, some additional preliminaries are needed.

▶ **Lemma 9.6** *If u, v are integers, $0 \leq f < 1$, $0 \leq g < 1$, and $u + f = v + g$, then $u = v$ and $f = g$.*

Proof: Exercise.

Theorem 9.13

Every positive irrational number has at most one representation as an infinite continued fraction with integer terms.

─────────────

Proof: Let x be a positive irrational number. Suppose that

$$x = [a_0, a_1, a_2, \cdots] = [b_0, b_1, b_2, \cdots]$$

Then we have

$$\lim_{t \to \infty} [a_0, a_1, a_2, \cdots, a_t] = \lim_{t \to \infty} [b_0, b_1, b_2, \cdots, b_t]$$

so that

$$\lim_{t \to \infty} \left(a_0 + \frac{1}{[a_1, a_2, \cdots, a_t]} \right) = \lim_{t \to \infty} \left(b_0 + \frac{1}{[b_1, b_2, \cdots, b_t]} \right)$$

that is

$$a_0 + \frac{1}{\lim_{t \to \infty} [a_1, a_2, \cdots, a_t]} = b_0 + \frac{1}{\lim_{t \to \infty} [b_1, b_2, \cdots, b_t]}$$

Since $a_1 \geq 1$ and $b_1 \geq 1$, it follows that each of the two limits in the immediately preceding equation exceed 1. Therefore Lemma 9.6 implies $a_0 = b_0$ and

$$\lim_{t \to \infty} [a_1, a_2, \cdots, a_t] = \lim_{t \to \infty} [b_1, b_2, \cdots, b_t]$$

Similarly, $a_1 = b_1, a_2 = b_2, \cdots, a_n = b_n, \cdots$. ■

Theorem 9.13 says that if a positive irrational number has a representation as an infinite continued fraction with integer terms, then that representation is unique. The following theorem guarantees the existence of such a representation and indicates how to compute it.

Theorem 9.14   If x_0 is a positive irrational number, let $a_0 = [x_0]$. If $n \geq 1$, let $x_n = 1/(x_{n-1} - a_{n-1})$, and let $a_n = [x_n]$. Then $x_0 = [a_0, a_1, a_2, \cdots]$.

Proof: First, we show that $x_0 = [a_0, a_1, a_2, \cdots, a_{n-1}, x_n]$ for all $n \geq 1$, using induction on n. By hypothesis, $x_1 = 1/(x_0 - a_0)$, so $x_0 = a_0 + 1/x_1 = [a_0, x_1]$. By induction hypothesis, we have

$$x_0 = [a_0, a_1, a_2, \cdots, a_{n-2}, x_{n-1}]$$

However, recall that $x_{n-1} = a_{n-1} + \frac{1}{x_n} = [a_{n-1}, x_n]$.

Therefore

$$x_0 = [a_0, a_1, a_2, \cdots, a_{n-2}, [a_{n-1}, x_n]] = [a_0, a_1, a_2, \cdots, a_{n-2}, a_{n-1}, x_n]$$

by Theorem 9.1. Now let

$$\frac{p_n}{q_n} = C_n = [a_0, a_1, a_2, \cdots, a_n]$$

Theorem 9.5 implies that

$$x_0 = \frac{x_{n+1} p_n + p_{n-1}}{x_{n+1} q_n + q_{n-1}}$$

Therefore

$$|x_0 - C_n| = |\frac{x_{n+1} p_n + p_{n-1}}{x_{n+1} q_n + q_{n-1}} - \frac{p_n}{q_n}|$$

$$= |\frac{p_{n-1} q_n - p_n q_{n-1}}{q_n (x_{n+1} q_n + q_{n-1})}|$$

$$= \frac{1}{q_n (x_{n+1} q_n + q_{n-1})} < \frac{1}{q_n (a_{n+1} q_n + q_{n-1})}$$

$$= \frac{1}{q_n q_{n+1}} < \frac{1}{\alpha^{2n-1}}$$

by Lemma 9.4 and Theorem 9.6(a). Therefore

$$\lim_{n \to \infty} (x_0 - C_n) = 0 \to x_0 = \lim_{n \to \infty} C_n$$

$$= \lim_{n \to \infty} [a_0, a_1, a_2, \cdots, a_n] =$$

$$[a_0, a_1, a_2, \cdots] \blacksquare$$

For example, let us compute the infinite continued fraction expansion of $\sqrt{22}$.

$$x_0 = \sqrt{22} = 4 + (\sqrt{22} - 4) \qquad\qquad \rightarrow a_0 = 4$$

$$x_1 = \frac{1}{\sqrt{22} - 4} = \frac{\sqrt{22} + 4}{6} = 1 + \frac{\sqrt{22} - 2}{6} \rightarrow a_1 = 1$$

$$x_2 = \frac{6}{\sqrt{22} - 2} = \frac{\sqrt{22} + 2}{3} = 2 + \frac{\sqrt{22} - 4}{3} \rightarrow a_2 = 2$$

$$x_3 = \frac{3}{\sqrt{22} - 4} = \frac{\sqrt{22} + 4}{2} = 4 + \frac{\sqrt{22} - 4}{2} \rightarrow a_3 = 4$$

$$x_4 = \frac{2}{\sqrt{22} - 4} = \frac{\sqrt{22} + 4}{3} = 2 + \frac{\sqrt{22} - 2}{3} \rightarrow a_4 = 2$$

$$x_5 = \frac{3}{\sqrt{22} - 2} = \frac{\sqrt{22} + 2}{6} = 1 + \frac{\sqrt{22} - 4}{6} \rightarrow a_5 = 1$$

$$x_6 = \frac{6}{\sqrt{22} - 4} = \sqrt{22} + 4 = 8 + \sqrt{22} - 4 \rightarrow a_6 = 8$$

At this point, repetition sets in. We have $x_6 - a_6 = \sqrt{22} - 4 = x_0 - a_0$. Now $x_7 = (x_6 - a_6)^{-1} = (x_0 - a_0)^{-1} = x_1$, so $a_7 = [x_7] = [x_1] = a_1$. More generally, for all $k \geq 1$, we have

$$x_{6+k} = \frac{1}{x_{5+k} - a_{5+k}} = \frac{1}{x_{k-1} - a_{k-1}} = x_k \rightarrow a_{6+k} = a_k$$

Therefore we have

$$\sqrt{22} = [4, 1, 2, 4, 2, 1, 8, 1, 2, 4, 2, 1, 8, \cdots]$$

As a consequence of Theorems 9.13 and 9.14, we are able to state the following theorem:

Theorem 9.15 Every positive irrational number x_0 has a unique representation as an infinite continued fraction with integer terms, namely, $x_0 = [a_0, a_1, a_2, \cdots]$ where $a_n = [x_n]$ for all $n \geq 0$ and $x_n = 1/(x_{n-1} - a_{n-1})$ for all $n \geq 1$.

Proof: This follows from Theorems 9.13 and 9.14. ∎

It appears from Theorem 9.15 that the infinite continued fraction representation of a positive irrational number may be obtained by an automatic procedure. In fact, the procedure is the same as that for obtaining the finite continued fraction expansion of a rational number, except that it never terminates. There is a catch, however. In the preceding example, x_1 through x_6 were presented in radical form, so no roundoff error was introduced in computing the x_i. Generally, however, this is not the case. If an x_i that is an irrational number is approximated by a rational number, then the error thus introduced will propagate larger errors in the x_k such that $k > i$. As a consequence, for sufficiently large k, the absolute value of the error in x_k will be 1 or greater, so that a_k will be incorrect, as well as all a_n such that $n \geq k$. For example, it is known that

$$\pi = [3, 7, 15, 1, 292, 1, 1, 1, 2, 1, 3, \cdots]$$

In fact the familiar rational approximation to π, namely, $22/7$, is just the first convergent: $[3, 7]$. If you try to verify the continued fraction expansion for π with a hand calculator, using the approximation $\pi = 3.141592654$, you will not obtain $a_8 = 2$.

In Section 9.4 we will see that the best rational approximations to an irrational number are obtained by the convergents to the infinite continued fraction representation of the irrational number. In general, however, the infinite continued fraction representation of an irrational number is hard to obtain. A notable exception is the case where the irrational number is quadratic, that is, the irrational number is a real, positive solution of a quadratic equation with integer coefficients. This case is investigated in Section 9.5.

Section 9.3   Exercises

1. Let $x = [1, 2, 3, \cdots]$. Find the least integer n such that $C_n = p_n/q_n$ estimates x correctly to 6 decimal places.

2. Use Theorem 9.14 to obtain the continued fraction expansion of $\sqrt{41}$. Using this result, estimate $\sqrt{41}$ to the nearest hundredth.

3. Prove that if x is the irrational number whose infinite continued fraction expansion is $[m, m, m, \cdots]$ where the integer $m \geq 1$, then $x = \frac{1}{2}(m + \sqrt{m^2 + 4})$.

4. By letting $m = 2n$ in the result from Exercise 3, plus some elementary manipulation, prove that $\sqrt{n^2 + 1} = [n, 2n, 2n, 2n, \cdots]$.

5. Using the result of Exercise 4, prove that $\sqrt{n^2 + 1} < n + \frac{1}{2n}$.

6. Prove Lemma 9.6.

7. Using the approximation, $e = 2.718281828$, find

    (a) the first 6 terms in the continued fraction expansion of $(e + 1)/(e - 1)$

    (b) the first 8 terms in the continued fraction expansion of $(e^2 + 1)/(e^2 - 1)$.

8. Prove that

(a) $[2, 1, 2, 1, \cdots] = 1 + \sqrt{3}$

(b) $[1, 2, 1, 2, \cdots] = \frac{1}{2}(1 + \sqrt{3})$

9.4 Approximation by Continued Fractions

We will now investigate just how well a positive irrational number is approximated by the convergents of its infinite continued fraction. First, we show that each convergent provides a better approximation than its immediate predecessor.

Theorem 9.16

If x_0 is a positive irrational number, $n \geq 1$, and p_n/q_n is the nth convergent in the infinite continued fraction for x_0, then

$$\left| x_0 - \frac{p_n}{q_n} \right| < \left| x_0 - \frac{p_{n-1}}{q_{n-1}} \right|$$

Proof: As in the proof of Theorem 9.14, we let $a_n = [x_n]$ for all $n \geq 0$ and $x_n = (x_{n-1} - a_{n-1})^{-1}$ for all $n \geq 1$ so that $x_0 = [a_0, a_1, a_2, \cdots, a_n, x_{n+1}]$. Now Theorem 9.5 implies

$$x_0 = \frac{x_{n+1}p_n + p_{n-1}}{x_{n+1}q_n + q_{n-1}} \quad \text{so} \quad x_{n+1}q_n x_0 + q_{n-1}x_0 = x_{n+1}p_n + p_{n-1}$$

Therefore

$$x_{n+1}(q_n x_0 - p_n) = -q_{n-1}x_0 + p_{n-1}$$

so

$$x_{n+1}(q_n x_0 - p_n) = -q_{n-1}\left(x_0 - \frac{p_{n-1}}{q_{n-1}}\right)$$

Dividing by $x_{n+1}q_n$, we obtain

$$x_0 - \frac{p_n}{q_n} = -\frac{q_{n-1}}{x_{n+1}q_n}\left(x_0 - \frac{p_{n-1}}{q_{n-1}}\right)$$

which implies

$$\left| x_0 - \frac{p_n}{q_n} \right| = \frac{q_{n-1}}{x_{n+1}q_n}\left| x_0 - \frac{p_{n-1}}{q_{n-1}} \right|$$

Now $0 < x_n - a_n < 1$, so $x_{n+1} = (x_n - a_n)^{-1} > 1$. Since $q_0 = 1 \leq q_1$, Lemma 9.5 implies $q_{n-1} \leq q_n$ for all $n \geq 1$. Therefore

$$\left|x_0 - \frac{p_n}{q_n}\right| < \left|x_0 - \frac{p_{n-1}}{q_{n-1}}\right| \quad \blacksquare$$

For example, we saw previously that $\sqrt{22} = [4, 1, 2, 4, 2, 1, 8, \cdots]$. Therefore we have

$$C_0 = [4] = 4 \qquad |\sqrt{22} - C_0| = |\sqrt{22} - 4| \qquad = 0.690$$

$$C_1 = [4, 1] = 5 \qquad |\sqrt{22} - C_1| = |\sqrt{22} - 5| \qquad = 0.310$$

$$C_2 = [4, 1, 2] = \frac{14}{3} \qquad |\sqrt{22} - C_2| = |\sqrt{22} - \frac{14}{3}| \qquad = 0.024$$

$$C_3 = [4, 1, 2, 4] = \frac{61}{13} \qquad |\sqrt{22} - C_3| = |\sqrt{22} - \frac{61}{13}| \qquad = 0.002$$

Let $y_n = |x_0 - C_n|$, where C_n is the nth convergent in the infinite continued fraction representation of x_0. By virtue of Theorems 9.14 and 9.15, we know that the sequence $\{y_n\}$ converges monotonically to zero; but how rapid is this convergence? This question is answered by the next several theorems.

Theorem 9.17　If p_n/q_n is the nth convergent in the infinite continued fraction representation of the positive irrational x_0, then

$$\frac{1}{2q_n q_{n+1}} < \left|x_0 - \frac{p_n}{q_n}\right| < \frac{1}{q_n q_{n+1}}$$

Proof: Since x_0 is between p_n/q_n and p_{n+1}/q_{n+1}, we have

$$\left|x_0 - \frac{p_n}{q_n}\right| + \left|x_0 - \frac{p_{n+1}}{q_{n+1}}\right| = \left|\frac{p_{n+!}}{q_{n+1}} - \frac{p_n}{q_n}\right| = \frac{1}{q_n q_{n+1}}$$

by Theorem 9.6(b). Now Theorem 9.15 implies

$$0 < \left|x_0 - \frac{p_{n+1}}{q_{n+1}}\right| < \left|x_0 - \frac{p_n}{q_n}\right|$$

Therefore

$$\left|x_0 - \frac{p_n}{q_n}\right| < \frac{1}{q_n q_{n+1}} < 2\left|x_0 - \frac{p_n}{q_n}\right|$$

from which the conclusion follows.　$\blacksquare$

For example, we saw previously that if $x_0 = \sqrt{22}$, then $C_2 = 14/3$, so $q_2 = 3$; also, $q_3 = 13$. Theorem 9.16 says that

$$\frac{1}{2q_2q_3} < |\sqrt{22} - C_2| < \frac{1}{q_2q_3}$$

that is,

$$\frac{1}{78} < |\sqrt{22} - \frac{14}{3}| < \frac{1}{39} \quad \text{or} \quad .0128 < |\sqrt{22} - \frac{14}{3}| < .0256|$$

(Actually, $|\sqrt{22} - \frac{14}{3}| = .0239$.)

If p_n/q_n is the nth convergent in the infinite continued fraction representation of x, then Theorem 9.17 and Lemma 9.5 imply $|x - p_n/q_n| < 1/q_n^2$. As a result, we see that for every positive irrational number x, there are infinitely many rationals a/b such that $|x - a/b| < 1/b^2$. This result may be improved as follows:

Theorem 9.18

If x is a positive irrational number and $n \geq 1$, then

(a) at least one of every two consecutive convergents to x satisfies

$$|x - \frac{p_n}{q_n}| < \frac{1}{2q_n^2}$$

and

(b) at least one of every three consecutive convergents satisfies

$$|x - \frac{p_n}{q_n}| < \frac{1}{\sqrt{5}q_n^2}.$$

Proof of (a): Recall from the proof of Theorem 9.16 that

$$|x_0 - \frac{p_n}{q_n}| + |x_0 - \frac{p_{n+1}}{q_{n+1}}| = \frac{1}{q_nq_{n+1}}$$

That is, if

$$|x - \frac{p_k}{q_k}| \geq \frac{1}{2q_k^2}$$

$$\text{for} \quad k = n \text{ and } k = n + 1$$

then

$$\frac{1}{2q_n^2} + \frac{1}{2q_{n+1}^2} \leq \frac{1}{q_nq_{n+1}} \rightarrow \frac{q_{n+1}}{q_n} + \frac{q_n}{q_{n+1}} \leq 2$$

If we let $r = q_{n+1}/q_n$, we have $r + r^{-1} \leq 2$, which implies $r = 1$, so $q_{n+1} = q_n$. Now Lemma 9.5 implies $n = 0$, contrary to the hypothesis. ∎

Proof of (b): Let $r = q_{n+1}/q_n$ and $s = q_{n+2}/q_{n+1}$. If

$$|x - \frac{p_k}{q_k}| \geq \frac{1}{\sqrt{5}q_k^2} \quad \text{for} \quad k = n, \ n+1, \ \text{and } n+2$$

then, as in the proof of (a), we get $r + r^{-1} \leq \sqrt{5}$, so $r \leq (\sqrt{5}+1)/2$. Since r is rational, we must have $r < (\sqrt{5}+1)/2$. Similarly, $s < (\sqrt{5}+1)/2$. However

$$q_{n+2} = a_{n+1}q_{n+1} + q_n \geq q_{n+1} + q_n \rightarrow \frac{q_{n+2}}{q_{n+1}} \geq 1 + \frac{q_n}{q_{n+1}} \rightarrow s \geq 1 + r^{-1}$$

But since $r < \frac{\sqrt{5}+1}{2}$, we have $r^{-1} > \frac{\sqrt{5}-1}{2}$, which implies $s > \frac{\sqrt{5}+1}{2}$, an impossibility. ∎

For example, let $\alpha = (1 + \sqrt{5})/2$. By the result of Section 9.2, Exercise 3, the nth convergent to α is F_{n+1}/F_n, where F_n denotes the nth Fibonacci number. One can show that

$$|\alpha - \frac{F_{n+1}}{F_n}| < \frac{1}{\sqrt{5}F_n^2}$$

if and only if n is even. We leave the proof of this assertion as an exercise.

As an immediate consequence of Theorem 9.18, we obtain the following:

Theorem 9.19

(Hurwitz)

If x be a positive irrational, then there are infinitely many rationals, a/b such that $|x - a/b| < 1/\sqrt{5}b^2$.

Proof: This follows directly from Theorem 9.18(b).

Theorem 9.18(a) has a "converse" of sorts that states that a sufficiently accurate rational approximation to an irrational number *must* be a convergent to the infinite continued fraction expansion of the irrational number. Before we can prove this surprising result, we need the following theorems.

Theorem 9.20

If p_n/q_n is the nth convergent in the infinite continued fraction expansion of the positive irrational number x, and $1 \leq b < q_{n+1}$, then $|q_n x - p_n| \leq |bx - a|$ for all a.

Proof: Let us seek integers u and v such that

$$\begin{cases} p_n u + p_{n+1} v = a \\ q_n u + q_{n+1} v = b \end{cases}$$

Since the determinant of this system of two linear equations in two unknowns is $p_n q_{n+1} - p_{n+1} q_n = (-1)^{n+1} \neq 0$, the system has a unique solution, namely,

$$u = (-1)^{n+1}(a q_{n+1} - b p_{n+1})$$

$$v = (-1)^{n+1}(b p_n - a q_n)$$

If $u = 0$, then $a q_{n+1} = b p_{n+1}$. Since $(p_{n+1}, q_{n+1}) = 1$ by Theorem 9.7, Euclid's Lemma implies that $q_{n+1} | b$, so $q_{n+1} \leq b$, contrary to hypothesis. If $u \neq 0$ but $v = 0$, then $b = q_n u$ and $a = p_n u$, so

$$|bx - a| = |q_n u x - p_n u|$$

$$= |u||q_n x - p_n| \geq |q_n x - p_n|$$

If $u \neq 0$ and $v \neq 0$, then we claim that u and v have opposite signs. Indeed, since $q_n u = b - q_{n+1} v$, if $v < 0$, then $q_n u > 0$, so $u > 0$. If $v > 0$, then $v \geq 1$, so $b - q_{n+1} v \leq b - q_{n+1}$. However $b - q_{n+1} < 0$ by hypothesis, so $b - q_{n+1} v < 0 \rightarrow q_n u < 0 \rightarrow u < 0$.

In addition, since x lies between the consecutive convergents p_n/q_n and p_{n+1}/q_{n+1}, the quantities $q_n x - p_n$ and $q_{n+1} x - p_{n+1}$ have opposite signs. Therefore $u(q_n x - p_n)$ and $v(q_{n+1} x - p_{n+1})$ have the same sign. Now

$$|bx - a| = |(q_n u + q_{n+1} v)x - (p_n u + p_{n+1} v)|$$

$$= |u(q_n x - p_n) + v(q_{n+1} x - p_{n+1})|$$

$$= |u(q_n x - p_n)| + |v(q_{n+1} x - p_{n+1})| \geq |u||q_n x - p_n| \geq |q_n x - p_n| \quad \blacksquare$$

Now we can show that the convergents of the infinite continued fraction expansion of a positive irrational number provide the best rational approximations with bounded denominators.

Theorem 9.21

If p_n/q_n is the nth convergent in the infinite continued fraction expansion of the positive irrational number x, $1 \leq b < q_n$, and a is an arbitrary natural number, then

$$\left| x - \frac{p_n}{q_n} \right| \leq \left| x - \frac{a}{b} \right|$$

Proof: If $|x - (a/b)| < |x - (p_n/q_n)|$, then

$$|q_n x - p_n| = q_n|x - \frac{p_n}{q_n}| > q_n|x - \frac{a}{b}| \geq b|x - \frac{a}{b}| = |bx - a|$$

so $|q_n x - p_n| > |bx - a|$, contrary to Theorem 9.20.   ∎

We are now ready to present the following theorem.

Theorem 9.22

If x is a positive irrational number, and if a, b are integers such that $(a, b) = 1$ and $|x - (a/b)| < 1/2b^2$, then a/b is a convergent in the infinite continued fraction expansion of x, that is $a = p_n$ and $b = q_n$ for some n.

Proof: Assume the contrary. Lemma 9.5 implies that we can find an integer n such that $q_n \leq b < q_{n+1}$. Now

$$|q_n x - p_n| \leq |bx - a| = b|x - \frac{a}{b}| \leq \frac{1}{2b}$$

by hypothesis. Therefore $|x - \frac{p_n}{q_n}| < \frac{1}{2bq_n}$. If $a/b \neq p_n/q_n$, then $|bp_n - aq_n| \geq 1$. Therefore

$$\frac{1}{bq_n} \leq \frac{|bp_n - aq_n|}{bq_n} = |\frac{p_n}{q_n} - \frac{a}{b}| \leq |\frac{p_n}{q_n} - x| + |x - \frac{a}{b}| < \frac{1}{2bq_n} + \frac{1}{2b^2}$$

This implies $b < q_n$, an impossibility. Therefore $a/b = p_n/q_n$. Since $(a, b) = 1$ by hypothesis and $(p_n, q_n) = 1$ by Lemma 9.7, it follows that $a = p_n$ and $b = q_n$.   ∎

Note that if x is a positive irrational number, then Theorem 9.22 gives conditions that are sufficient, but not necessary, for a rational approximation to x to be a convergent.

For example, if $x = x_0 = \sqrt{13}$, then $a_0 = [x_0] = [\sqrt{13}] = 3$, and

$$x_1 = \frac{1}{x_0 - a_0} = \frac{1}{\sqrt{13} - 3} = \frac{\sqrt{13} + 3}{4} \rightarrow a_1 = [x_1] = 1$$

$$x_2 = \frac{1}{x_1 - a_1} = \frac{4}{\sqrt{13} - 1} = \frac{\sqrt{13} + 1}{3} \rightarrow a_2 = [x_2] = 1$$

$$x_3 = \frac{1}{x_2 - a_2} = \frac{3}{\sqrt{13} - 2} = \frac{\sqrt{13} + 2}{3} \rightarrow a_3 = [x_3] = 1$$

The following table lists a_n, p_n, q_n for $0 \le n \le 3$.

n	a_n	p_n	q_n
0	3	3	1
1	1	4	1
2	1	7	2
3	1	11	3

Now $|x - (p_3/q_3)| = |\sqrt{13} - (11/3)| = .0611 > .0556 = 1/18 = 1/2q_3^2$, so the inequality in the hypothesis of Theorem 9.22 is not satisfied, even though $11/3$ is a convergent. The following theorem, however, gives conditions that are both necessary and sufficient for a rational approximation of an irrational number to be a convergent.

Theorem 9.23

Let a and b be natural numbers such that $a/b = [a_0, a_1, a_2, \cdots, a_n] = p_n/q_n$. If x is a positive irrational number, then a/b is convergent in the continued fraction expansion for x if and only if $|x - (a/b)| < 1/q_n(q_n + q_{n-1})$.

Proof: We omit the proof, but refer the interested reader to [10], pp. 42–45.

Theorem 9.23 enables us to obtain the following improvement to Theorem 9.22.

Theorem 9.24

If a and b are natural numbers such that $(a, b) = 1$ and $b > 1$, and if x is a positive irrational number such that $|x - (a/b)| < 1/b(2b - 1)$, then a/b is a convergent in the continued fraction representation of x.

Proof: Let $a/b = [a_0, a_1, a_2, \cdots, a_n] = p_n/q_n$ with $a_n > 1$ if $n > 0$. If $n = 0$, then $a/b = [a_0] = a_0/1$, so $b = 1$, contrary to hypothesis. If $n \ge 1$, then Lemma 9.5 implies $q_n \ge q_{n-1} + 1$, so that $2q_n - 1 \ge q_n + q_{n-1}$. Now

$$|x - \frac{a}{b}| < \frac{1}{b(2b - 1)} \rightarrow |x - \frac{a}{b}| < \frac{1}{q_n(2q_n - 1)} \le \frac{1}{q_n(q_n + q_{n-1})}$$

The conclusion now follows from Theorem 9.23. ∎

Section 9.4   Exercises

1. Find the least n such that $\sqrt{6}$ can be approximated to three decimal places by the nth convergent of its continued fraction.

2. Let C_n be the nth convergent in the continued fraction expansion of $\alpha = (1 + \sqrt{5})/2$. Prove that $|\alpha - C_n| < 1/\sqrt{5}q_n^2$ if and only if n is even.

3. Given that $(e + 1)/(e - 1) = [2, 6, 10, 14, \cdots]$, use a convergent to estimate $(e + 1)/(e - 1)$ to 4 decimal places.

4. Use the properties of Fibonacci numbers to prove that if $|\alpha - (a/b)| < 1/hb^2$, then $h \leq \sqrt{5}$.

Section 9.4 Computer Exercises

5. Let p_n/q_n be the nth convergent in the infinite continued fraction representation of θ, a positive irrational number. Write a computer program to evaluate $|\theta - (p_n/q_n)|q_n^2$ for $n = 1, 2, 3, \cdots$. (You might take $\theta = \sqrt{2}$ or $\theta = \sqrt{5}$, for example.) What does the numerical evidence suggest?

9.5 Periodic Continued Fractions: I

We saw in Section 9.3 that $\sqrt{22} = [4, 1, 2, 4, 2, 1, 8, 1, 2, 4, 2, 1, 8, \cdots]$. That is, the continued fraction representation of $\sqrt{22}$ consists of an initial term, namely, 4, followed by a sequence of 6 integers that repeats indefinitely, namely, 1, 2, 4, 2, 1, 8. A more convenient notation for this infinite continued fraction is

$$\sqrt{22} = [4, \overline{1, 2, 4, 2, 1, 8}]$$

We say that the continued fraction representation for $\sqrt{22}$ is *periodic*. In general, an infinite continued fraction is called periodic if it ends with an indefinitely repeating finite sequence of integers. Some examples are

$$[2, 1, 3, 4, 5, 3, 4, 5, 3, 4, 5, \cdots] = [2, 1, \overline{3, 4, 5}]$$

and

$$[1, 4, 2, 3, 1, 4, 2, 3, 1, 4, 2, 3, \cdots] = [\overline{1, 2, 4, 3}]$$

The second example, where the continued fraction contains no nonperiodic terms, is called *purely periodic*.

It can be shown that if x is a positive irrational number, then the infinite continued fraction representation of x is periodic if and only if x is a quadratic irrational, that is, x is the real root of an irreducible quadratic equation with integer coefficients. For example, we saw earlier, in Section 9.3, that if $\alpha = (1 + \sqrt{5})/2$, then $\alpha = [1, 1, 1, \cdots] = [\overline{1}]$. Now α is the positive real root of the equation, $t^2 - t - 1 = 0$.

We will prove the slightly narrower result: If C and D are integers such that $0 < C < D$ and D/C is not a square, then

$$\sqrt{D/C} = [a_0, \overline{a_1, a_2, \cdots, a_t}] \tag{9.1}$$

for some $t \geq 1$. The smallest t for which equation (9.1) holds is called the *length of the period* of the continued fraction. The integer sequence $a_1, a_2, \cdots, a_t$ is called the *period* of the continued fraction. The periodic infinite continued fractions presented above have periods of length 6, 3, 4, 1 respectively. In the work ahead, $[y]$ denotes the integer part of y, *not* a continued fraction with a single term.

Theorem 9.25 If C and D are integers such that $0 < C < D$ and D/C is not a square, then there exists $t \geq 1$ such that $\sqrt{D/C} = [a_0, \overline{a_1, a_2, \cdots, a_t}]$.

Proof: Let $b_0 = 0, r_0 = C, x_0 = \sqrt{D/C}, a_0 = [x_0]$. For $k \geq 1$, let

$$b_k = a_{k-1}b_{k-1} - r_{k-1}$$

$$r_k = \frac{CD - b_k^2}{r_{k-1}}$$

$$x_k = \frac{\sqrt{CD} + b_k}{r_k}$$

$$a_k = [x_k]$$

In particular, $b_1 = C[\sqrt{D/C}]$. Then

$$x_k = a_k + \frac{(\sqrt{CD} + b_k - a_k r_k)}{r_k}$$

$$= a_k + r_{k+1}(\sqrt{CD} + b_{k+1})$$

$$= a_k + \frac{1}{x_{k+1}}$$

We will show by induction on k that each of b_k, r_k, a_k is an integer. From this, and from Theorem 9.15, it follows that

$$\sqrt{D/C} = [a_0, a_1, a_2, \cdots]$$

Periodicity is a consequence of the fact (to be demonstrated below) that b_k and the r_k are positive integers bounded from above. First, we need several lemmas.

▶ **Lemma 9.7** r_k and b_k are integers for all $k \geq 0$.

Proof: (Induction on k) $b_0 = 0$, $r_0 = C$, $b_1 = a_0 r_0 - b_0 = C[\sqrt{D/C}]$, $r_1 = (CD - b_1^2)/r_0 = (CD - C^2[\sqrt{D/C}]^2)/C = D - C[\sqrt{D/C}]^2$. By definition,

a_k is an integer for all k. If $k \geq 1$, then since $b_k = a_{k-1}r_{k-1} - b_{k-1}$, it follows by induction hypothesis that b_k is an integer. If $k \geq 2$, then

$$r_k = \frac{CD - b_k^2}{r_{k-1}}$$

$$= \frac{CD - (a_{k-1}r_{k-1} - b_{k-1})^2}{r_{k-1}}$$

$$= \frac{CD - b_{k-1}^2}{r_{k-1}} + a_{k-1}(2b_{k-1} - a_{k-1}r_{k-1})$$

$$= r_{k-2} + a_{k-1}(b_{k-1} - b_k)$$

By induction hypothesis, each of r_{k-2}, b_{k-1} is an integer. We have just proved that b_k is an integer. Therefore r_k is an integer. ∎

▶ **Lemma 9.8** *For all $k \geq 0$, we have*

$$1 \leq r_k \leq 2[\sqrt{CD}] \ , \ b_k \leq [\sqrt{CD}] \ , \ a_k \geq 1$$

Proof: (Induction on k) The conclusion holds for $k = 0$, since $r_0 = C$, $b_0 = 0$, $a_0 = [\sqrt{D/C}] \geq 1$. Let $k \geq 1$. Since b_{k-1} and r_{k-1} are integers by Lemma 9.7, it follows that $x_{k-1} = (\sqrt{CD} + b_{k-1})/r_{k-1}$ is irrational. Now

$$x_{k-1} - [x_{k-1}] = \frac{\sqrt{CD} - b_k}{r_{k-1}} = \frac{r_k}{\sqrt{CD} + b_k}$$

so we have

$$0 < \frac{\sqrt{CD} - b_k}{r_{k-1}} < 1 \tag{9.2}$$

and

$$0 < \frac{r_k}{\sqrt{CD} + b_k} < 1 \tag{9.3}$$

Since $r_{k-1} \geq 1$ by induction hypothesis, (9.2) implies $b_k < \sqrt{CD}$, so $b_k \leq [\sqrt{CD}]$ and $CD - b_k^2 > 0$. Since $a_k = [x_k] = [\frac{\sqrt{CD}+b_k}{r_k}]$, (9.3) implies $a_k \geq 1$. Now

$$r_k = \frac{CD - b_k^2}{r_{k-1}} > 0 \rightarrow r_k \geq 1$$

Finally, (3) implies $r_k < \sqrt{CD} + b_k \leq \sqrt{DC} + [\sqrt{CD}]$. Since r_k is an integer, we have $r_k \leq 2[\sqrt{CD}]$. ∎

▶ **Lemma 9.9** $b_k \geq 1$ *for all $k \geq 1$.*

Proof: Recall that $b_1 = C[\sqrt{D/C}] \geq 1$. Now assume that $k \geq 1$ and $b_{k+1} \leq 0$. This implies $a_k r_k \leq b_k$. Since $a_k \geq 1$, we have $r_k \leq b_k$, and also $a_k \leq b_k/r_k$, hence $a_k \leq [b_k/r_k]$, that is $[(\sqrt{CD} + b_k)/r_k] \leq [b_k/r_k]$. Since $CD > 0$, the reverse inequality also holds. Therefore $[(\sqrt{CD} + b_k)/r_k] = [b_k/r_k]$. This implies $\sqrt{CD} < r_k$. But then Lemma 9.8 implies $b_k < r_k$, an impossibility. The conclusion now follows. ∎

Proof: (Theorem 9.25, continued) Resuming the proof of Theorem 9.25, we have seen via Lemmas 9.8 and 9.9 that $1 \leq b_k \leq [\sqrt{CD}]$ and $1 \leq r_k \leq 2[\sqrt{CD}]$ for all $k \geq 1$. Therefore, if $m = 2[\sqrt{CD}]^2$, there exist at most m distinct ordered pairs (b_i, r_j). This implies there exists an integer t such that $1 \leq t \leq m$, $b_{t+1} = b_1$, $r_{t+1} = r_1$. The least such t is called the *length of the period* of the continued fraction representation of $\sqrt{D/C}$. Now

$$x_{t+1} = \frac{\sqrt{CD} + b_{t+1}}{r_{t+1}} = \frac{\sqrt{CD} + b_1}{r_1} = x_1$$

so $a_{t+1} = [x_{t+1}] = [x_1] = a_1$. By induction on k, we get $b_{t+k} = b_k$, $r_{t+k} = r_k$, $a_{t+k} = a_k$, $x_{t+k} = x_k$ for all $k \geq 1$. Therefore

$$\sqrt{D/C} = [a_0, a_1, a_2, \cdots, a_t, a_1, a_2, \cdots, a_t, \cdots]$$

which we write more conveniently as

$$\sqrt{D/C} = [a_0, \overline{a_1, a_2, \cdots, a_t}] \quad ∎$$

For example, we saw in Section 9.3 that $\sqrt{22} = [4, \overline{1, 2, 4, 2, 1, 8}]$. We could obtain this result without performing algebraic operations on irrational fractions, provided that we make use of the inductive definitions of b_k, r_k, a_k. If we wish to obtain the continued fraction representation of $\sqrt{22}$, we note that $D = 22$, $C = 1$ and construct the following table:

k	b_k	r_k	a_k
0	0	1	4
1	4	6	1
2	2	3	2
3	4	2	4
4	4	3	2
5	2	6	1
6	4	1	8
7	4	6	1

Now $b_7 = 4 = b_1$, $r_7 = 6 = r_1$, $a_7 = 1 = a_1$. It is clear from the table that $t = 6$ is the least positive integer such that $b_{t+1} = b_1$, $r_{t+1} = r_1$, $a_{t+1} = a_1$. Therefore $\sqrt{22} = [4, \overline{1, 2, 4, 2, 1, 8}]$.

The reader may note that (i) the last term in the period of the continued fraction, namely, 8, is double the initial nonperiodic term, 4, and (ii) except for the last term, the period is symmetric, that is, $a_{t-k} = a_k$ for all k such that $1 \leq k \leq t - 1$. We shall see that these phenomena are not isolated, but occur in the general case.

In the several lemmas that follow, b_k, r_k, a_k are defined as in the proof of Theorem 9.25.

▶ **Lemma 9.10** $b_j < b_k + r_k$ for all j, k.

Proof: Since $b_k = a_{k-1}r_{k-1} - b_{k-1}$, we have $b_k + b_{k-1} = a_{k-1}r_{k-1}$. Lemma 9.8 implies $a_{k-1} \geq 1$, so $b_k + b_{k-1} \geq r_{k-1}$. Lemma 9.8 also implies $b_{k-1} < \sqrt{CD}$, so $\sqrt{CD} + b_k > b_{k-1} + b_k$, which in turn implies $\sqrt{CD} + b_k > r_{k-1}$. However $CD - b_k^2 = r_k r_{k-1}$, so we must have $\sqrt{CD} - b_k < r_k$, that is, $\sqrt{CD} < b_k + r_k$. Lemma 9.8 implies $b_j < \sqrt{CD}$, hence $b_j < b_k + r_k$. ■

▶ **Lemma 9.11** $b_t = b_1$ and $a_t = 2a_0$.

Proof:

$$r_t = \frac{CD - b_{t+1}^2}{r_{t+1}} = \frac{CD - b_1^2}{r_1} = r_0 = C$$

Recall that $b = C[\sqrt{D/C}] = Ca_0$. Now

$$b_t = a_t r_t - b_{t+1} = Ca_t - b_1 = Ca_t - Ca_0 = C(a_t - a_0)$$

Lemma 9.10 implies $b_1 < b_t + r_t$, that is, $b_1 < b_t + C$, so $b_1/C < b_t/C + 1$. Since b_1/C and b_t/C are integers, it follows that $b_1/C \leq b_t/C$, so $b_1 \leq b_t$. Now

$$\sqrt{CD} = b_1 + (\sqrt{CD} - b_1) = b_1 + (\sqrt{CD} - Ca_0)$$

$$= b_1 + C(\sqrt{D/C} - [\sqrt{D/C}])$$

Let $r = \sqrt{D/C}$, so $\sqrt{CD} = b_1 + C(r - [r])$. Lemma 9.8 implies $b_t < \sqrt{CD}$, so $b_t < b_1 + C(r - [r])$. Since b_1/C and b_t/C are integers and $0 < r - [r] < 1$, it follows that $b_t/C \leq b_1/C$, hence $b_t \leq b_1$. Therefore $b_t = b_1$. Also, $a_t = (b_t + b_{t+1})/r_t = 2b_1/C = 2a_0$. ■

▶ **Lemma 9.12** If $1 \leq t \leq k - 1$, then $r_{t-k} = r_k$, $a_{t-k} = a_k$, and $b_{t-k} = b_{k+1}$.

Proof: (Induction on k) We have seen in the proof of Lemma 9.11 that $r_t = r_0$ and $b_t = b_1$, so the conclusion holds for $k = 0$. By induction hypothesis, we have $r_{t-k+1} = r_{k-1}$ and $b_{t-k+1} = b_k$. Now

$$r_{t-k} = \frac{CD - b_{t-k+1}^2}{r_{t-k+1}} = \frac{CD - b_k^2}{r_{k-1}} = r_k$$

Also, $b_{t-k} + b_{t-k+1} = a_{t-k} r_{t-k}$, so $b_{t-k} + b_k = a_{t-k} r_{t-k}$. Moreover, $b_k + b_{k+1} = a_k r_k$, so $b_k + b_{k+1} = a_k r_{t-k}$. Subtracting, we get $b_{t-k} - b_{k+1} = r_{t-k}(a_k - a_{t-k})$, so $a_k - a_{t-k} = (b_{k+1} - b_{t-k})/r_{t-k}$. If $1 \leq k \leq t - 1$, then Lemma 9.10 implies $(b_{k+1} - b_{t-k})/r_{t-k} < 1$. Therefore $a_k - a_{t-k} \leq 0$, so $a_k \leq a_{t-k}$. But simultaneously, we have $1 \leq t - k \leq t - 1$, so we similarly obtain $a_{t-k} \leq a_k$. Therefore $a_{t-k} = a_k$, from which it follows that $b_{t-k} = b_{k+1}$. ∎

We summarize our results in the following theorem:

Theorem 9.26

If C and D are integers such that $0 < C < D$ and $D/C \neq s^2$, then the infinite continued fraction representation of $\sqrt{D/C}$ is given by

$$\sqrt{D/C} = [a_0, \overline{a_1, a_2, a_3, \cdots, a_3, a_2, a_1, 2a_0}]$$

Proof: This follows from Theorem 9.25 and Lemmas 9.11 and 9.12. ∎

Remarks

The first proof of Theorem 9.25 was published in 1770 by Lagrange (see [7]). The first proof of Theorem 9.26 was published in 1828 by Galois (see [3]). These distinguished mathematicians used more complex methods to obtain their results.

For example, let us develop the continued fraction representation for $\sqrt{21/13}$. Following the notation used in the proof of Theorem 9.25, we have $b_0 = 0$, $r_0 = C = 13$, $a_0 = [\sqrt{21/13}] = 1$. For $k \geq 1$, we use the formulas

$$b_k = a_{k-1} r_{k-1} - b_{k-1}$$

$$r_k = \frac{CD - b_k^2}{r_{k-1}} = \frac{273 - b_k^2}{r_{k-1}}$$

$$a_k = \left[\frac{[\sqrt{CD}] + b_k}{r_k}\right] = \left[\frac{[\sqrt{273}] + b_k}{r_k}\right] = \left[\frac{16 + b_k}{r_k}\right]$$

The resulting data are listed in the following table:

k	b_k	r_k	a_k
0	0	13	1
1	13	8	3
2	11	19	1
3	8	11	2
4	14	7	4
5	14	11	2
6	8	19	1
7	11	8	3
8	13	13	2
9	13	8	3

Since $b_9 = b_1 = 13$, $r_9 = r_1 = 8$, and $a_9 = a_1 = 3$, we know that $t = 8$, so we have

$$\sqrt{21/13} = [1, \overline{3, 1, 2, 4, 2, 1, 3, 2}]$$

Section 9.5 Exercises

1. Find the infinite continued fraction representations for each of the following:

(a) $\sqrt{2}$

(b) $\sqrt{3}$

(c) $\sqrt{11}$

(d) $\sqrt{22}$

(e) $\sqrt{31}$

(f) $\sqrt{74}$

(g) $\sqrt{94}$

(h) $\sqrt{75}$

(i) $\sqrt{8/5}$

(j) $\sqrt{22/3}$

2. Prove that if $\sqrt{N} = [a, \overline{b, b, 2a}]$, then $(b^2 + 1)|(2ab + 1)$.

3. Prove that $\sqrt{m^2 + 1} = [m, \overline{2m}]$.

4. Prove that $\sqrt{m^2 - 1} = [m - 1, \overline{1, 2m - 2}]$.

5. Prove that $\sqrt{m^2 + 2} = [m, \overline{m, 2m}]$.

6. Prove that $\sqrt{m^2 - 2} = [m - 1, \overline{1, m - 2, 1, 2m - 2}]$ if $m \geq 3$.

7. Prove that $\sqrt{4m^2 + 4} = [2m, \overline{m, 4m}]$.

8. Prove that $\sqrt{4m^2 + 4m + 5} = [2m + 1, \overline{m, 1, 1, m, 4m + 2}]$ if $m \geq 1$.

9. Prove that $\sqrt{4m^2 - 4} = [2m - 1, \overline{1, m - 2, 1, 4m - 2}]$ if $m \geq 3$.

10. Prove that $\sqrt{4m^2 + 4m - 3} = [2m, \overline{1, m - 1, 2, m - 1, 1, 4m}]$ if $m \geq 2$.

11. Prove that $\sqrt{9m^2 + 3} = [3m, \overline{2m, 6m}]$.

12. Prove that $\sqrt{9m^2 - 3} = [3m - 1, \overline{1, 2m - 2, 1, 6m - 2}]$ if $m \geq 2$.

13. Prove that $\sqrt{9m^2 + 6} = [3m, \overline{m, 6m}]$.

14. Prove that $\sqrt{9m^2 - 6} = [3m - 1, \overline{1, m - 2, 1, 6m - 2}]$ if $m \geq 3$.

★15. Prove that if $\sqrt{N}$ has an infinite continued fraction representation whose period has odd length, then N is a sum of two squares.

Section 9.5 Computer Exercises

16. Write a computer program to find the infinite continued fraction expansion of $\sqrt{D/C}$, where C and D are integers such that $0 < C < D$ and $D/C \neq s^2$.

9.6 Periodic Continued Fractions: II

The next chapter deals with nonlinear Diophantine equations, among them the *generalized Pell's equation*

$$Cx^2 - Dy^2 = E$$

where C, D, E are given integers such that $0 < C < D$ and $D/C \neq s^2$. We now present some additional properties of periodic infinite continued fractions that will be useful in obtaining the solutions of the generalized Pell's equation. First, some additional terminology is needed.

Definition 9.4 n^{th} complete quotient

Let $x_0 = [a_0, a_1, a_2, \cdots] = \lim_{t \to \infty}[a_0, a_1, a_2, \cdots, a_t]$. If $n \geq 1$, let $x_n = [a_n, a_{n+1}, a_{n+2}, \cdots] = \lim_{t \to \infty}[a_n, a_{n+1}, a_{n+2}, \cdots, a_t]$. We call x_n the nth *complete quotient* of x_0.

For example, let $x_0 = [0, 2, 4, 6, 8, 10, \cdots]$. Then $x_3 = [6, 8, 10, 12, 14, \cdots]$.

Actually, these n^{th} complete quotients are not new; they are generated in the process of finding the terms a_n of the continued fraction representation of x_0, as the following theorem asserts.

Theorem 9.27   Let the positive irrational number x_0 have the continued fraction representation $x_0 = [a_0, a_1, a_2, \cdots]$ where the a_i are integers such that $a_0 \geq 0$ and $a_i \leq 1$ for all $i \geq 1$. Then the following are equivalent:

1. x_n is defined inductively by $a_n = [x_n]$ for all $n \geq 0$; also, $x_n = 1/(x_{n-1} - a_{n-1})$ for all $n \geq 1$.

2. $x_n = [a_n, a_{n+1}, a_{n+2}, \cdots]$ for all $n \geq 0$.

Proof: (1) implies (2): (Induction on n).

By hypothesis, (2) holds for $n = 0$, and we have $a_n + 1/x_{n+1} = x_n$ for all $n \geq 0$. By induction hypothesis,

$$x_n = [a_n, a_{n+1}, a_{n+2}, \cdots]$$

$$= \lim_{t \to \infty} [a_n, a_{n+1}, a_{n+2}, \cdots, a_{n+t}]$$

$$= \lim_{t \to \infty} \left(a_n + \frac{1}{[a_{n+1}, a_{n+2}, \cdots, a_{n+t}]} \right)$$

$$= a_n + \frac{1}{\lim_{t \to \infty} [a_{n+1}, a_{n+2}, \cdots, a_{n+t}]}$$

$$= a_n + 1/[a_{n+1}, a_{n+2}, \cdots]$$

Therefore $x_{n+1} = [a_{n+1}, a_{n+2}, \cdots]$.

(2) implies (1):

If $n \geq 0$, then $x_n = [a_n, a_{n+1}, a_{n+2}, \cdots]$ by hypothesis. The same set of equations that appear above in the proof of (1) show that $x_n = a_n + 1/x_{n+1}$, hence $x_{n+1} = 1/(x_n - a_n)$ for all $n \geq 0$. Referring to the continued fraction expansion for x_n, we have $C_0 = a_n < x_n < a_n + 1/a_{n+1} = C_1$. Since $a_{n+1} \geq 1$, we must have $a_n = [x_n]$. ∎

For example, referring again to the continued fraction expansion of $\sqrt{22}$ from Section 9.5, we have

$$x_0 = \sqrt{22} = [4, \overline{1, 2, 4, 2, 1, 8}]$$

$$x_1 = (\sqrt{22} + 4)/6 = [\overline{1, 2, 4, 2, 1, 8}]$$

$$x_2 = (\sqrt{22} + 2)/3 = [\overline{2, 4, 2, 1, 8, 1}]$$

$$x_3 = (\sqrt{22} + 4)/2 = [\overline{4, 2, 1, 8, 1, 2}]$$

$$x_4 = (\sqrt{22} + 4)/3 = [\overline{2, 1, 8, 1, 2, 4}]$$

$$x_5 = (\sqrt{22} + 2)/6 = [\overline{1, 8, 1, 2, 4, 2}]$$

$$x_6 = \sqrt{22} + 4 = [\overline{8, 1, 2, 4, 2, 1}]$$

We need not compute any further complete quotients here because $x_{k+6} = x_k$ for all $k \geq 0$.

Next, we show that an infinite continued fraction can also be presented as a finite continued fraction whose last term is an nth complete quotient.

Theorem 9.28 Let the positive irrational number x_0 have the continued fraction representation, $x_0 = [a_0, a_1, a_2, \cdots]$ where the a_i are integers such that $a_0 \geq 0$ and $a_i \leq 1$ for all $i \geq 1$. Let $x_n = [a_n, a_{n+1}, a_{n+2}, \cdots]$ where $n \geq 1$. Then $x_0 = [a_0, a_1, a_2, \cdots, a_{n-1}, x_n]$.

Proof: (Induction on n)

By hypothesis and Theorem 9.27, we have $x_n = a_n + 1/x_{n+1}$ for all $n \geq 0$. In particular, $x_0 = a_0 + 1/x_1 = [a_0, a_1]$. By induction hypothesis, $x_0 = [a_0, a_1, a_2, \cdots, a_{n-1}, x_n]$. Thus $x_0 = [a_0, a_1, a_2, \cdots, a_{n-1}, a_n + 1/x_{n+1}]$. Now Theorem 9.1 implies $x_0 = [a_0, a_1, a_2, \cdots, a_n, x_{n+1}]$. ∎

We can now derive the following identity, which will be useful in the next chapter.

Theorem 9.29 Let C, D be integers such that $0 < C < D$ and $D/C \neq s^2$. Let $\sqrt{D/C} = [a_0, \overline{a_1, a_2, \cdots, a_t}]$ for some $t \geq 1$. If $k \geq 0$, let r_k be defined as in Theorem 9.25, and let $p_k/q_k = [a_0, a_1, a_2, \cdots, a_k]$. Then

$$CP_{k-1}^2 - Dq_{k-1}^2 = (-1)^k r_k$$

Proof: By hypothesis and by Theorem 9.25, we have

$$b_0 = 0$$

$$r_0 = C$$

$$x_k = \frac{\sqrt{CD} + b_k}{r_k}$$

$$a_k = [x_k]$$

for all $k \geq 0$, also

$$b_k = a_{k-1}r_{k-1} - b_{k-1}$$

$$r_k = \frac{CD - b_k^2}{r_{k-1}}$$

for all $k \geq 1$. Now Theorem 9.27 implies $x_k = a_k + 1/x_{k+1} = [a_k, a_{k+1}, a_{k+2}, \cdots]$ for all $k \geq 0$. Therefore Theorem 9.28 implies that $x_0 = [a_0, a_1, a_2, \cdots, a_{k-1}, x_k]$. Now Theorem 9.5 implies

$$x_0 = \frac{x_k p_{k-1} + p_{k-2}}{x_k q_{k-1} + q_{k-2}}$$

that is,

$$\sqrt{D/C} = (\frac{\sqrt{CD} + b_k)/r_k)p_{k-1} + p_{k-2}}{\sqrt{CD} + b_k)/r_k)q_{k-1} + q_{k-2}} = \frac{(\sqrt{CD} + b_k)p_{k-1} + r_k p_{k-2}}{(\sqrt{CD} + b_k)q_{k-1} + r_k q_{k-2}}$$

so

$$Dq_{k-1} + \sqrt{D/C}(b_k q_{k-1} + r_k q_{k-2} = (\sqrt{CD})p_{k-1} + (b_k p_{k-1} + r k p_{k-2})$$

Equating the irrational terms, and also the rational terms, and then simplifying, we obtain

$$b_k q_{k-1} + r_k q_{k-2} = C p_{k-1}$$

$$b_k p_{k-1} + r_k p_{k-2} = D q_{k-1}$$

If we multiply the first equation by p_{k-1}, the second equation by q_{k-1} and then subtract, we get

$$C p_{k-1}^2 - D q_{k-1}^2 = r_k(p_{k-1}q_{k-2} - p_{k-2}q_{k-1} = (-1)^{k-2}r_k = (-1)^k r_k$$

by Theorem 9.6(a).   ∎

For example, we previously computed the terms in the period of the continued fraction expansion of $\sqrt{21/13}$. Let us reproduce this data in the table below, but let us also compute the p_k and the q_k such that p_k/q_k is the kth convergent.

k	b_k	r_k	a_k	p_k	q_k
-2				0	1
-1				1	0
0	0	13	1	1	1
1	13	8	3	4	3
2	11	19	1	5	4
3	8	11	2	14	11
4	14	7	4	61	48
5	14	11	2	136	107
6	8	19	1	197	155
7	11	8	3	727	572
8	13	13	2	1651	1299

Let us verify Theorem 9.29 using $C = 13$ and $D = 21$. We have

$$13p_{-1}^2 - 21q_{-1}^2 = 13(1^2) - 21(0^2) = 13 = r_0$$

$$13p_0^2 - 21q_0^2 = 13(1^2) - 21(1^2) = -8 = -r_1$$

$$13p_1^2 - 21q_1^2 = 13(4^2) - 21(3^2) = 3 = r_2$$

$$13p_2^2 - 21q_2^2 = 13(5^2) - 21(4^2) = -11 = -r_3$$

In the proof of Theorem 9.25, we saw that if x_k denotes the kth complete quotient in the continued fraction expansion of $\sqrt{D/C}$, where C and D are integers such that $0 < C < D$ and $D/C \neq s^2$, then $x_{t+k} = x_k$ for all $k \geq 1$, where t denotes the length of the period of the continued fraction expansion. We now prove that the complete quotients, which are periodic, starting with x_1, also have period of length t.

Theorem 9.30

Let C, D be integers such that $0 < C < D$ and $D/C \neq s^2$. Let $\sqrt{D/C} = [a_0, \overline{a_1, a_2, \cdots, a_t}]$ for some $t \geq 1$. Let x_k be the kth complete quotient, that is, $x_k = (\sqrt{CD} + b_k)/r_k$ where b_k, r_k are defined as in Theorem 9.25. Then the x_k are periodic with period of length t, that is, $x_{t+k} = x_k$ for all $k \geq 1$, and t is the least such positive integer.

Proof: We have seen in the proof of Theorem 9.25 that $x_{t+k} = x_k$ for all $k \geq 1$. Therefore it suffices to show that if $0 < j < k$ and $x_j = x_k$, then $k > t$. Suppose that $0 < j < k$ and $x_j = x_k$. Then

$$\frac{\sqrt{CD} + b_j}{r_j} = \frac{\sqrt{CD} + b_k}{r_k}$$

so

$$r_k\sqrt{CD} + b_j r_k = r_j\sqrt{CD} + b_k r_j$$

This implies

$$((r_k - r_j)\sqrt{CD} = b_k r_j - b_j r_k$$

Since the left side of the equation is an integer, it follows that

$$r_k - r_j = 0$$

and

$$b_k = b_j$$

We can now say more about the complete quotients.

Theorem 9.31

If x_k and t are defined as in the statement of Theorem 9.30, $1 \leq j \leq t$, and $j < k$, then $x_k = x_j$ if and only if $k = j + th$ for some $h \geq 1$.

Proof: (Sufficiency) By Theorem 9.25 and the definition of t, we have $b_{t+j} = b_j$ and $r_{t+j} = r_j$ for all $j \geq 1$. By induction on h, we can show $b_{th+j} = b_j$ and $r_{th+j} = r_j$ for all $h \geq 1$. Since $x_k = (\sqrt{CD} + b_k)/r_k$, it follows that $x_{th+j} = x_j$ for all $h, j \geq 1$.

(Necessity) Let $k = th + r$ where $0 \leq r < t$. By hypothesis, $x_k = x_j$, that is, $x_{th+r} = x_j$. But Theorem 9.31 implies $x_{th+r} = x_r$, so $x_r = x_j$. If $r = 0$, then $x_{th} = x_0$, so

$$\frac{\sqrt{CD}}{r_0} = \frac{\sqrt{CD} + b_{th}}{r_{th}} \rightarrow r_{th}\sqrt{CD} = r_0\sqrt{CD} + r_0 b_{th} \rightarrow r_{th} = r_0 = C$$

Therefore $Cb_{th} = 0 \rightarrow b_{th} = 0$, an impossibility. Consequently, we have $1 \leq r < t$. The proof of Theorem 9.30 implies that if j, r are integers such that $0 < j$, $r < t$ and $x_r = x_j$, then $r = j$. Thus we are done. ∎

Referring to the continued fraction expansion of $\sqrt{D/C}$, where C and D are as in the statement of Theorem 9.30, we have seen that r_k is periodic, that is, $r_{t+k} = r_k$ for all $k \geq 0$. We now show that if $C = r_0$ is square-free, then r_k assumes the value C only at the end of each period.

Theorem 9.32

Let C, D, T, b_k, r_k, and a_k be defined as in Theorem 9.30. If C is square-free and $r_k = C$ where $k > 0$, then $k = th$ for some $h \geq 1$.

Proof: Since $r_k r_{k-1} = CD - b_k^2$ and $r_k = C$ by hypothesis, we have $C|b_k^2$. Since C is square-free by hypothesis, it follows that $C|b_k$, so b_k/C is an integer. Now

$$a_k = [x_k] = \left[\frac{\sqrt{CD} + b_k}{r_k}\right] = \left[\sqrt{\frac{D}{C}} + \frac{b_k}{C}\right] = \left[\sqrt{\frac{D}{C}}\right] + \frac{b_k}{C} = \frac{b_1}{C} + \frac{b_k}{C} = \frac{b_1 + b_k}{C}$$

so $Ca_k - b_k = b_1$. However $b_{k+1} = a_k r_k - b_k = Ca_k - b_k$, so $b_{k+1} = b_1$. Also, $r_{k+1} = (CD - b_{k+1}^2)/r_k = (CD - b_1^2)/C = (CD - b_1^2)/r_0 = r_1$. Therefore $x_{k+1} = x_1$. Now Theorem 9.31 implies $k + 1 = th + 1$, so that $k = th$ for some $h \geq 1$. ∎

For example, if we refer to the table on page 246, which lists the r_k associated with the continued fraction expansion of $\sqrt{21/13}$, we may note that $t = 8$ and $r_0 = r_8 = 13$, but $r_k \neq 13$ if $0 < k < 8$.

By virtue of Theorem 9.29, if C, D, p_k, q_k, r_k are as in Theorem 9.29, it is not surprising that the variable $Cp_{k-1}^2 - Dq_{k-1}^2$ is periodic. Let T be the length of the period. The following theorem relates T to t, the length of the period of the continued fraction expansion of $\sqrt{D/C}$.

Theorem 9.33   If C and D are integers such that $0 < C < D$ and $D/C \neq s^2$, and p_k/q_k is the kth convergent in the continued fraction expansion of $\sqrt{D/C}$, then $Cp_{k-1}^2 - Dq_{k-1}^2$ is periodic with period of length T, where

$$T = \begin{cases} t & \text{if } t \text{ is even} \\ 2t & \text{if } t \text{ is odd} \end{cases}$$

Proof: The identity $Cp_{k-1}^2 - Dq_{k-1}^2 = (-1)^k r_k$ arose in the proof of Theorem 9.29, so it suffices to investigate the behavior of $(-1)^k r_k$. This variable is periodic, with a period whose length is at most $2t$, since if $k = 2t + j$ where $0 \leq j < 2t$, then $(-1)^k r_k = (-1)^{2t+j} r_{2t+j} = (-1)^j r_j$. If T is the length of the period of $(-1)^k r_k$, we have $T \leq 2t$. Since $(-1)^T r_T = (-1)^0 r_0 = C$, we must have $2|T$. Since $r_T = C$, Theorem 9.32 implies that $t|T$. Therefore $LCM[2, t]|T$. If t is odd, then $LCM[2, t] = 2t$, so $2t|T$. However, recall that $T \leq 2t$, so $T = 2t$. If t is even, then $LCM[2, t] = t$, so $t|T$, hence $t \leq T$. If j is any non-negative integer, then $(-1)^{t+j} r_{t+j} = (-1)^j r_j$, so $T \leq t$. Therefore $T = t$.   ■

For example, let us develop the continued fraction expansion of $\sqrt{41}$, listing the relevant data in the following table:

k	b_k	r_k	a_k	p_k	q_k	$p_{k-1}^2 - 41q_{k-1}^2$
-2				0	1	
-1				1	0	
0	0	1	6	6	1	1
1	6	5	2	13	2	-5
2	4	5	2	32	5	5
3	6	1	12	397	62	-1
4	6	5	2	826	129	5
5	4	5	2	2049	320	-5
6	6	1	12	25414	3969	1

We have $\sqrt{41} = [6, \overline{2, 2, 12}]$ and $t = 3$, which is odd, so $T = 6$.

We have devoted some attention to the infinite continued fraction expansion of $\sqrt{D/C}$, where C and D are integers such that $0 < C < D$ and $D/C \neq s^2$. We have seen that these continued fractions are periodic (with initial nonperiodic term). More generally, let x be a positive irrational number. It can be shown that $x = [b_0, b_1, b_2, \cdots, b_{m-1}, \overline{a_1, a_2, \cdots, a_t}]$ if and only if x is a quadratic irrational number; that is, the continued fraction expansion of x is eventually periodic if and only if x satisfies a quadratic equation with integer coefficients. We conclude this chapter by introducing a generalization of periodic continued fractions.

Definition 9.5   Arithmetic Progression of Degree m

If the integer $m \geq 0$, let $g(x)$ be a function such that (1) $g(n)$ is a natural number for all $n \geq 0$, and (2) $g(x) = f(x)/m!$ where $f(x)$ is a polynomial of degree m with integer coefficients. Then the sequence $\{g(0), g(1), g(2), \cdots\}$ is called an *arithmetic progression of degree m*.

For example, if $g(x) = (x + 1)(x + 2)(x + 3)/6$, then we obtain the sequence $\{1, 4, 10, 20, 35, 56, \cdots\}$, an arithmetic progression of order three. If $g(x)$ is constant, then we obtain a constant sequence, which is an arithmetic progression of order zero.

Now suppose that we have k functions, $g_1(x), g_2(x), \cdots, g_k(x)$, all of which satisfy the conditions of Definition 9.5.

Definition 9.6   Hurwitz-Type Continued Fraction

If

$$x = [a_0, a_1, a_2, \cdots, a_{h-1}, g_1(0), g_2(0), \cdots, g_k(0), g_1(1), g_2(1), \cdots, g_k(1),$$

$$g_1(2), g_2(2), \cdots, g_k(2), \cdots]$$

then we say x is a *Hurwitz-type continued fraction*. We denote x more conveniently as

$$x = [a_0, a_1, a_2, \cdots, a_{h-1}, \overline{g_1(n), g_2(n), \cdots, g_k(n)}|_{n=0}^{\infty}]$$

The functions $g_i(x)$ are assumed to satisfy the conditions given in Definition 9.5. If each $g_i(x)$ is constant, then x is simply a periodic continued fraction. The following results, which we present without proof, are given in terms of Hurwitz-type continued fractions.

Theorem 9.34   If n is a natural number, then

$$\coth \frac{1}{n} = \frac{e^{2/n} + 1}{e^{2/n} - 1} = [n, 3n, 5n, 7n, 9n, \cdots] = [\overline{(2m + 1)n}|_{m=0}^{\infty}]$$

In particular, we have

$$\coth 1 = \frac{e^2 + 1}{e^2 - 1} = [1, 3, 5, 7, 9, \cdots] = [\overline{(2m+1)}|_{m=0}^{\infty}]$$

Theorem 9.35

$$e = [2, \overline{1, 2+2n, 1}|_{n=0}^{\infty}] = [2, 1, 2, 1, 1, 4, 1, 1, 6, 1, 1, 8, 1, \cdots]$$

Theorem 9.36   If $m \geq 2$, then

$$e^{1/m} = [\overline{1, (1+2n)m - 1, 1}|_{n=0}^{\infty}]$$
$$= [1, m-1, 1, 1, 3m-1, 1, 1, 5m-1, 1, 1, \cdots]$$

For the proofs of the last three theorems, we refer the interested reader to [10].

Section 9.6   Exercises

1. Given the periodic infinite continued fraction representation of each of the following irrationals, find the corresponding closed-form representation. (For example, we have seen that if $x = [\overline{1}]$, then $x = (1 + \sqrt{5})/2$.)

   (a) $[\overline{1, 2}]$

   (b) $[\overline{2, 1}]$

   (c) $[\overline{1, 2, 3}]$

   (d) $[1, \overline{1, 2}]$

2. Let x be a positive irrational number whose infinite continued fraction representation is purely periodic, that is, $x = [\overline{a_0, a_1, a_2, \cdots, a_{m-1}}]$ for some $m \geq 1$. Prove that x is a quadratic irrational number.

Review Exercises

1. Find continued fraction representations with integer terms (and $a_n > 1$) for each of the following:

   (a) $\frac{8}{9}$

   (b) $\frac{55}{23}$

   (c) $\frac{100}{43}$

(d) $\frac{123}{76}$

(e) $\frac{1000}{901}$

2. Find rational numbers having the following continued fraction expansions:

(a) [1,4,7,10,13]

(b) [2,4,6,8,10]

(c) [5,4,3,2]

(d) [8,6,4,2]

3. (a) Obtain the continued fraction expansion of $\sqrt{43}$, and (b) use your result from part (a) to estimate $\sqrt{43}$ to 3 decimals.

4. Prove that $[\overline{1,1,2}] = \frac{2+\sqrt{10}}{3}$.

5. Given that $e = [2, 1, 2, 1, 1, 4, 1, 1, 6, \cdots]$, estimate e to 3 decimals.

6. Find the infinite continued fraction expansions for each of the following:

(a) $\sqrt{7}$

(b) $\sqrt{12}$

(c) $\sqrt{20}$

(d) $\sqrt{28}$

(e) $\sqrt{34}$

(f) $\sqrt{127}$

7. Find the infinite continued fraction expansions for each of the following:

(a) $\frac{1+\sqrt{7}}{2}$

(b) $\frac{\sqrt{3}-1}{2}$

(c) $2 - \sqrt{3}$

8. Find the infinite continued fraction expansions for each of the following:

(a) $\sqrt{\frac{3}{2}}$

(b) $\sqrt{\frac{5}{3}}$

(c) $\sqrt{\frac{8}{5}}$

Chapter 10

Nonlinear Diophantine Equations

10.1 Introduction

A *Diophantine equation* is an equation in two or more integer unknowns. Such equations are named after Diophantus of Alexandria, who lived about 250 A.D. In Section 4.5, we learned how to solve linear Diophantine equations in two unknowns, that is, equations of the form $ax + by = c$ where a, b, c are given integers and x and y are unknown. In Section 4.9, we learned how to find all solutions of the Pythagorean equation

$$x^2 + y^2 = z^2$$

In this chapter, we will consider several additional types of nonlinear Diophantine equations, that is, Diophantine equations in which at least one unknown term has a degree of at least two.

10.2 Fermat's Last Theorem

In a letter to Mersenne that he wrote in 1637, Fermat claimed that the equation

$$x^n + y^n = z^n \tag{10.1}$$

has no solution in positive integers for any $n \geq 3$. Fermat did not provide a proof of this assertion, which became known as *Fermat's Last Theorem*. (Fermat claimed that he had found a proof while reading Bachet's *Diophantus* but had lacked sufficient space to write the proof in the margin of the book.)

The case $n = 2^k$ where $k \geq 2$ can be excluded as a result of Section 4.9, Exercise 7. Therefore, let $n = mp$, where p is an odd prime. If Equation 10.1 holds, then we have

$$(x^m)^p + (y^m)^p = (z^m)^p \tag{10.2}$$

Therefore we need only consider the case where $n = p$, an odd prime.

Many of the world's most accomplished mathematicians attempted to prove Fermat's Last Theorem, achieving only partial results. Finally, in 1995, Andrew Wiles of the Institute for Advanced Study in Princeton published a valid proof, based on the techniques of arithmetic algebraic geometry. Wiles's proof is 110 pages long and is far beyond the scope of this text.

10.3 Pell's Equation: $x^2 - Dy^2 = 1$

John Pell (1611–1685)
John Pell was an English mathematician who in the 1640s taught mathematics in Holland, at the universities of Amsterdam and Breda.

In a letter to Goldbach dated August 10, 1732, Euler attributed the equation

$$x^2 - Dy^2 = 1 \tag{10.3}$$

to Pell. The name has stuck, although most mathematicians consider this attribution to be erroneous.

Note that Equation 10.3 is really a one-parameter family of equations, the parameter being D. Without loss of generality, assume that x and y are non-negative. It is easily seen that for any D, $(x, y) = (1, 0)$ is a solution of 10.3, which we call the *trivial* solution. Furthermore, if D is a square, then the trivial solution is the unique solution. We therefore assume henceforth that D is a positive, nonsquare integer. We shall see that if $(x, y) = (a, b)$ is a solution of 10.3, then a/b is an approximation to $\sqrt{D}$.

Throughout the ages, many mathematicians have taken an interest in Pell's equation. Baudhayana (India, fourth century B.C.) approximated $\sqrt{2}$ with $17/12$ and $577/408$. These ratios correspond to solutions to Pell's equation with $D = 2$. Archimedes (third century B.C.) approximated $\sqrt{3}$ by $1351/780$, which corresponds to a solution of 10.3 with $D = 3$. Pell's equation is discussed at length in both Brahmagupta's *Brahma-Sputa-Siddhanta* (seventh century A.D.) and Bhaskara's *Bija-Ganita* (twelfth century A.D.).

In more recent times, Fermat sparked an interest in Pell's equation among European mathematicians. In a letter of February 1657 written to Frenicle de Bessy, Fermat claimed that equation 10.3 has infinitely many solutions. He challenged the mathematicians of England, Holland, and Germany to provide the proof.

At first, the English mathematician John Wallis (Professor at Oxford) played down the importance of the problem. Later in 1657, Wallis' colleague, Lord William Brouncker, found a constructive method to find infinitely many solutions of equation 10.3. His method, which depends on the continued fraction representation of $\sqrt{D}$, was published by Wallis. At this point, Wallis changed his attitude about the significance of the problem. In a congratulatory letter to Brouncker, he wrote: "You have shown that England's champions of wisdom are as strong as those of war!"

Brouncker's method was subsequently rediscovered by Euler. Lagrange noted that the earlier efforts of Wallis and Euler were incomplete. In 1766, Lagrange managed to fill the gaps by providing a completely rigorous proof of Fermat's assertion that (for any positive, nonsquare D), Pell's equation has infinitely many solutions. (For further details, see [8].)

We shall see that if C and D are square-free integers such that $0 < C < D$, and if E is an integer such that $|E| < \sqrt{CD}$, then all solutions, if any, of the generalized Pell's equation

$$Cx^2 - Dy^2 = E \qquad (10.4)$$

can be obtained from the convergents in the infinite continued fraction representation of $\sqrt{D/C}$.

Theorem 10.1

Let C, D be square-free integers such that $0 < C < D$. Let E be an integer such that $|E| < \sqrt{CD}$. Let r_k be defined as in Theorem 9.25. Let T be the length of the period of $(-1)^k r_k$. Then

(a) equation 10.4 has solutions if and only if there exists k such that $0 \leq k < T$ and $E = (-1)^k r_k$

(b) for each such E and each such k, all solutions of equation 10.4 are given by

$$x_{n,k} = p_{k-1+nT}$$

$$y_{n,k} = q_{k-1+nT}$$

where $n = 0, 1, 2, 3$, etc. and p_j/q_j is the jth convergent in the continued fraction representation of $\sqrt{D/C}$.

Proof of (a): (Sufficiency) If $E = (-1)^k r_k$ for some k such that $0 \le k < T$, then Theorem 9.29 implies that $(x, y) = (p_{k-1}, q_{k-1})$ is a solution of Equation 10.4.

(Necessity) If Equation 10.4 holds, then $|Cx^2 - Dy^2| = |E|$. Now $|E| < \sqrt{CD}$ by hypothesis, so $|Cx^2 - Dy^2| < \sqrt{CD}$; thus

$$|x^2 - \frac{D}{C}y^2| < \sqrt{\frac{D}{C}}$$

Let $r = \sqrt{\frac{D}{C}}$, so $|x^2 - r^2 y^2| < r$. Now we have

$$|\frac{x}{y} - r||x + yr| < \frac{r}{y} \rightarrow |\frac{x}{y} - r| < \frac{r}{y(ry + x)} \rightarrow |\frac{x}{y} - r| < \frac{1}{y(y + x/r)}$$

By Theorem 9.24, it suffices to prove that $y + \frac{x}{r} > 2y - 1$, that is, $\frac{x}{r} > y - 1$. This inequality is trivially true if $y \le 1$. By hypothesis, we have

$$\frac{Cx^2}{D} = y^2 + \frac{E}{D} > y^2 - \frac{\sqrt{CD}}{D}$$

that is, $(x/r)^2 > y^2 - (1/r)$. Since $0 < C < D$ by hypothesis, we have $r > 1$, so $y^2 - (1/r) > y^2 - 1 > (y-1)^2$. Therefore $(x/r)^2 > (y-1)^2$, so $x/r > y-1$, and we are done. Therefore, there exists k such that $(x, y) = (p_{k-1}, q_{k-1})$. Equation 10.4 and Theorem 9.29 imply that $E = (-1)^k r_k$ for some k. Since $Cp_{k-1}^2 - Dq_{k-1}^2$ is periodic with period of length T, according to Theorem 9.23, we may further specify that $0 \le k < T$. We call this solution a *fundamental solution*. ■

Proof of (b): In the proof of (a), we saw that if Equation 10.4 holds, then $(x, y) = (p_{j-1}, q_{j-1})$ for some $j \ge 1$. Let $k = j - T[j/T]$. Then $0 \le k < T$ and $Cp_{k-1}^2 - Dq_{k-1}^2 = (-1)^k r_k = (-1)^j r_j = E$ by hypothesis and by definition of T, so $(x, y) = (p_{k-1}, q_{k-1})$ is a fundamental solution of Equation 10.4 associated with the solution $(x, y) = (p_{j-1}, q_{j-1})$. By virtue of Theorems 9.29 and 9.33, each fundamental solution, which we now denote $(x_{0,k}, y_{0,k}) = (p_{k-1}, q_{k-1})$, generates an infinite family of solutions: $(x_{n,k}, y_{n,k}) = (p_{k-1+nT}, q_{k-1+nT})$ where $n = 0, 1, 2, 3, \cdots$. Also, there are no additional solutions to Equation 10.4. ■

For example, consider the equation

$$2x^2 - 5y^2 = -3 \tag{10.5}$$

Since $|-3| = 3 < \sqrt{10}$, Theorem 10.1 is applicable. We therefore develop the continued fraction expansion of $\sqrt{5/2}$, following the notation of Theorem 9.25. We list the results in Table 10.1 on the next page. Note that $t = 3$ and $T = 6$.

Table 10.1

k	b_k	r_k	a_k	p_k	q_k	$(-1)^k r_k$
-2				0	1	
-1				1	0	
0	0	2	1	1	1	2
1	2	3	1	2	1	-3
2	1	3	1	3	2	3
3	2	2	2	8	5	-2
4	2	3	1	11	7	3
5	1	3	1	19	12	-3
6	2	2	2	49	31	2
7	2	3	1	68	43	-3
8	1	3	1	117	74	3
9	2	2	2	302	191	-2
10	2	3	1	419	265	3
11	1	3	1	721	456	-3

We obtain two fundamental solutions of equation 10.5, namely, for $k = 1$, we get $(x_{0,1}, y_{0,1}) = (p_0, q_0) = (1, 1)$; for $k = 5$, we get $(x_{0,5}, y_{0,5}) = (p_4, q_4) = (11, 7)$. Therefore, all solutions of equation 10.5 are given by the two families, $(x_{n,1}, y_{n,1}) = (p_{6n}, q_{6n})$ and $(x_{5,n}, y_{5,n}) = (p_{5+6n}, q_{5+6n})$ where the integer $n \geq 0$.

In particular, we may use Theorem 10.1 to obtain all solutions of Pell's equation, that is, Equation 10.3.

Theorem 10.2

Let D be a positive nonsquare integer. Let t be the length of the period of the continued fraction expansion of $\sqrt{D}$. Then Pell's equation has infinitely many solutions, all given as follows:

(i) if t is even, then $(x_n, y_n) = (p_{nt-1}, q_{nt-1})$ for $n \geq 0$

(ii) if t is odd, then $(x_n, y_n) = (p_{2nt-1}, q_{2nt-1})$ for $n \geq 0$.

Proof: We apply Theorem 10.1 with $C = E = 1$. Therefore if T is the length of the period of $(-1)^k r_k$, there exists k such that $0 \leq k < T$ and $(-1)^k r_k = 1$. This implies that $r_k = 1 = C$ and $2|k$. In general, if $(-1)^j r_j = 1$, then Theorem 9.32 implies $t|j$. Therefore $t|k$. Also, since $2|k$, we have $[2, t]|k$. If t is odd, then $[2, t] = 2t = T$, so $T|k$. This implies that $k = 0$. If t is even, then Theorem 9.33 implies that $T = t$, so again $T|k$, hence $k = 0$. Since we obtain a single family of solutions, following Theorem 10.1, we may denote these solutions as $(x_n, y_n) = (p_{nT-1}, q_{nT-1})$ where $n \geq 0$. The conclusion now follows by appeal to Theorem 9.33. ■

For example, suppose we wish to solve the Pell's equation

$$x^2 - 7y^2 = 1 \qquad (10.6)$$

Now $\sqrt{7} = [2, \overline{1, 1, 1, 4}]$, so $t = T = 4$, and our solutions are $(x_n, y_n) = (p_{4n-1}, q_{4n-1})$ where $n \geq 0$. Let us compute some convergents to $\sqrt{7}$ using the recursion relations of Theorem 9.5. We list our results in Table 10.2.

Table 10.2

k	a_k	p_k	q_k	k	a_k	p_k	q_k
-2		0	1	5	1	45	17
-1		1	0	6	1	82	31
0	2	2	1	7	1	127	48
1	1	3	1	8	4	590	223
2	1	5	2	9	1	717	271
3	1	8	3	10	1	1307	494
4	4	37	14	11	1	2024	765

From Table 10.2, we see that the first few solutions of equation 10.6 are

$$x_0 = p_{-1} = 1 \qquad\qquad y_0 = q_{-1} = 0$$

$$x_1 = p_3 = 8 \qquad\qquad y_1 = q_3 = 3$$

$$x_2 = p_7 = 127 \qquad\qquad y_2 = q_7 = 48$$

$$x_3 = p_{11} = 2024 \qquad\qquad y_3 = q_{11} = 765$$

As a second example, let us solve the Pell's equation

$$x^2 - 41y^2 = 1 \qquad (10.7)$$

Now $\sqrt{41} = [6, \overline{2, 2, 12}]$, so $t = 3$, $T = 6$ and our solutions are $(x_n, y_n) = (p_{6n-1}, q_{6n-1})$ where $n \geq 0$. Let us compute some convergents to $\sqrt{41}$. We list our results in Table 10.3.

From this table, we see that the first few solutions of Equation 10.7 are

$$x_0 = p_{-1} = 1 \qquad\qquad y_0 = q_{-1} = 0$$

$$x_1 = p_5 = 2049 \qquad\qquad y_1 = q_5 = 320$$

$$x_2 = p_{11} = 8396801 \qquad\qquad y_2 = q_{11} = 1311360$$

Table 10.3

k	a_k	p_k	q_k
-2		0	1
-1		1	0
0	6	6	1
1	2	13	2
2	2	32	5
3	12	397	62
4	2	826	129
5	2	2049	320
6	12	25414	3969
7	2	52877	8258
8	2	131168	20485
9	12	1626893	254078
10	2	3384954	528641
11	2	8396801	1311360

The equation

$$x^2 - Dy^2 = -1 \qquad (10.8)$$

is called the *associated Pell's equation*. The following theorem, whose proof is left as an exercise, concerns the solutions of Equation 10.8.

Theorem 10.3

Let D be a positive nonsquare integer. Let t be the length of the period of the continued fraction expansion of $\sqrt{D}$. If t is even, then Equation 10.8 has no solutions. If t is odd, then Equation 10.8 has infinitely many solutions, all given by $(x_n, y_n) = (p_{(2n-1)t-1}, q_{(2n-1)t-1})$.

Proof: Exercise.

For example, the equation $x^2 - 7y^2 = -1$ has no solutions, since $\sqrt{7} = [2, \overline{1, 1, 1, 4}]$, so $t = 4$. (We could also reach this conclusion by considering quadratic residues (mod 7).)

Again, the equation $x^2 - 41y^2 = -1$ has solutions since $\sqrt{41} = [6, \overline{2, 2, 12}]$, so $t = 3$. The first two such solutions are

$$(x_1, y_1) = (p_2, q_2) = (32, 5) \quad \text{and} \quad (x_2, y_2) = (p_8, q_8) = (131168, 20485)$$

Next, we define the fundamental solution of Pell's equation, and we develop a way to obtain all solutions from the fundamental solution. This method will make use of two-by-two matrices and will use continued fractions to a lesser extent.

Definition 10.1

If D is a positive nonsquare integer, let T be the length of the period of $(-1)^k r_k$ in the continued fraction expansion of $\sqrt{D}$. Then $(x_1, y_1) = (p_{T-1}, q_{T-1})$ is called the *fundamental solution* of Equation 10.4.

For example, we have just seen that the fundamental solution of Equation 10.6 is $(x_1, y_1) = (p_3, q_3) = (8, 3)$. Also, the fundamental solution of Equation 10.7 is $(x_1, y_1) = (p_5, q_5) = (2049, 320)$.

In order to show how all solutions of Equation 10.4 can be obtained from the fundamental solution, we must make a brief excursion into algebraic number theory.

Definition 10.2

If D is a positive nonsquare integer, let $S(D)$ denote the set of all real numbers $a + b\sqrt{D}$, where a and b are integers.

For example, $S(7) = \{a + b\sqrt{7} : a, b \in Z\}$. We note that $S(D)$ is closed under multiplication. That is, we have

Theorem 10.4   If $z_1 \in S(D)$ and $z_2 \in S(D)$, then $z_1 z_2 \in S(D)$.

Proof: Let $z_i = a_i + b_i \sqrt{D}$ where $i \in \{1, 2\}$. Then

$$z_1 z_2 = (a_1 + b_1 \sqrt{D})(a_2 + b_2 \sqrt{D})$$
$$= (a_1 a_2 + D b_1 b_2) + (a_1 b_2 + a_2 b_1)\sqrt{D} \in S(D). \quad \blacksquare$$

For example, if $z_1 = 2 + 3\sqrt{7}$ and $z_2 = 3 + 5\sqrt{7}$, then $z_1 z_2 = 111 + 19\sqrt{7}$. If $z \in S(D)$, it is convenient to assign to z an integer that we call its *norm*, which we define as follows:

Definition 10.3

If $z = a + b\sqrt{D}$, let $N(z) = a^2 - Db^2$. The integer $N(z)$ is called the *norm* of z.

For example, $N(2+3\sqrt{7}) = 2^2 - 7(3^2) = 4 - 63 = -59$. The norm of z is a completely multiplicative function of z, as in the next theorem.

Theorem 10.5 If $z_i \in S(D)$, where $i \in \{1, 2\}$, then $N(z_1 z_2) = N(z_1)N(z_2)$

Proof:

$$N(z_1)N(z_2) = (a_1^2 - Db_1^2)(a_2^2 - Db_2^2)$$

$$= a_1^2 a_2^2 - Da_1^2 b_2^2 - Da_2^2 b_1^2 + D^2 b_1^2 b_2^2$$

$$= (a_1^2 a_2^2 + D^2 b_1^2 b_2^2 + 2Da_1 a_2 b_1 b_2) - D(a_1^2 b_2^2 + a_2^2 b_1^2 + 2a_1 a_2 b_1 b_2)$$

$$= (a_1 a_2 + Db_1 b_2)^2 - D(a_1 b_2 + a_2 b_1)^2$$

$$= N(z_1 z_2) \quad \blacksquare$$

In order to establish the connection between Pell's equation and the set $S(D)$, we need to define what are known as *units*.

Definition 10.4

Let $u = a + b\sqrt{D} \in S(D)$. We say that u is a *unit* if $N(u) = 1$, that is, if $a^2 - Db^2 = 1$.

For example, $8 + 3\sqrt{7}$ is a unit in $S(7)$, since $N(8+3\sqrt{7}) = 8^2 - 7(3^2) = 64 - 63 = 1$. Note that $u = a + b\sqrt{D}$ is a unit in $S(D)$ if and only if $(x, y) = (a, b)$ is a solution of the Pell's equation $x^2 - Dy^2 = 1$.

We now develop some properties of units. Let $u^{-1} = 1/u$.

Theorem 10.6 If u is a unit in $S(D)$, then so is u^{-1}.

Proof: First, we must verify that $u^{-1} \in S(D)$. If $u = a + b\sqrt{D}$, then

$$u^{-1} = \frac{1}{u} = \frac{1}{a + b\sqrt{D}} = \frac{a - b\sqrt{D}}{a^2 - Db^2} = \frac{a - b\sqrt{D}}{1} = a - b\sqrt{D}$$

so $u^{-1} \in S(D)$. Now $N(u^{-1}) = N(u^{-1})1 = N(u^{-1})N(u) = N(u^{-1}u) = N(1) = 1$ so u^{-1} is a unit in $S(D)$. $\blacksquare$

For example, if $u + 8 + 3\sqrt{7}$, then $u^{-1} = 8 - 3\sqrt{7}$. We have just seen that the inverse of a unit is a unit. We are about to see that the product of two units is also a unit.

Theorem 10.7   If u_1 and u_2 are units in $S(D)$, then so is u_1u_2.

Proof: By hypothesis and Theorem 10.4, $u_1u_2 \in S(D)$. By Theorem 10.5, $N(u_1u_2) = N(u_1)N(u_2) = 1^2 = 1$, so u_1u_2 is a unit in $S(D)$. ∎

For example, $u_1 = 3 + 2\sqrt{2}$ and $u_2 = 17 + 12\sqrt{2}$ are both units in $S(2)$, since $N(u_1) = 3^2 - 2(2^2) = 1$ and $N(u_2) = 17^2 - 2(12^2) = 1$. Now $u_1u_2 = 99 + 70\sqrt{2}$, which is also a unit in $S(2)$, since $N(u_1u_2) = 99^2 - 2(70^2) = 1$.

If $u = r + s\sqrt{D}$ is a unit in $S(D)$, we say that r is the *rational part* and s is the *irrational part* of u. The units of $S(D)$ are real numbers and therefore may be compared in size. If we look at units whose rational and irrational parts are both nonnegative, we can say more.

Theorem 10.8   Let $u_1 = a_1 + b_1\sqrt{D}$ and $u_2 = a_2 + b_2\sqrt{D}$ be units in $S(D)$ such that a_1, a_2, b_1, b_2 are all nonnegative. Then the following are equivalent: (i) $a_1 < a_2$, (ii) $b_1 < b_2$, and (iii) $u_1 < u_2$.

Proof: Since

$$a_1^2 - Db_1^2 = N(u_1) = 1 = N(u_2) = a_2^2 - Db_2^2 \to a_1^2 - a_2^2 = D(b_1^2 - b_2^2)$$

it follows that (i) and (ii) are equivalent. If $a_1 < a_2$, then also since $b_1 < b_2$, it follows that $u_1 < u_2$. Therefore $u_2 \geq u_1 \to a_2 \geq a_1$. But if $u_1 = u_2$, then $a_1 + b_1\sqrt{D} = a_2 + b_2\sqrt{D}$, so $a_1 - a_2 = (b_2 - b_1)\sqrt{D}$, hence $a_1 = a_2$ and $b_1 = b_2$. It now follows that $u_1 > u_2 \to a_1 > a_2$, or equivalently, $u_1 < u_2 \to a_1 < a_2$. Therefore (i) and (iii) are equivalent, so we are done. ∎

We will soon see that all units in $S(D)$ with nonnegative rational and irrational parts are powers of a single generating unit. First, however, we need the following lemma.

▶ **Lemma 10.1** *f $u = a + b\sqrt{D}$ is a unit in $S(D)$ such that $u > 1$, then $a \geq 2$ and $b \geq 1$.*

Proof: By hypothesis, we have

$$(a - b\sqrt{D})(a + b\sqrt{D}) = a^2 - Db^2 = N(u) = 1$$

and $a + b\sqrt{D} = u > 1$, so $0 < a - b\sqrt{D} < 1$. Therefore $a - 1 < b\sqrt{D} < a$. If $a \leq 0$, then $b < 0$, so $u = a + b\sqrt{D} < 0$, contrary to hypothesis. If

$a = 1$, then $1 - Db^2 = 1$, so $b = 0$, and hence $u = 1$, contrary to hypothesis. Therefore $a \geq 2$. This implies that $1 < b\sqrt{D}$, so $b \geq 1$. ∎

Theorem 10.9 Let U be the set of units u of $S(D)$ such that $u > 1$. Then there exists $\theta \in U$ such that for all $v \in U$ there exists $k \geq 1$ such that $v = \theta^k$.

Proof: Let t be the length of the period of the continued fraction expansion of $\sqrt{D}$. Let

$$T = \begin{cases} t & \text{if } t \text{ is even} \\ 2t & \text{if } t \text{ is odd} \end{cases}$$

Theorem 10.2 implies $u = p_{T-1} + q_{T-1}\sqrt{D} \in U$, so U is nonempty. Let A be the set of all natural numbers r such that $r + s\sqrt{D} = u \in U$. Since U is non empty, Lemma 10.1 implies that A is also non empty. By the well-ordering principle, A has a least element, a. Let $\theta = a + b\sqrt{D}$ be the corresponding element of U. Now let v belong to U. Since $v > 1$ and $\theta > 1$, there exists $k \geq 1$ such that $\theta^k \leq v < \theta^{k+1}$. If $\theta^k < v$, then $1 < \theta^{-k}v < \theta$. Let $w = \theta^{-k}v$. Theorems 10.6 and 10.7 imply $w \in U$. However, $w < \theta$, which contradicts the definition of θ. Therefore $v = \theta^k$. ∎

For example, we saw earlier that $u_1 = 3 + 2\sqrt{2}$ and $u_2 = 17 + 12\sqrt{2}$ are units in $S(2)$. One can show that if $a + b\sqrt{D}$ is a unit in $S(D)$, then $a \not\equiv D$ (mod 2). Therefore, if U is the set of units u in $S(2)$ such that $u > 1$ and $\theta = a + b\sqrt{2}$ generates U, it follows that a is odd. Now Lemma 10.1 implies $a \geq 3$, so Theorem 10.8 implies $\theta = u_1$. One can easily verify that $\theta^2 = u_2$.

The following theorem enables us to obtain all solutions of Pell's equation from the fundamental solution.

Theorem 10.10 If D is a positive nonsquare integer, and if (x_1, y_1) is the fundamental solution to Equation 10.3, $x^2 - Dy^2 = 1$, then all solutions with $x > 0$ and $y > 0$ are given by (x_n, y_n), where $x_n + y_n\sqrt{D} = (x_1 + y_1\sqrt{D})^n$ with $n \geq 1$.

Proof:

$$x_n^2 - Dy_n^2 = N(x_n + y_n\sqrt{D}) = N((x_1 + y_1\sqrt{D})^n)$$

$$= (N(x_1 + y_1\sqrt{D}))^n = 1^n = 1$$

so $(x, y) = (x_n, y_n)$ is a solution to Equation 10.4. If $(x, y) = (a, b)$ is a solution to Equation 10.4 such that $a > 0$ and $b > 0$, then $u = a + b\sqrt{D}$ is a unit in $S(D)$ and $u > 1$. Therefore Theorem 10.9 implies that $a + b\sqrt{D} = (x_1 + y_1\sqrt{D})^n = x_n + y_n\sqrt{d}$ for some $n \geq 1$. ∎

For example, we have seen that in $S(2)$, we have $(x_1, y_1) = (3, 2)$. Now

$$x_2 + y_2\sqrt{2} = (x_1 + y_1\sqrt{2})^2 = (3 + 2\sqrt{2})^2 = 17 + 12\sqrt{2}$$

so $(x_2, y_2) = (17, 12)$. Likewise, we have

$$x_3 + y_3\sqrt{2} = (x_1 + y_1\sqrt{2})^3 = (3 + 2\sqrt{2})^3 = 99 + 70\sqrt{2}$$

so $(x_3, y_3) = (99, 70)$. In practice, using Theorem 10.10 to compute the solutions to Equation 10.3 would be laborious. The following theorem, which makes use of two-by-two matrices, offers a shortcut, though.

Theorem 10.11　Let D be a positive nonsquare integer. Let $(x_0, y_0) = (1, 0)$ be the trivial solution to equation 10.3. Then all solutions to Equation 10.3 in nonnegative integers may be expressed by each of the two following forms, where the integer $n \geq 1$:

(a)
$$\begin{bmatrix} x_n \\ y_n \end{bmatrix} = \begin{bmatrix} x_1 & Dy_1 \\ y_1 & x_1 \end{bmatrix} \begin{bmatrix} x_{n-1} \\ y_{n-1} \end{bmatrix}$$

(b)
$$\begin{bmatrix} x_n \\ y_n \end{bmatrix} = \begin{bmatrix} x_1 & Dy_1 \\ y_1 & x_1 \end{bmatrix}^n \begin{bmatrix} 1 \\ 0 \end{bmatrix}$$

Proof: Since (b) follows from (a), it suffices to prove (a). Theorem 10.10 implies that

$$x_n + y_n\sqrt{D} = (x_1 + y_1\sqrt{D})^n$$
$$= (x_1 + y_1\sqrt{D})(x_1 + y_1\sqrt{D})^{n-1}$$
$$= (x_1 + y_1\sqrt{D})(x_{n-1} + y_{n-1}\sqrt{D})$$
$$= (x_1x_{n-1} + Dy_1y_{n-1}) + (x_1y_{n-1} + x_{n-1}y_1)\sqrt{D}$$

Therefore, we have

$$x_n = x_1x_{n-1} + Dy_1y_{n-1} \quad \text{and} \quad y_n = x_1y_{n-1} + x_{n-1}y_1$$

which implies (a). ∎

For example, if $D = 7$, we saw previously that the fundamental solution to Equation 10.4 is $(x_1, y_1) = (8, 3)$. Using Theorem 10.11(a), we have

$$\begin{bmatrix} x_2 \\ y_2 \end{bmatrix} = \begin{bmatrix} x_1 & Dy_1 \\ y_1 & x_1 \end{bmatrix} \begin{bmatrix} x_1 \\ y_1 \end{bmatrix} = \begin{bmatrix} 8 & 21 \\ 3 & 8 \end{bmatrix} \begin{bmatrix} 8 \\ 3 \end{bmatrix} = \begin{bmatrix} 127 \\ 48 \end{bmatrix}$$

$$\begin{bmatrix} x_3 \\ y_3 \end{bmatrix} = \begin{bmatrix} x_1 & Dy_1 \\ y_1 & x_1 \end{bmatrix} \begin{bmatrix} x_2 \\ y_2 \end{bmatrix} = \begin{bmatrix} 8 & 21 \\ 3 & 8 \end{bmatrix} \begin{bmatrix} 127 \\ 48 \end{bmatrix} = \begin{bmatrix} 2024 \\ 765 \end{bmatrix}$$

In general, to solve a Pell's equation, one must first use the continued fraction expansion of $\sqrt{D}$ to obtain the fundamental solution and then obtain all other solutions via Theorem 10.11.

For example, in order to solve $x^2 - 6y^2 = 1$, we begin by obtaining the continued fraction expansion of $\sqrt{6}$. (See Table 10.4.)

Table 10.4

k	b_k	r_k	a_k	p_k	q_k
-2				0	1
-1				1	0
0	0	1	2	2	1
1	2	2	2	5	2
2	2	1	4	22	9
3	2	2	2	49	20

We see that $\sqrt{6} = [2, \overline{2, 4}]$, so $T = t = 2$. Now the fundamental solution to this Pell's equation is $(x_1, y_1) = (p_{T-1}, q_{T-1}) = (p_1, q_1) = (5, 2)$. Therefore, by Theorem 10.11, all nonnegative solutions of $x^2 - 6y^2 = 1$ are given by

$$\begin{bmatrix} x_n \\ y_n \end{bmatrix} = \begin{bmatrix} 5 & 12 \\ 2 & 5 \end{bmatrix}^n \begin{bmatrix} 1 \\ 0 \end{bmatrix}$$

Section 10.3 Exercises

1. For each of the following values of D, find the first 2 nontrivial solutions to the corresponding Pell's equation.

   (a) 5

   (b) 10

   (c) 13

   (d) 39

2. Prove Theorem 10.3.

3. Does the equation $x^2 - 15y^2 = -1$ have any solutions? Explain using

   (a) continued fractions

   (b) quadratic residues

4. Prove that if (x_1, y_1) is the fundamental solution to the associated Pell's equation $x^2 - Dy^2 = -1$, then all solutions are given by

$$\begin{bmatrix} x_n \\ y_n \end{bmatrix} = \begin{bmatrix} x_1 & Dy_1 \\ y_1 & x_1 \end{bmatrix}^{2n-1} \begin{bmatrix} 1 \\ 0 \end{bmatrix}$$

5. Prove that the equation: $x^4 - 2y^2 = 1$ has only the trivial solutions $(x, y) = (\pm 1, 0)$ using

   (a) continued fractions

   (b) other considerations

6. Let t be the length of the period of the continued fraction expansion of $\sqrt{D}$, where D is a positive nonsquare integer. Prove that if D has a prime factor p such that $p \equiv 1 \pmod 4$, then $2 | t$.

7. Find the first 3 solutions in positive integers to the generalized Pell's equation $3x^2 - 5y^2 = -2$.

8. Prove that there are infinitely many integers n such that each of n, $n+1$, $n+2$ is a sum of two squares.

Section 10.3   Computer Exercises

9. Write a computer program to find the first 20 solutions to the equation $x^2 - 11y^2 = 1$ using Theorem 10.11(a). (The fundamental solution is $(x_1, y_1) = (10, 3)$.)

10.4   Mordell's Equation: $x^3 = y^2 + k$

In 1621, the French mathematician Bachet de Meziriac stated that if $k = 2$, then the equation

$$x^3 = y^2 + k \tag{10.9}$$

has the unique positive integer solution $(x, y) = (3, 5)$. In 1657 Fermat stated that if $k = 4$, then equation 10.9 has only two solutions: $(x, y) = (2, 2)$ and $(5, 11)$. Two centuries elapsed before it was possible to verify the validity of these assertions. In 1738 Euler proved that if $k = -1$, then $(x, y) = (2, 3)$ is the unique solution to Equation 10.9. In 1869 Lebesgue proved that if $k = -7$, then Equation 10.9 has no solution. We give this proof in Theorem 10.12. In 1930 the Norwegian mathematician Nagell proved that if $k = -17$, then Equation 10.9 has 8 solutions. Louis J. Mordell (1888–1972) was Sadlerian Professor of Mathematics at Cambridge University in England. He wrote extensively on Diophantine equations in general, and on Equation 10.9 in particular. His name is now associated with Equation 10.9. Note that Equation 10.9 is a one-parameter family of equations, k being the parameter.

Mordell's equation is inhomogeneous, that is to say, the variable terms in the equation do not all have the same degree. As a consequence, Mordell's equation is more difficult to solve for most values of k than the other Diophantine equations considered earlier in this chapter. Usually, the solution requires the use of algebraic number theory, which is not treated in this text. The cases $k = -7$ and $k = 16$, however, can be handled by elementary methods, as we show below.

Theorem 10.12 The equation

$$x^3 = y^2 - 7 \tag{10.10}$$

has no solution in integers.

Proof: Assume the contrary. If $2|x$, then $y^2 \equiv 7 \pmod 8$, an impossibility. If $x \equiv 3 \pmod 4$, then $x^3 \equiv 3 \pmod 4$, so $y^2 \equiv 2 \pmod 4$, an impossibility. Therefore $x \equiv 1 \pmod 4$, so $x + 2 \equiv 3 \pmod 4$. Now Equation 10.10 implies $x^3 + 8 = y^2 + 1$, so $(x + 2)(x^2 - 2x + 4) = y^2 + 1$. Since $x + 2 \equiv 3 \pmod 4$, there exists a prime, p, such that $p \equiv 3 \pmod 4$ and $p|(x + 2)$. However, then $p|(y^2 + 1)$, which contradicts Theorem 4.12. ∎

Theorem 10.13 The equation

$$x^3 = y^2 + 16 \tag{10.11}$$

has no solution in integers.

Proof: Assume the contrary. Then x and y have the same parity. If x and y are both even, then $x = 2u$ and $y = 2v$ for some integers u, v. If we substitute these values into Equation 10.11, we obtain $8u^3 = 4v^2 + 16$. Therefore $8|4v^2$, so $v = 2t$. Thus $8u^3 = 16t^2 + 16$. Therefore $16|8u^3$, so $u = 2w$. Thus $64w^3 = 16t^2 + 16$. This implies $4w^3 = t^2 + 1$, so $t^2 \equiv -1 \pmod 4$, an impossibility.

If x and y are both odd, then $x^2 \equiv y^2 \equiv 1 \pmod{8}$, so Equation 10.11 implies $x \equiv 1 \pmod{8}$. Also, Equation 10.11 implies

$$y^2 + 8 = x^3 - 8 = (x-2)(x^2 + 2x + 4)$$

It is easily seen that Equation 10.11 has no solution with $x = 3$. Therefore, $x - 2$ has an odd prime factor p. Since $p|(y^2 + 8)$, we have $1 = (\frac{-8}{p}) = (\frac{-2}{p})$. This implies that $p \equiv 1 \pmod{8}$ or $p \equiv 3 \pmod{8}$. Since $x - 2$ is the product of such primes, it follows that $x - 2 \equiv 1 \pmod{8}$ or $x - 2 \equiv 3 \pmod{8}$, hence $x \equiv 3 \pmod{8}$ or $x \equiv 5 \pmod{8}$, an impossibility.   ■

Section 10.4   Exercises

1. Prove that equation 10.9 has no solutions for the following cases:

   (a) $k = 96$

   (b) $k = -215$

Review Exercises

1. Show why it is true that if $n \geq 2$, then the equation $x^{2^n} + y^{2^n} = z^{2^n}$ has no solution in positive integers.

2. For each of the following values of D, find the first two nontrivial solutions to the Pell's equation: $x^2 - Dy^2 = 1$.

   (a) 7

   (b) 14

   (c) 19

   (d) 23

3. For each of the following values of D, either (i) find the first two solutions to the associated Pell's Equation, $x^2 - Dy^2 = -1$; or (ii) explain why no such solutions exist.

   (a) 5

   (b) 10

   (c) 23

   (d) 29

4. Find four solutions in positive integers to the equation $5x^2 - 7y^2 = -2$.

5. Show that the equation $x^3 = y^2 - 999$ has no solution.

6. Show that the equation $x^3 = y^2 + 80$ has no solution.

7. Show that the equation $x^5 = y^2 + 40$ has no solution.
   (Hint: Re-examine the proof of Theorem 10.13.)

Chapter 11

Computational
Number Theory

11.1 Introduction

Computational number theory is the newest branch of number theory. It has developed rapidly in the last several decades, as ever-faster computers have become available to scientists and engineers. Computational number theory is *not* pure mathematics, although it makes ample use thereof. It is a blend of mathematics with computer science, with some of the flavor of engineering thrown in.

An *algorithm* is a specific procedure that accepts a given input and produces a desired output. Computational number theory takes a hard look at the algorithms of elementary number theory. Some of these algorithms require too much computer time or computer memory for a given sufficiently large input. Such algorithms are called *computationally infeasible* and are therefore rejected. An example of a computationally infeasible algorithm is trial division as a test of primality for large numbers. Computational number theory seeks ever more efficient algorithms, that is, algorithms that require less time or memory than algorithms currently in use. Computational number theory is closely related to that branch of computer science known as *complexity theory*.

It is worth mentioning that computational number theory is an *experimental* science with a very tolerant viewpoint such as one finds in engineering. That is to say, an algorithm may be considered useful that works for some, but not all, inputs of a given size. Furthermore, many of the algorithms of computational number theory are *nondeterministic*, that is, they require the use of random (actually pseudorandom) numbers. The emphasis is on what works. It is not unusual in the literature of computational number theory to find the phrase "the cost of factoring an integer." Computational number theory is also subject to transient influences, such as the cost and availability of certain computer hardware items.

Computational number theory has its theorems, but it is largely concerned with algorithms. Just as today's computers have rendered obsolete those of yesterday, and will surely themselves be rendered obsolete by those of tomorrow, it is reasonable to expect that many of the algorithms yet to come will make museum pieces of those presently in use.

The achievements of computational number theory include fast primality tests for large integers and less rapid but often workable methods for the factorization of large composite integers. An important application of computational number theory has been to *cryptology*, which is discussed in Chapter 12. We begin the study of computational number theory by introducing the concept of *computational feasibility*.

Definition 11.1

An algorithm is said to be *computationally feasible* for a given input if it can be executed in a reasonable amount of time and requires no more than a reasonable amount of computer memory.

"A reasonable amount of time" means seconds, minutes, hours, days, weeks, or even months, but *not* years or centuries. An algorithm which, for a given input, requires an excessive amount of time or space is called *computationally infeasible*. Some algorithms are computationally feasible for small inputs, but computationally infeasible for large inputs. This is the case for trial division as a test of primality. Other algorithms, such as Euclid's algorithm for finding the greatest common divisor of two natural numbers, are computationally feasible for all inputs.

In Section 2.4, we presented the binary number system. This is the system by which numbers are represented in computers. Recall that each natural number has a binary representation that consists of a string of 0's and 1's. For example, $34_{10} = 100010_2$. The 0's and 1's that appear are called the *binary digits* or *bits*. We define a *bit operation* as an addition, subtraction, or multiplication of two 1-bit integers, the division of a 2-bit integer by a 1-bit integer, or a one-place shift of a binary integer. Computer time is usually estimated in terms of the number of bit operations needed to execute an algorithm. In the 1970s, the noted computer scientist Donald

Knuth introduced the custom of using "big oh" notation in making such estimates. "Big oh" notation had previously been used in analytic number theory.

Definition 11.2   "**Big Oh**" **notation.**

Let $f(x)$ be a function defined on the reals. Suppose there exists a constant $k > 0$ and a function $g(x) > 0$ such that $|f(x)| < kg(x)$ for all sufficiently large x. Then we say "$f(x)$ is big oh of $g(x)$" and we write $f(x) = O(g(x))$.

For example, $x + \sin x = O(x)$, since $|x + \sin x| \le |x| + |\sin x| \le |x| + 1 \le 2x$ if $x \ge 1$. Generally, if $f(x) = O(x^n)$ for some n, we seek the least n for which this is true. Note that if $f(x) = O(1)$, then $f(x)$ is bounded. We leave as exercises the proofs of the two following theorems.

Theorem 11.1   If $f(x) = O(g(x))$ and c is any constant, then $cf(x) = O(g(x))$.

Proof: Exercise.

For example, let $f(x) = x^2 + x$. Then $f(x) = O(x^2)$ since $x^2 + x \le 2x^2$ if $x \ge 1$. Let $c = 2$. Now $2f(x) = 2(x^2 + x) \le 4x^2$ if $x \ge 1$, so $2f(x) = O(x^2)$.

Theorem 11.2   If $f_1(x) = O(g_1(x))$ and $f_2(x) = O(g_2(x))$, then

(a) $f_1(x) + f_2(x) = O(g_1(x) + g_2(x))$ and

(b) $f_1(x)f_2(x) = O(g_1(x)g_2(x))$.

Proof: Exercise.

For example, $\sqrt{x+1} = O(x^{1/2})$ and $\sqrt{x^2 - x + 1} = O(x)$, so $\sqrt{x^3 + 1} = \sqrt{(x+1)(x^2 - x + 1)} = O(x^{3/2})$.

As a consequence of Theorems 11.1 and 11.2, it can be shown that (1) adding or subtracting two n-bit integers requires $O(n)$ bit operations and (2) a conventional multiplication of a m-bit integer by an n-bit integer requires $O(mn)$ bit operations. In particular, the multiplication of two n-bit integers requires $O(n^2)$ bit operations. (By means of a technique known as the *fast Fourier transform*, the multiplication of two n-bit integers can be performed in $O(n \log n)$ bit operations.) To conventionally compute the integer part of the quotient of a $2n$-bit integer by an n-bit integer requires $O(n^2)$ bit operations.

Next, we analyze Euclid's algorithm for finding the greatest common divisor of two given natural numbers.

Theorem 11.3

If a and b are integers such that $a > b > 0$, then the number of bit operations needed to compute (a, b) using Euclid's algorithm is $O((\log a)^3)$.

Proof: If Euclid's algorithm requires n iterations to compute (a, b), then Lamé's Theorem (Theorem 2.4) implies that $n < 1 + \log_\alpha b$, where $\alpha = \frac{1}{2}(1 + \sqrt{5})$; hence $n = O(\log a)$. Each iteration is a division that requires $O((\log a)^2)$ bit operations. Therefore Theorem 11.2b implies that the total number of bit operations needed is $O((\log a)^3)$. ∎

For example, suppose that a and b are 100-digit integers, while c and d are 200-digit integers. Theorem 11.3 implies that computing (c, d) will take about 8 times longer than computing (a, b).

Definition 11.3

An algorithm that accepts n as an input is said to be *polynomial time* if it requires $O((\log n)^c)$ bit operations for some $c > 0$.

For example, the preceding discussion shows that Euclid's algorithm is polynomial time, with $c = 3$.

As a contrast, suppose that n is a given integer such that $(6, n) = 1$, and we wish to determine whether n is prime or composite. Trial division consists of dividing n by every integer d such that $(6, d) = 1$ and $5 \leq d \leq \sqrt{n}$. If n/d is an integer for such a d, then n is composite and d is a factor of n. If no quotient n/d is an integer for $5 \leq d \leq \sqrt{n}$, then n is prime.

It would be sufficient, in theory, to consider only those potential divisors d that are prime. However, this would require either (1) testing each d itself for primality or (2) storing all the primes up to $\sqrt{n}$. Such an endeavor would use more computer time and memory than simply trial division by all integers $6k \pm 1$ not exceeding $\sqrt{n}$, even though many of these divisions are redundant.

In trial division, the number of divisors is $O(\sqrt{n})$ and each division requires $O((\log_2(n))^2)$ bit operations, so the total number of bit operations needed to test n for primality (or to factor n) is $O(\sqrt{n}(\log_2(n))^2)$. Therefore trial division is *not* a polynomial time algorithm and is computationally infeasible for large n. Nevertheless, in several current factorization techniques, trial division is still used to find *small* factors of a large integer n, that is to say, factors d such that $d < (\log_2(n))^2$.

In the work ahead, we will discuss computationally feasible tests of compositeness and primality, as well as methods of factorization of composite integers. How significant is this enterprise? Let us quote Gauss: "The problem of distinguishing prime numbers from composite numbers and resolving the latter into their prime factors is known to be one of the most important and useful in arithmetic ... The dignity of the science itself seems to require that every possible means be explored for the solution of a problem so elegant and so celebrated." ([4], Art. 329)

The reader who wishes to learn more about computational number theory would do well to consult the monograph [2] by R. Crandall and C. Pomerance.

11.2 Pseudoprimes and Carmichael Numbers

Given a large odd integer n, we wish to determine whether n is prime or composite. Recall that the prime number theorem implies that the proportion of primes below x is approximately $1/\log x$. Therefore, if n is chosen at random, n is far more likely to be composite than prime.

If b is an integer such that $(b, n) = 1$ and $b^{n-1} \not\equiv 1 \pmod{n}$, then Fermat's Little Theorem (Theorem 4.8) implies that n is composite. For example, if $n = 221$, let $b = 2$. Now $(2, 221) = 1$ and $2^{220} \equiv 16 \not\equiv 1 \pmod{221}$, so we may conclude that 221 is composite. Unfortunately, this test for compositeness does not always work, since it may occur that $(b, n) = 1$ and $b^{n-1} \equiv 1 \pmod{n}$ even though n is composite. For example, $341 = 11 * 31$ is composite and $(2, 341) = 1$, yet $2^{340} \equiv 1 \pmod{341}$. This leads us to the following definition.

Definition 11.4

n is a *pseudoprime to base b* if n is composite, $(b, n) = 1$, and $b^{n-1} \equiv 1 \pmod{n}$. If so, we write n is PSP(b).

The preceding example shows that 341 is a pseudoprime to base 2; the reader may verify that so are 561 and 645. Pseudoprimes to base 2 are rare. Of all the odd integers below 10^{10}, only 14,884 are pseudoprimes to base 2, whereas 455,051,511 of them are prime. Nevertheless, as we shall see shortly, there are infinitely many pseudoprimes to base 2. First, we need the following lemma.

▶ **Lemma 11.1** *If $b \neq 1$ and $m|n$, then $(b^m - 1)|(b^n - 1)$.*

Proof: By hypothesis, we have $n = km$. Therefore

$$b^n - 1 = (b^m)^k - 1 = (b^m - 1)((b^m)^{k-1} + (b^m)^{k-2} + \cdots + b^m + 1)$$

from which the conclusion follows. ∎

Theorem 11.4   There are infinitely many pseudoprimes to base 2.

Proof: Let n be a PSP(2), that is, n is an odd composite number such that $2^{n-1} \equiv 1 \pmod{n}$. If $m = 2^n - 1$, then m is odd. Lemma 11.1 implies m is composite. Now $2^n \equiv 2 \pmod{n}$, so $m - 1 = 2^n - 2 = kn$ for some integer k. Now Lemma 11.1 implies $m | (2^{kn} - 1)$, that is, $2^{kn} \equiv 1 \pmod{m}$. Therefore $2^{m-1} \equiv 1 \pmod{m}$, so m is also a PSP(2). Since $n_1 = 341$ is a PSP(2), we can generate an infinite sequence of PSP(2)'s: $\{n_1, n_2, n_3, \cdots\}$ via the formula

$$n_k = 2^{n_{k-1}} - 1 \qquad (k \geq 2) \quad \blacksquare$$

Remarks

$n_2 = 2^{n_1} - 1 = 2^{341} - 1$ is an integer with more than 100 decimal digits!

$\blacksquare$

If n is pseudoprime to base b, we may yet unmask its compositeness. Very few composite numbers are pseudoprime to several different bases simultaneously. We may therefore proceed as follows. Choose integers $b_1, b_2, \cdots, b_m$ such that $(b_i, n) = 1$ for each index, i. Compute $b_i^{n-1} \pmod{n}$ for $i = 1, 2, 3, \cdots, m$. If for some index i we get $b_i^{n-1} \not\equiv 1 \pmod{n}$, then n is composite.

For example, $(2, 341) = 1$ and $2^{340} \equiv 1 \pmod{341}$, but $(3, 341) = 1$ and $3^{340} \equiv 56 \not\equiv 1 \pmod{341}$, so 341 is composite.

This test for compositeness has its limitations, however, since even if $b^{n-1} \equiv 1 \pmod{n}$ for *all* b such that $(b, n) = 1$, n may still be composite. If this is the case, we call n a *Carmichael number*.

Definition 11.5

n is a *Carmichael number* if n is composite and $b^{n-1} \equiv 1 \pmod{n}$ for all b such that $(b, n) = 1$.

The smallest Carmichael number is $561 = 3 * 11 * 17$. Let us verify that 561 is indeed a Carmichael number. Suppose that b is an integer such that $(b, 561) = 1$. Therefore $(b, 3) = (b, 11) = (b, 17) = 1$. Fermat's Little Theorem implies $b^2 \equiv 1 \pmod{3}$, so $b^{560} \equiv (b^2)^{280} \equiv 1^{280} \equiv 1 \pmod{3}$. Similarly, $b^{10} \equiv 1 \pmod{11} \rightarrow b^{560} \equiv (b^{10})^{56} \equiv 1^{56} \equiv 1 \pmod{11}$. Finally $b^{16} \equiv 1 \pmod{17} \rightarrow b^{560} \equiv (b^{16})^{35} \equiv 1^{35} \equiv 1 \pmod{17}$. Therefore $b^{560} \equiv 1 \pmod{[3, 11, 17]}$; that is, $b^{560} \equiv 1 \pmod{561}$.

The several theorems that follow give a characterization of Carmichael numbers.

Theorem 11.5

If $n = \prod_{i=1}^{r} p_i$ where the p_i are distinct odd primes such that $(p_i - 1)|(n-1)$ for each i, then n is a Carmichael number.

Proof: By hypothesis, n is composite. If b is an integer such that $(b, n) = 1$, then $(b, p_i) = 1$ for all i, so $b^{p_i-1} \equiv 1 \pmod{p_i}$ for all i. By hypothesis, for each i, we have $n - 1 = k_i(p_i - 1)$. Therefore $b^{n-1} = b^{k_i(p_i-1)} = (b^{p_i-1})^{k_i} \equiv 1^{k_i} \equiv 1 \pmod{p_i}$. Therefore $b^{n-1} \equiv 1 \pmod{\prod_{i=1}^{r} p_i}$, that is $b^{n-1} \equiv 1 \pmod{n}$, so n is a Carmichael number. $\blacksquare$

We will soon see that every Carmichael number is of the type mentioned in Theorem 11.5, but first we need the following lemma.

▶ **Lemma 11.2** *If p is an odd prime factor of n and $h \geq 1$, then there exists an integer r such that $(r, n) = 1$ and r is a primitive root $\pmod{p^h}$.*

Proof: Let g be a primitive root $\pmod{p^h}$ such that $0 < g < p^h$. If $(g, n) = 1$, then $r = g$ and we are done. Otherwise, let $t_j = g + jp^h$ where $j = 1, 2, 3, \cdots$. Since $(g, p) = 1$, Dirichlet's Theorem (Theorem 3.13) implies that there are infinitely many j such that t_j is prime. Pick the least j such that t_j is prime and t_j exceeds all the prime factors of n. Then $(t_j, n) = 1$. Also, t_j is a primitive root $\pmod{p^h}$, so let $r = t_j$. $\blacksquare$

For example, let $n = 900 = 2^2 3^2 5^2$, $p = 3$, and $h = 2$. Now $g = 2$ is a primitive root $\pmod{3^2}$, but $(g, n) = (2, 900) = 2 > 1$. However $t_1 = 2 + 3^2 = 11$, which is prime, and $11 > 5$, so $r = 11$.

Theorem 11.6

Let n be the product of two or more distinct primes p_i. If n is a Carmichael number, then $(p_i - 1)|(n-1)$ for each index i.

Proof: For each index i, Lemma 11.2 implies that there exists an integer r_i such that $(r_i, n) = 1$ and r_i is a primitive root $\pmod{p_i}$. Since n is a Carmichael number by hypothesis, we have $r^{n-1} \equiv 1 \pmod{n}$, hence $r^{n-1} \equiv 1 \pmod{p_i}$. Now Theorem 6.1 implies $(p_i - 1)|(n-1)$. $\blacksquare$

Theorem 11.7

Every Carmichael number is square-free.

Proof: Suppose that n is a Carmichael number that is not square-free, so $n = p^h m$ where p is prime, $h \geq 2$, and $p \nmid m$. If $p = 2$, let q be a prime such that $q > m$ and $q \equiv 3 \pmod{2^h}$. (Dirichlet's Theorem guarantees the existence of such a q.) Therefore $(q, n) = 1$. Since n is a Carmichael number by hypothesis, we have $q^{n-1} \equiv 1 \pmod{n}$, from which it follows that $q^{2^h m - 1} \equiv 1 \pmod{2^h}$. By Euler's Theorem (Theorem 5.25), we have $q^{2^{h-1}} \equiv 1 \pmod{2^h}$. Since $2^{h-1} | 2^h m$, it follows that $q^{2^h m} \equiv 1 \pmod{2^h}$. This implies $q \equiv 1 \pmod{2^h}$, that is, $3 \equiv 1 \pmod{2^h}$, an impossibility, since $h \geq 2$.

If p is odd, then Lemma 11.2 implies there exists r such that $(r, n) = 1$ and r is a primitive root $\pmod{p^h}$. Euler's Theorem implies $r^{\phi(p^h)} \equiv 1 \pmod{p^h}$. Since n is a Carmichael number by hypothesis, we have $r^{n-1} \equiv 1 \pmod{n}$, hence $r^{n-1} \equiv 1 \pmod{p^h}$. Now Theorem 6.1 implies $\phi(p^h) | (n-1)$, that is, $p^{h-1}(p-1) | (p^h m - 1)$. This is impossible, since $h \geq 2$ by hypothesis. ∎

Theorem 11.8   Every Carmichael number is odd.

Proof: Exercise.

Theorem 11.9   Every Carmichael number has at least 3 distinct prime factors.

Proof: Exercise.

We can now state the following:

Theorem 11.10   Every Carmichael number is the product of three or more distinct odd prime factors. Furthermore, if n is a Carmichael number and if p is an odd prime, then $p | n$ if and only if $(p-1) | (n-1)$.

Proof: This follows from Theorems 11.5 through 11.9. ∎

For example, 561 is a Carmichael number since $561 = 3 * 11 * 17$, and 560 is divisible by each of 2, 10, 16. Also, 8911 is a Carmichael number, since $8911 = 7 * 19 * 67$, and 8910 is divisible by each of 6, 18, 66.

Remarks

There are only 1547 Carmichael numbers below 10^{10}. The existence of infinitely many Carmichael numbers was proved in 1994. In fact, if x is a positive real number, let $C(x)$ denote the number of Carmichael numbers not exceeding x. For x sufficiently large, we have $C(x) > x^{2/7}$. (see [1]) It is not known whether there exist infinitely many Carmichael numbers with a fixed number of prime factors.

Section 11.2 Exercises

1. Prove

   (a) Theorem 11.1

   (b) Theorem 11.2

2. If $f(x)$ is a polynomial of degree n, prove that $f(x) = O(x^n)$.

3. Use Fermat's Little Theorem to prove that 221 is composite.

4. Prove that if $m \geq 1$ and if each of $6m + 1$, $12m + 1$, $18m + 1$ is prime, then their product is a Carmichael number.

5. Find all Carmichael numbers, if any, of the form $3pq$ such that p and q are primes with $3 < p < q$.

6. Prove

   (a) Theorem 11.8

   (b) Theorem 11.9

7. Find all Carmichael numbers, if any, of the form $5pq$ such that p and q are primes with $5 < p < q$.

Section 11.2 Computer Exercises

8. Write a computer program to find the smallest pseudoprime to base 3 that is not a Carmichael number.

9. Write a computer program, that, given an odd prime p, finds all Carmichael numbers, if any, of the form pqr, where q and r are primes such that $p < q < r$.

10. Write a computer program to find all Carmichael numbers below 10^5 that have four distinct prime factors.

11.3 Miller's Test and Strong Pseudoprimes

Let n be a large odd integer. We have seen that we may use Fermat's Little Theorem to prove that n is composite, unless n is a pseudoprime or a Carmichael Number. Miller's test, which we describe below, was devised to cope with the obstacle posed by pseudoprimes and Carmichael numbers.

Suppose that x is an integer such that $x^2 \equiv 1 \pmod{n}$. If n is prime, then Theorem 4.11 implies that $x \equiv \pm 1 \pmod{n}$. Therefore, if $x^2 \equiv 1 \pmod{n}$ and $x \not\equiv \pm 1 \pmod{n}$, we may conclude that n is composite. In particular, if b is an integer such that $(b, n) = 1$ and $b^{n-1} \equiv 1 \pmod{n}$ but $b^{1/2(n-1)} \not\equiv \pm 1 \pmod{n}$, then n is composite.

For example, let $n = 561$ and $b = 5$. Now $(5, 561) = 1$ and $5^{560} \equiv 1 \pmod{561}$ but $5^{280} \equiv 67 \not\equiv \pm 1 \pmod{561}$, so 561 is composite.

More generally, let n be odd, and let b be an integer such that $(b, n) = 1$ and $b^{n-1} \equiv 1 \pmod{n}$. Let $n - 1 = 2^k m$ where $k \geq 1$ and m is odd. Let j be the least non-negative integer such that $b^{2^j} \equiv 1 \pmod{n}$. (By hypothesis, j exists and $j \leq k$.) If n is prime and $j > 0$, then Theorem 4.11 implies that $b^{2^{j-1}} \equiv -1 \pmod{n}$. Therefore, if $j > 0$ and $b^{2^{j-1}} \not\equiv -1 \pmod{n}$, we know that n is composite. These considerations lead to what is known as *Miller's test*.

Definition 11.6   Miller's test

Let $n - 1 = 2^k m$, where $k \geq 1$ and m is odd. Let b be an integer such that $(b, n) = 1$. We say that **n passes Miller's test to base b** if either

$$\text{(i)} \qquad\qquad b^m \equiv \pm 1 \pmod{n}$$

or

$$\text{(ii)} \qquad\qquad b^{2^j m} \equiv -1 \pmod{n}$$

for some j such that $1 \leq j \leq k - 1$.

The following theorem establishes the utility of Miller's test.

Theorem 11.11   If p is an odd prime and $(b, p) = 1$, then p passes Miller's test to base b.

Proof: Let $p - 1 = 2^k m$ where $k \geq 1$ and m is odd. It suffices to show that either (i) $b^m \equiv 1 \pmod{p}$ or (ii) $b^{2^j m} \equiv -1 \pmod{p}$ for some j such that $0 \leq j \leq k - 1$. Let b have order $h \pmod{p}$. By Theorems 4.8 and 6.1, we have $h | (p - 1)$. If h is odd, then $h | m$, so $m = ht$ for some $t \geq 1$. Now $b^m = b^{ht} = (b^h)^t \equiv 1^t \equiv 1 \pmod{p}$. If h is even, let $h = 2^{j+1} r$ where $j \geq 0$ and r is odd. Since $h | (p - 1)$, we have $2^{j+1} r | 2^k m$, which implies that

$j \leq k-1$ and $r|m$. Since $b^{2^{j+1}r} \equiv 1 \pmod{p}$, it follows from Theorem 4.11 that $b^{2^j r} \equiv \pm 1 \pmod{p}$. Since $2^j r = h/2 < h$, it follows from the definition of h that $b^{2^j r} \not\equiv 1 \pmod{p}$. Therefore $b^{2^j r} \equiv -1 \pmod{p}$. Now $m = rt$ with t odd, so $b^{2^j m} = (b^{2^j r})^t \equiv (-1)^t \equiv -1 \pmod{p}$. ∎

As a consequence of Theorem 11.11, if $(b,n) = 1$ and n fails Miller's test to base b, then we know that n is composite.

For example, let $n = 6601$. Then $n - 1 = 6600 = 2^3(825)$. Now

$$2^{825} \equiv 2738 \pmod{6601}$$

$$2^{1650} \equiv 4509 \pmod{6601}$$

$$2^{3300} \equiv 1 \pmod{6601}$$

Therefore 6601 fails Miller's test to base 2, so 6601 is composite.

If n passes Miller's test to base b, then we suspect that n is prime. We cannot be certain of this, since some composite numbers pass Miller's test. For example, $2047 - 1 = 2(1023)$ and $2^{1023} \equiv 1 \pmod{2047}$, yet $2047 = 23 * 89$, so 2047 is composite. This leads to the following definition.

Definition 11.7

If n is composite, yet n passes Miller's test to base b, then we say n is a ***strong pseudoprime to base b***, and we write n is SPSP(b).

The preceding example shows that 2047 is a strong pseudoprime to base 2. Although strong pseudoprimes are rare, we shall demonstrate that there are infinitely many strong pseudoprimes to base 2.

Theorem 11.12   There are infinitely many strong pseudoprimes to base 2.

Proof: Since there are infinitely many PSP(2)'s, it suffices to show that if n is a PSP(2) and $m = 2^n - 1$, then m is a SPSP(2). By hypothesis, n is odd and composite; consequently, so is m. By hypothesis, $2^{n-1} \equiv 1 \pmod{n}$, so $2^{n-1} - 1 = km$ for some odd k. Now $m - 1 = 2^n - 2 = 2(2^{n-1} - 1) = 2kn$. Since $m = 2^n - 1$, we have $2^n \equiv 1 \pmod{m}$. Therefore $2^{\frac{m-1}{2}} = 2^{kn} = (2^n)^k \equiv 1^k \equiv 1 \pmod{m}$. Therefore m is a SPSP(2). ∎

A Carmichael number may be considered to be a universal pseudoprime. There is no analogy for strong pseudoprimes. It can be shown (see [15]) that if n is odd and composite, then n passes Miller's test for at most $\frac{1}{4}(n-1)$ bases b with $0 < b < n$. As a result, if n passes Miller's test for more than $\frac{1}{4}(n-1)$ different such bases, then n must be prime. Such a primality test would be computationally infeasible, even worse than trial division. Still, something can be salvaged here, providing that one is willing to sacrifice certainty for computational feasibility.

Theorem 11.13

Rabin's Probabilistic Primality Test

Let n be an odd positive integer. Pick k distinct bases b such that $0 < b < n$ and $(b, n) = 1$. Perform Miller's test for each base. If n is composite, then the probability that n passes all k tests is less than $(\frac{1}{4})^k$.

Proof: If n is composite and p is the probability that n passes Miller's test to base b, where b is a randomly chosen integer such that $0 < b < n$ and $(b, n) = 1$, then $p \le (n - 1)/4n < \frac{1}{4}$. If the k different bases are chosen independently, then the probability that n passes all k tests is $p^k < (\frac{1}{4})^k$. ∎

For example, given n, let us pick 30 distinct integers b at random such that $0 < b < n$ and $(b, n) = 1$. Let us perform Miller's test for each of these 30 bases. If n is composite, then the probability that n passes all 30 tests is less than 10^{-18}.

If we suspect that n is prime, and if we are able to obtain at least a partial factorization of $n - 1$, then we may be able to prove that n is prime by means of Pocklington's theorem, which follows.

Theorem 11.14

Pocklington's Theorem

Let $s|(n - 1)$. Let p be a prime divisor of n. If there exists c such that $c^{n-1} \equiv 1 \pmod{n}$ and $(c^{(n-1)/q} - 1, n) = 1$ for all primes q such that $q|s$, then $p \equiv 1 \pmod{s}$. In particular, if $s \ge \sqrt{n}$, then n is prime.

Proof: Let p be a prime divisor of n. Let $b \equiv c^{(n-1)/s} \pmod{n}$. Then $b^s \equiv c^{n-1} \equiv 1 \pmod{n}$, so $b^s \equiv 1 \pmod{p}$. Let b have order $h \pmod{p}$. Then $h|s$ and $h|(p-1)$. We will show that in fact $h = s$. Now $b^{s/q} \equiv c^{(n-1)/q} \pmod{p}$. By hypothesis, $(c^{(n-1)/q} - 1, n) = 1$, so $(c^{(n-1)/q} - 1, p) = 1$. Therefore $c^{(n-1)/q} \not\equiv 1 \pmod{p}$, so $b^{s/q} \not\equiv 1 \pmod{p}$. Therefore $h \ne s/q$ for all primes q such that $q|s$. This implies that $h = s$, so $s|(p - 1)$; that is, $p \equiv 1 \pmod{s}$. If also $s \ge \sqrt{n}$, then $p > \sqrt{n}$, so $p = n$. ∎

For example, let $n = 3851$. Then $n - 1 = 3850 = 2 * 5^2 * 7 * 11$. Let $s = 77 = 7 * 11 > \sqrt{3851}$. Let $c = 2$. Now $2^{3850} \equiv 1 \pmod{3851}$. Also,

$$2^{3850/7} = 2^{550} \equiv 1192 \pmod{3851} \qquad (2^{550} - 1, 3851) = (1191, 3850) = 1$$

$$2^{3850/11} = 2^{350} \equiv 9 \pmod{3851} \qquad (2^{350} - 1, 3851) = 98, 3851) \qquad = 1$$

Therefore 3851 is prime.

To prove that n is prime using Pocklington's theorem, we may proceed as follows. Find an integer s such that $s|(n-1)$ and $s \geq \sqrt{n}$. Randomly select integers c such that $0 < c < n$ and $(c, n) = 1$ until one is found that satisfies the hypothesis of Pocklington's theorem. If such a c is found, then n is prime. If numerous attempts fail to find such a c, then we suspect that n is composite.

Section 11.3 Exercises

1. Verify that none of the following is a SPSP(2):

    (a) 561

    (b) 1105

    (c) 2465

    (d) 1729

    (e) 8911

2. Determine whether 15841 is

    (a) a SPSP(2)

    (b) a SPSP(3)

3. Use Pocklington's theorem to verify the primality of each of the following:

    (a) 997

    (b) 5003

    (c) 10007

    (d) 10009

    (e) 99991

    (f) 100003

Section 11.3 Computer Exercise

4. Write a computer program to test the primality of $\frac{1}{2}(3^8 + 1)$ using Pocklington's theorem.

11.4 Factoring: Fermat's Method and the Continued Fraction Method

Let n be an odd composite positive integer. We know that n is a square if and only if $n = [\sqrt{n}]^2$. If n is not a square, then $n = pq$ where $n > p > q > 1$. The case of greatest interest is where p and q are primes. We shall consider several methods of factorization, namely, trial division, Fermat's method, the continued fraction method, the quadratic sieve method, and the Pollard $p - 1$ method.

Trial Division

As we saw at the beginning of this chapter (see p. 282), the number of bit operations needed to factor n using trial division is $O(\sqrt{n}(\log_2(n))^2)$. Therefore trial division is computationally infeasible for large n. Nevertheless, trial division is a practical method for obtaining prime factors of n that are less than a suitable bound, such as $(\log_2(n))^2)$.

Fermat's Method

Let $n = pq$, where p and q are odd primes such that $p > q$. Also, then $n = x^2 - y^2$, where $x = \frac{1}{2}(p + q)$ and $y = \frac{1}{2}(p - q)$. To factor n, we seek a positive integer y such that $n + y^2$ is a square. We therefore compute $n + y^2$ for $y = 1, 2, 3$, etc. We succeed when we try $y = \frac{1}{2}(p - q)$, but not sooner. Then $p = x + y$ and $q = x - y$.

For example, let us use Fermat's method to factor 377.

$$377 + 1^2 = 378 \neq x^2$$

$$377 + 2^2 = 381 \neq x^2$$

$$377 + 3^2 = 386 \neq x^2$$

$$377 + 4^2 = 393 \neq x^2$$

$$377 + 5^2 = 402 \neq x^2$$

$$377 + 6^2 = 413 \neq x^2$$

$$377 + 7^2 = 426 \neq x^2$$

$$377 + 8^2 = 441 = 21^2$$

Therefore $x = 21$ and $y = 8$, so $p = 21 + 8 = 29$ and $q = 21 - 8 = 13$. Our result is $377 = 29 * 13$.

Note that the number of iterations needed to factor by Fermat's method is $\frac{1}{2}(p - q)$, which is $O(p)$, thus greater than $O(\sqrt{n})$. Therefore, unless $p - q$ is small, Fermat's method is computationally infeasible.

Continued Fraction Method

Fermat's method of factoring an odd integer n consists of finding integers x, y such that $n = x^2 - y^2$. Instead, let us try to find integers x, y such that

$n|(x^2 - y^2)$, that is, $x^2 \equiv y^2 \pmod{n}$, or $x^2 - y^2 = mn$ for some positive integer m. Without loss of generality, we may specify that $0 < y < x < n$. Once we obtain such a pair of integers x, y, we compute $d = (x - y, n)$. If d is a nontrivial divisor of n, that is, if $1 < d < n$, then $d = p$ or q, and $n/d = q$ or p. We have therefore succeeded in factoring n.

Let p_k/q_k denote the kth convergent of the infinite fraction expansion of $\sqrt{n}$. Recall that in order to obtain p_k and q_k, we need to compute b_k, r_k, a_k. (see Theorem 9.25) Furthermore, recall the identity

$$p_{k-1}^2 - nq_{k-1}^2 = (-1)^k r_k$$

If we can find an *even* k such that $r_k = y^2$, then we have

$$p_{k-1}^2 - nq_{k-1}^2 = y^2$$

so $n|(p_{k-1}^2 - y^2)$. We then compute $(p_{k-1} - y, n)$, hoping to find a nontrivial factor of n. (It suffices to compute the $p_k \pmod{n}$.)

For example, let us factor 481 by the continued fraction method. We list our results in computing the continued fraction expansion of $\sqrt{481}$ as follows:

k	b_k	r_k	a_k	p_k
-2				0
-1				1
0	0	1	21	21
1	21	40	1	22
2	19	3	13	307
3	20	27	1	329
4	7	16		

Since $r_4 = 16 = 4^2$, we compute $d = (p_3 - 4, n) = (329 - 4, 481) = (325, 481) = 13$. Therefore, $n/d = 481/13 = 37$. Our result is $481 = 13 * 37$.

As a second example, let us factor 161 by the continued fraction method. We compute the continued fraction expansion of $\sqrt{161}$ and list our results in tabular form.

k	b_k	r_k	a_k	p_k
-2				0
-1				1
0	0	1	12	12
1	12	17	1	13
2	5	8	2	38
3	11	5	4	4
4	9	16	1	42
5	7	7	2	88
6	7	16		

Now $r_4 = 16 = 4^2$, so $d = (p_3 - 4, 161) = (4 - 4, 161) = (0, 161) = 161$, so we have not obtained a nontrivial factor of 161. Continuing, we find that $r_6 = 16 = 4^2$ and $d = (p_5 - 4, 161) = (88 - 4, 161) = (84, 161) = 7$. Therefore, $161 = 7 * 23$.

If numerous iterations in the continued fraction expansion of $\sqrt{n}$ fail to yield nontrivial factors of n, then it may pay to start over by computing the continued fraction of $\sqrt{jn}$, where $j = 2, 6, 30$, etc. (In general, j is the product of the first few primes.) This variation of the continued fraction method corresponds to finding integers x, y such that $x^2 \equiv y^2 \pmod{jn}$ and then computing $(x - y, jn)$, hoping to obtain a nontrivial factor of n.

For example, let us try to factor 66043 by the continued fraction method. Using the continued fraction expansion of $\sqrt{66043}$, we have

k	b_k	r_k	a_k	p_k	k	b_k	r_k	a_k	p_k
-2				0	17	217	234	2	51424
-1				1	18	251	13	39	16392
0	0	1	256	256	19	256	39	13	7368
1	256	507	1	257	20	251	78	6	61140
2	251	6	84	21844	21	217	243	1	2465
3	253	339	1	22101	22	26	269	1	63605
4	86	173	1	43945	23	243	26	19	22186
5	87	338	1	3	24	251	117	4	20263
6	251	9	56	44113	25	217	162	2	62712
7	253	226	2	22186	26	107	337	1	16932
8	199	117	3	44628	27	230	39	12	1724
9	152	367	1	771	28	238	241	2	20380
10	215	54	8	50796	29	244	27	18	38349
11	217	351	1	51567	30	242	277	1	58729
12	134	137	2	21844	31	35	234	1	31035
13	140	339	1	7368	32	199	113	4	50783
14	199	78	5	58684	33	253	18	28	13
15	191	379	1	9	34	251	169		
16	188	81	5	58729					

Our efforts to obtain a nontrivial factor of 66023 have been fruitless so far. We had $r_6 = 9 = p_5^2$, and $r_{16} = 81 = p_{15}^2$, $r_{34} = 169 = p_{33}^2$, all of which led to trivial factorizations.

Let us shift to the continued fraction expansion of $\sqrt{2 * 66043} = \sqrt{132086}$.

We have

k	b_k	r_k	a_k	p_k	k	b_k	r_k	a_k	p_k
-2				0	10	228	331	1	77755
-1				1	11	103	367	1	4466
0	0	1	363	363	12	264	170	3	91153
1	363	317	2	727	13	246	421	1	95619
2	271	185	3	2544	14	175	241	2	18219
3	284	278	2	5815	15	307	157	4	36409
4	272	209	3	19989	16	321	185	3	127446
5	355	29	24	89293	17	234	418	1	31769
6	341	545	1	109282	18	184	235	2	58898
7	204	166	3	20881	19	286	214	3	76377
8	294	275	2	18958	20	356	25		
9	256	242	2	58797					

Now $r_{20} = 25 = 5^2$, and $(p_{19}-5, 132066) = (76377-5, 132066) = (76372, 132066) = 313$. Therefore, $66043 = 313 * 211$.

Section 11.4 Exercises

1. Use Fermat's method to factor each of the following:

   (a) 299

   (b) 629

   (c) 5141

   (d) 7493

   (e) 17063

2. Use the continued fraction method to factor each of the following:

   (a) 10511

   (b) 17711

   (c) 32639

   (d) 121393

   (e) 141191

   (f) 196423

Section 11.4 Computer Exercise

3. Write a computer program to factor composite integers using the continued fraction method.

11.5 Quadratic Sieve Method

As in the continued fraction method, given an odd composite integer n, we seek integers x, y such that

$$x^2 \equiv y^2 \pmod{n} \tag{11.1}$$

Once we find x and y, we compute $d = (x - y, n)$. If d is nontrivial, that is, if $1 < d < n$, then we have found a factor of n, namely, d. In the quadratic sieve method, Congruence 11.1 is obtained by multiplying together several preliminary congruences, as we shall explain.

We begin by choosing a factor base F whose k elements are 2 and the first $k - 1$ odd primes p such that the Legendre symbol $\left(\frac{n}{p}\right) = 1$. It has been suggested that the optimal k for a given n is approximately $\sqrt{L(n)}$, where $L(n) = exp(\sqrt{\log n \log \log n})$. Here $exp(t) = e^t$. Now define the polynomial $g(x) = x^2 - n$. It follows that $x^2 \equiv g(x) \pmod{n}$. Now let $x_i = i + [\sqrt{n}]$ where $i = 1, 2, 3$, etc. We are interested in finding values x_i such that all prime factors of $g(x_i)$ belong to the factor base F. For each of these x_i, we obtain a congruence of the form

$$x_i^2 \equiv y_i \pmod{n} \tag{11.2}$$

where

$$y_i = \prod_{j=1}^{k} p_{ij}^{a_{ij}} \tag{11.3}$$

and where each prime factor $p_{ij} \in F$ and each exponent $a_{ij} \geq 0$. Suppose we find r such congruences, where

$$\prod_{i=1}^{r} y_i = y^2 \tag{11.4}$$

Then, letting $x = \prod_{i=1}^{r} x_i$, we obtain the desired Congruence 11.1. Note that for a particular value of i, $g(x_i)$ may well fail to factor completely in F. If so, we discard this x_i.

Each congruence of type 11.2 gives rise to a k-dimensional vector:

$$\vec{v_i} = \langle v_{i1}, v_{i2}, v_{i3}, \cdots, v_{ik} \rangle \tag{11.5}$$

where each component v_{ij} is the least positive residue of the exponent a_{ij} (mod 2). Therefore $v_{ij} \in \{0, 1\}$ for all i, j.

Suppose we generate m congruences of type 11.2, where $m > k$. Then certainly $m - k$ dependency relations exist among the corresponding m vectors of type 11.5. (Vector addition is performed (mod 2).) A typical dependency relation might be $\vec{v_2} + \vec{v_3} + \vec{v_7} = \vec{0}$. (Here $\vec{0}$ denotes the k-dimensional zero vector.) For each subset of dependent vectors, multiplying together the corresponding congruences of type 11.2 leads to the desired congruence of type 11.1.

The congruence of type 11.1 that we have obtained may yield a trivial d; that is, $d = 1$ or n. It can be shown that the probability of such an event is at most $\frac{1}{2}$. If this occurs, then we must generate an additional congruence of type 11.1 via other dependency relations among our vectors. If we let $m = k + 10$, then the probability that all 10 congruences of type 11.1 that we obtain will yield a trivial d is at most 2^{-10}. In other words, we are better than 99.9% certain of obtaining a nontrivial factor of n.

The following remarks are pertinent:

1. We exclude from the factor base all primes p such that $\left(\frac{n}{p}\right) = -1$, because for such primes, $x_i^2 \not\equiv n \pmod{p}$, so $p \nmid g(x_i)$.

2. A dependency relation among the vectors is certain to arise once we have at least $k + 1$ vectors, but may arise among fewer vectors.

3. It is useful to note that if $p | g(x_i)$, then $p | g(x_i + jp)$ for all integers j.

For example, let us use the quadratic sieve method to factor 28421. Note that $[\sqrt{28421}] = 168$. The criteria given above indicate that the factor base F should contain 11 primes, namely, 2, and the first 10 odd primes p such that $\left(\frac{28421}{p}\right) = 1$. This yields

$$F = \{2, 5, 7, 13, 19, 23, 29, 31, 41, 43, 53\}$$

In Table 11.1, for each i such that $1 \leq i \leq 18$, we list the following: i, $x_i = i + 168$, $g(x_i) = x_i^2 - 28421$, its factorization over F (where possible), and the corresponding vector.

At this point, although we have only generated 7 (not 12) vectors, we may note the dependency relation $\vec{v_3} + \vec{v_5} + \vec{v_7} = \vec{0}$. The corresponding congruences of type 11.2 are

$$173^2 \equiv 2^2 * 13 * 29 \pmod{28421}$$

$$175^2 \equiv 2^2 * 19 * 29 \pmod{28421}$$

$$186^2 \equiv 5^2 * 13 * 19 \pmod{28421}$$

Multiplying them together, we get

$$(173 * 175 * 186)^2 \equiv (2^2 * 5 * 13 * 29)^2 \pmod{28421}$$

that is, $3792^2 \equiv 1155^2 \pmod{28421}$. Finally, $d = (3792 - 1155, 28421) = 293$, and $28421 = 293 * 97$.

Table 11.1

i	x_i	$x_i^2 - 28421$	v_i
1	169	$140 = 2^2 * 5 * 7$	$< 0, 1, 1, 0, 0, 0, 0, 0, 0, 0, 0 > = \vec{v_1}$
2	170	479	None
3	171	$820 = 2^2 * 5 * 41$	$< 0, 1, 0, 0, 0, 0, 0, 0, 1, 0, 0 > = \vec{v_2}$
4	172	1163	None
5	173	$1508 = 2^2 * 13 * 29$	$< 0, 0, 0, 1, 0, 0, 1, 0, 0, 0, 0 > = \vec{v_3}$
6	174	$1855 = 5 * 7 * 53$	$< 0, 1, 1, 0, 0, 0, 0, 0, 0, 0, 1 > = \vec{v_4}$
7	175	$2204 = 2^2 * 19 * 29$	$< 0, 0, 0, 0, 1, 0, 1, 0, 0, 0, 0 > = \vec{v_5}$
8	176	$2555 = 5 * 7 * 73$	None
9	177	$2908 = 2^2 * 727$	None
10	178	$3263 = 13 * 251$	None
11	179	$3620 = 2^2 * 5 * 181$	None
12	180	$3979 = 23 * 173$	None
13	181	$4340 = 2^2 * 5 * 7 * 31$	$< 0, 1, 1, 0, 0, 0, 0, 1, 0, 0, 0 > = \vec{v_6}$
14	182	4703	None
15	183	$5068 = 2^2 * 7 * 181$	None
16	184	$5435 = 5 * 1087$	None
17	185	$5804 = 2^2 * 1451$	None
18	186	$6175 = 5^2 * 13 * 19$	$< 0, 0, 0, 1, 1, 0, 0, 0, 0, 0, 0 > = \vec{v_7}$

Section 11.5   Exercises

1. For each of the following integers

    i. determine whether it is prime or composite

    ii. if the integer is composite, factor it using the quadratic sieve method

(a) 30941

(b) 33919

(c) 68569

(d) 838861

(e) 1016801

(f) 1149847

(g) 1346269

11.6 Pollard $p-1$ Method

Let p be a prime factor of the odd composite integer N. Let q be an integer such that $(p-1)|q$. Then Fermat's Little Theorem and Theorem 2.12 imply $p|(2^q - 1)$. Let $d = (2^q - 1, N)$. Therefore $p|d$. If also $N \nmid (2^q - 1)$, then $1 < d < N$, so we have obtained a nontrivial factor of N.

We can obtain such a q as follows: If k is a natural number, let $M(k)$ be the least common multiple of all integers from 1 to k, that is, $M(k) = [1, 2, 3, \cdots, k]$. For sufficiently large k, such as $k = p-1$, we know that $(p-1)|M(k)$, so we can let $q = M(k)$. In order for this method of factorization to work efficiently, it is necessary that the least k such that $(p-1)|M(k)$ be small. This in turn requires that all prime factors of $p-1$ be small. Incidentally, if k is not a prime or a power of a prime, then $M(k) = M(k-1)$. Therefore it suffices to generate $M(k)$ only for primes and their powers.

We begin by listing all primes and prime powers not exceeding a preassigned bound, B. In Table 11.2, for $B = 16$, we list the following:

(a) in the top row, the integers from 1 to 10

(b) in the middle row, the first 10 primes and prime powers

(c) in the bottom row, the corresponding primes, denoted p_n

n	1	2	3	4	5	6	7	8	9	10
	2	3	4	5	7	8	9	11	13	16
p_n	2	3	2	5	7	2	3	11	13	2

Given N, we start by letting $b_1 = 2$. For $n \geq 2$, we let $b_n \equiv b_{n-1}^{p_{n-1}}$ (mod N). We compute $d_n = (b_n - 1, N)$. If $1 < d_n < N$, then we have found a nontrivial factor of N. Otherwise, we compute b_{n+1}, d_{n+1}, etc.

For example, let us factor $N = 10001$ by the $p-1$ method. Our computations, listed in Table 11.3, indicate that $73|10001$. Indeed, $10001 = 73 * 137$.

Let N have the prime factor p. If q is a prime factor of $p-1$, suppose that $q^k|(p-1)$ but $q^{k+1} \nmid (p-1)$. If for each such q and k it is true that $q^k \leq B$, then the $p-1$ method will succeed in identifying p as a factor of N. In fact, if q^k is the largest prime power factor of $p-1$, and if q^k corresponds to p_n (as shown in Table 11.2, extended if necessary), then at most $n+1$ iterations are needed to obtain the factor p.

In the preceding example, $73 - 1 = 72 = 2^3 3^2$. The largest prime power factor of 72 is $3^2 = 9$, which corresponds to $n = 7$. Therefore 8 iterations were needed to obtain the factor 73. Note that at most, 11 iterations would be necessary to obtain any prime factor p such that $p - 1 = 2^a 3^b 5^c 7^d 11^e 13^f$ where $1 \leq a \leq 4$, $0 \leq b \leq 2$, and each of the other exponents is 0 or 1.

Table 11.3

n	p_n	$b_n \ (mod N)$	$b_n - 1 \ (mod N)$	d_n
1	2	2	1	1
2	3	4	3	1
3	2	64	63	1
4	5	4096	4095	1
5	7	8532	8531	1
6	2	5466	5455	1
7	3	4169	4168	1
8	11	3578	3577	73

We conclude by briefly mentioning some additional methods of factorization. These include the *Pollard rho method,* the *p+1 method,* the *elliptic curve method,* and the *number field sieve.* All methods of factorization currently in use have in common that the factorization of a given odd integer N requires approximately $exp\sqrt{\log N \log \log N}$ bit operations. This means that it takes 3.9 hours to factor a 50-digit integer, 104 days for a 75-digit integer, and 74 years for a 200-digit integer. It is not known whether factoring large composite integers is intrinsically time-consuming. The apparent difficulty in factoring large integers is put to good use in public-key cryptosystems, which are discussed in the concluding chapter.

Section 11.6   Exercises

1. Extend the table of consecutive prime powers to accommodate the bound $B = 30$.

2. Use the $p - 1$ method to factor each of the following:

   (a) 30607

   (b) 76201

   (c) 145921

   (d) 175717

   (e) 876437

Section 11.6   Computer Exercise

3. Write a computer program to implement factorization using the $p - 1$ method.

Review Exercises

1. Use Fermat's Little Theorem to show that 91 is composite.

2. Find a Carmichael number of the form $7pq$, where p and q are primes such that $7 < p < q$.

3. Determine whether 341 is a SPSP(2).

4. Use Pocklington's Theorem to verify the primality of 1003.

5. Factor each integer listed below using

    i. Fermat's method

    ii. the continued fraction method

    iii. the quadratic sieve method

    iv. the Pollard $p - 1$ method

    (a) 16799

    (b) 19109

    (c) 20303

    (d) 20453

Chapter 12

Cryptology

12.1 Introduction

The ability to transmit confidential information securely is of great importance in diplomatic, military, and commercial affairs. The success of a given enterprise may well require that a transmitted message remain secret to all but the intended recipient. The process of encoding a message in order to protect its secrecy is known as *enciphering, encryption, or cryptography*. Having received an encoded message, the intended recipient uses a key to decode it. The key must of course be kept secret.

If an encoded message is sent by an adversary or competitor, it is desirable to be able to decode the message so as to render it comprehensible. The latter process is called *deciphering, decryption, or cryptanalysis*. The science of *cryptology*, which includes the complementary operations of cryptography and cryptanalysis, relies heavily on the use of numbers, and therefore, on number theory. The military use of cryptology dates back at least as far as Julius Caesar (100–44 B.C.), to whom are attributed the "Caesar ciphers" which are discussed below. First, we mention some historical instances where cryptology played a significant role.

In the 1580s, England and Spain were at odds. England was ruled by Queen Elizabeth I, a Protestant, while Spain was governed by Philip II, a powerful Catholic ruler who also controlled Portugal, Naples, Sicily, and parts of Holland. Mary Stuart (also known as Mary, Queen of Scots), a Catholic cousin of Elizabeth, was confined in a castle. She plotted with Philip to overthrow Elizabeth and replace her as Queen of England. Mary's enciphered messages were intercepted and deciphered by the British authorities. She doomed herself when she wrote "Let this enterprise begin." Mary was convicted of treason by a Star Chamber, and was beheaded in 1587. The following year, Philip sent the Spanish Armada in an unsuccessful attempt to conquer England.

In the 1590s, Philip was at odds with French king, Henry IV. A French royal privy councilor, Francois Vieta (1540–1603) succeeded in breaking the Spanish code, which had more than 400 characters. As a result, the French were able for some time to thwart the designs of their Spanish adversaries. Philip even complained to the Pope that the French were using black magic against him. Vieta was a lawyer who at one time had Mary Stuart as a client. When not occupied with legal and court matters, Vieta busied himself with mathematics. More than anyone else, he is responsible for the introduction of modern algebraic symbolism.

In 1642, in the early days of the English Civil War, the mathematician John Wallis (1616–1703) deciphered encoded messages for the Parliamentarians. In recognition of his services and his abilities, he was named Savilian Professor of Geometry in 1649. Much later, in the 1690s, Wallis rejected a request from Leibniz to train some German students in cryptology.

In August 1914, in the early days of World War I, the failure of two Russian armies to encipher their communications led to a humiliating defeat at the hands of the Germans at Tannenberg. In 1917, British cryptologists deciphered the so-called Zimmermann telegram, in which the German foreign minister proposed an alliance with Mexico against America. Soon after this was made public, America, which had until then remained neutral, entered the war against Germany.

Before America entered World War II in 1941, American cryptologists had broken several Japanese codes. In the first few months after attacking Pearl Harbor, the Japanese had a free hand in southeast Asia and the Pacific, conquering and/or occupying the Phillipines, Malaysia, Thailand, Burma, Indo-China, the northern portion of New Guinea, and many islands. Thanks to continued cryptological access to Japanese communications, the numerically inferior American navy was able to defeat the Japanese at the Battle of the Coral Sea in May, 1942, and shortly afterwards in the Battle of Midway Island. A planned Japanese invasion of Australia was thereby averted.

In April, 1943, American forces learned of a planned visit by Admiral Isoroku Yamamoto to some Japanese-held islands in the Pacific. (Yamamoto, the Japanese chief of naval operations, had planned the attack on Pearl Harbor.) Yamamoto's plane, although escorted by six fighters, was intercepted and shot down. It has been estimated that because of American use of cryptology, World War II was shortened by at least a year.

Before presenting the details of cryptology, we introduce some notation. The term *literal plaintext* will denote the original message, expressed in the English language. The term *numerical plaintext* will denote a numerical equivalent of the literal plaintext, obtained as follows. Let f be the bijective function whose domain is the alphabet in its usual order and whose range is the set of integers from 00 to 25. Table 12.1 defines f specifically.

A	B	C	D	E	F	G	H	I	J	K	L	M
00	01	02	03	04	05	06	07	08	09	10	11	12
N	O	P	Q	R	S	T	U	V	W	X	Y	Z
13	14	15	16	17	18	19	20	21	22	23	24	25

For example, the literal plaintext MATH corresponds to the numerical plaintext 12001907. Also, the numerical plaintext 180402201781924 corresponds to the literal plaintext SECURITY. Note that an m-letter literal plaintext can be converted via the above table to a $2m$-digit numerical plaintext and vice versa.

Similarly, the term *literal ciphertext* denotes an encoded message in literal form, whereas the term *numerical ciphertext* denotes an encoded message in numerical form.

12.2 Character Ciphers

In a *substitution* or *character cipher*, each letter in the literal plaintext is transformed to another letter in the literal ciphertext. One type of character cipher is the *linear cipher*. For example, if n is a two-digit unit of numerical plaintext, we might encipher by the key

$$E(n) \equiv an + b \pmod{26}$$

where $1 \le a \le 25$, $(a, 26) = 1$, $0 \le b \le 25, 0 \le E(n) \le 25$. A linear cipher with $a = 1$ is known as a Caesar cipher.

Suppose our literal plaintext is the proverb SPEAK SOFTLY BUT CARRY A BIG STICK, rewritten in 4-letter blocks as SPEA KSOF TLYB UTCA RRYA BIBG TICK. The corresponding numerical plaintext is 18140400 10181405 19112401 20190200 171721400 01080618 19080210. Suppose that we encipher using a linear cipher with $a = 7$ and $b = 3$. Enciphering 18, we have $E(18) \equiv 7(18) + 3 \equiv 129 \equiv 25 \pmod{26}$. Continuing in like fashion, we obtain the numerical ciphertext: 25230503 21252132 06021510 13061703 18181503 10071925 06071721, which corresponds to the literal ciphertext

ZXFD VZXM GCPK NGRD SSPD KHTZ GHRV.

Decryption is achieved by converting the literal ciphertext back to numerical ciphertext and then employing the inverse transformation:

$$D(n) \equiv 15n + 7 \pmod{26}$$

For example, to decipher Z, we apply the decryption key to its numerical equivalent, 25:

$$D(25) \equiv 15(25) + 7 \equiv 382 \equiv 18 \pmod{26}$$

Therefore Z deciphers as the literal equivalent of 18, namely, S. The decryption key for a linear cipher has the form

$$D(n) \equiv cn + d \pmod{26}$$

where $1 \leq c \leq 25$, $(c, 26) = 1$, $0 \leq d \leq 25$. There are 12 possible values for c and 26 possible values for d. Therefore there are $12 * 26 = 312$ possible decryption keys. Using a computer, one might try all these possibilities until the right one appeared. This method of decryption is known as *exhaustive cryptanalysis*.

Another method of decryption is *frequency analysis*. It is claimed that the ranking of the letters of the alphabet by relative frequency in the English language is given by the list below.

ETAOINSRHDLCUMFPGWYBVKXJQZ

Assume that the following message has been encoded using a linear cipher:

DWALQ OWLYE RTQYF YRUYE LQEIW WALDW LYDEV NYLQJ LWAIN VYWLV NWRVN SQOWR VSRYR VERYK REVSW ROWRO YSFYT SRHSJ YLVUE RTTYT SOEVY TVWVN YBLBW WQSVS WRVNE VEHHM YRELY OLYEV YTYGA EH.

The letters appearing most frequently are Y (19), V (15), W (15). Apparently Y deciphers as E, and V or W deciphers as T. Assume that the former is true. We can now find the values of c and d in the decryption key by passing to numerical equivalents. We have $D(24) \equiv 4 \pmod{26}$ and $D(21) \equiv 19 \pmod{26}$. This yields the system of simultaneous linear congruences

$$24c + d \equiv 4 \pmod{26} \; ; \; 21c + d \equiv 19 \pmod{26}$$

To solve the system, we first subtract, obtaining $3c \equiv -15 \pmod{26}$, so that $c \equiv -5 \equiv 21 \pmod{26}$. Also, $d \equiv 4 - 24c \equiv 4 - 24(-5) \equiv 124 \equiv 20 \pmod{26}$. We have succeeded in obtaining the decryption key: $D(n) \equiv 21n + 20 \pmod{26}$. The ciphertext may now be transformed to the following plaintext message: FOURS COREA NDSEV ENYEA RSAGO OURFO

REFAT HERSB ROUGH TFORT HONTH ISCON TINEN TANEW
NATIO NCONC EIVED INLIB ERTYA NDDED ICATE DTOTH
EPROP OSITI ONTHA TALLM ENARE CREAT EDEQU AL.

Finally, converting from 5-letter blocks to ordinary English, we have the first sentence of Lincoln's Gettysburg Address.

More generally, a character cipher might consist of a permutation of the alphabet, as given by the following encryption key.

Plaintext: A B C D E F G H I J K L M N O P Q R S T U V W X Y Z

Ciphertext: T H E Q U I C K B R O W N F X J M P S V L A Z Y D G

There are 26!, or about $4(10^{26})$ possible encryption keys for a character cipher, so exhaustive cryptanalysis is not computationally feasible. Frequency analysis, however, provides an effective means of decrypting character ciphers.

Section 12.2 Exercises

1. Each of the following ciphertexts has been enciphered with a Caesar cipher. Decipher them by exhaustive cryptanalysis.

   (a) NDJGC JBQTG XHJE

   (b) ZKPJN KYGPD AXKWP

2. Encipher the plaintext HIT THE ROAD JACK with the linear cipher $E(n) \equiv 5n + 8 \pmod{26}$.

3. The ciphertext VKYAQ VAKEC has been enciphered with the linear cipher $E(n) \equiv 17n + 10 \pmod{26}$. Decipher it.

4. Encipher the plaintext WHAT A SURPRISE three times in succession with the linear cipher $E(n) \equiv 3n + 2 \pmod{26}$.

5. Each of the following ciphertexts have been enciphered with a linear cipher. Decipher them using frequency analysis.

   (a) RLOI NRLO PZHP HOIN TVNN YFUO INHZ PNYZ
       NHLY GYVR SNAO INFA XPHO INTV NNYF URLO
       INRL OPZH

   (b) TGVE DQVA KQGT HLGV AQTY QGDQ VRQV RDQD IDQG
       VLRO MAYA GTM

6. If any linear cipher is applied to a given plaintext 312 times in succession, what is the result? Why is this so?

7. A linear cipher $E(n) \equiv an + b \pmod{26}$ leaves the letter N unchanged. What value or values must b have?

8. A linear cipher interchanges N and S. What does it do to X?

12.3 Block Ciphers

In a block or polygraph cipher, every n-letter block of literal plaintext is transformed to an n-letter block of ciphertext. For example, a *digraph* or *two-character block cipher* might be encoded with the following encryption key: $E(P_1P_2) = C_1C_2$, where $C_1 \equiv 3P_1 + 4P_2$ (mod 26) and $C_2 \equiv 5P_1 + 7P_2$ (mod 26). Here P_1 and P_2 denote the numerical equivalents of plaintext letters, while C_1 and C_2 denote the numerical equivalent of ciphertext letters. The plaintext message SPEAK SOFTL YBUTC ARRYA BIGST ICK would be transformed to the ciphertext

KNMUY UKBXQ YXZGZ PPWUQ MFMAL UNZ

Our encryption key can be presented conveniently in matrix form

$$\begin{bmatrix} C_1 \\ C_2 \end{bmatrix} \equiv \begin{bmatrix} 3 & 4 \\ 5 & 7 \end{bmatrix} \begin{bmatrix} P_1 \\ P_2 \end{bmatrix} \quad (\text{mod } 26)$$

The matrix that appears, which we designate M, is called the *encryption matrix*. In order to decipher, we need M^{-1}, the inverse of M (mod 26). It can be shown that M^{-1} always exists, provided that $(|M|, 26) = 1$, where $|M|$ is the determinant of M. In the preceding example, we have

$$M = \begin{bmatrix} 3 & 4 \\ 5 & 7 \end{bmatrix} \qquad M^{-1} = \begin{bmatrix} 7 & -4 \\ -5 & 3 \end{bmatrix} \equiv \begin{bmatrix} 7 & 22 \\ 21 & 3 \end{bmatrix} \quad (\text{mod } 26)$$

In general, let $\vec{P}$ represent the column vector whose components are P_1 and P_2, while $\vec{C}$ represents the column vector whose components are C_1 and C_2. Then encryption is performed via the matrix congruence $C \equiv MP$ (mod 26), while decryption is performed via $P \equiv M^{-1}C$ (mod 26). To decipher a digraph, one needs to know the four elements of M^{-1}. One might compute the frequencies of the various digraphs that appear in the ciphertext, place these digraphs in descending order of frequency, and then make use of a digraph table. In general, the most frequently occurring digraph is TH.

The procedure described above can be generalized to the n-block cipher (also known as the *Hill cipher*). Again, encryption is performed by the matrix congruence $C \equiv MP$ (mod 26). Here M denotes an n by n matrix with integer entries such that $(|M|, 26) = 1$, while P and C denote n-dimensional column vectors corresponding to blocks of plaintext and ciphertext respectively. For n sufficiently large, decryption by frequency analysis is no longer computationally feasible.

Although computers play a large role in modern cryptology, this chapter contains no exercises specifically designated as computer exercises. The exercises in this chapter have been written so as to be capable of solutions either with or without computers, according to the reader's preference.

Section 12.3 Exercises

1. Encipher the plaintext PRACTICE MAKES PERFECT using the block cipher $C_1 \equiv 9P_1 + 5P_2 \pmod{26}$, $C_2 \equiv 3P_1 + 4P_2 \pmod{26}$.

2. The ciphertext TX NT QM TW HT YB SK ER AT UT QE has been enciphered with the block cipher $C_1 \equiv 5P_1 + 2P_2 \pmod{26}$, $C_2 \equiv 7P_1 + 3P_2 \pmod{26}$. Decipher it.

3. It is known that the two most frequently occurring digraphs in the English language are TH and HE. In a ciphertext, the two most frequently occurring digraphs are XY and BT. Assuming that a plaintext was enciphered with a block cipher, so that $P_1 \equiv aC_1 + bC_2 \pmod{26}$, $P_2 \equiv cC_1 + dC_2 \pmod{26}$, determine a, b, c, d.

12.4 One-Time Pads: Exponential Ciphers

Suppose that we wish to encipher an n-character plaintext message. The corresponding numerical plaintext may be considered as a vector: $\vec{P} =< P_1, P_2, P_3, \cdots, P_n >$, where each component P_i satisfies: $0 \le P_i \le 25$.

Let the encryption vector $\vec{K} =< K_1, K_2, K_3, \cdots, K_n >$ satisfy the same conditions as the P_i.

The corresponding numerical ciphertext is a vector: $\vec{C} =< C_1, C_2, C_3, \cdots, C_n >$ obtained via $C_i \equiv P_i + K_i \pmod{26}$ for all i.

For example, suppose our literal plaintext message is SWIMMING IS FUN. Then $\vec{P} =< 18, 22, 08, 12, 12, 08, 13, 06, 08, 18, 05, 20, 13 >$. Let $\vec{K} =< 19, 04, 03, 04, 06, 07, 04, 00, 09, 10, 02, 09, 21 >$. Then $\vec{C} =< 11, 00, 11, 16, 18, 15, 17, 06, 17, 02, 07, 03, 08 >$. This corresponds to the literal ciphertext L A L Q S P P R G R C H D I .

On the other hand, suppose that our literal plaintext message is G O T Y O U R N U M B E R . Then
$\vec{P} =< 06, 14, 19, 24, 14, 20, 17, 13, 20, 12, 01, 04, 17 >$
Let $\vec{K} =< 05, 12, 18, 18, 04, 21, 00, 19, 23, 16, 06, 25, 08 >$.
Then we obtain the same $\vec{C}$, and hence the same literal ciphertext as before.

An unintended recipient without knowledge of the encryption vector $\vec{K}$ has no way of obtaining the plaintext from the cipher text, *provided that $\vec{K}$ is used only once*. In general, there are 26^n possible values of $\vec{K}$, so exhaustive cryptanalysis is not computationally feasible. This cryptosystem, known as the *one-time pad*, was invented by Joseph O. Mauborgne of the U.S. Army Signal Corps during World War I. It is an unbreakable code and has been used in the "hot line" from Washington to Moscow.

Nevertheless, the one-time pad has its disadvantages. The key, which must be sent in advance to the intended recipient, must be changed very frequently. Furthermore, since the key is as long as the message, this code is not practical for the transmission of long messages.

Let a literal plaintext be split into m-letter blocks that are converted to $2m$-digit blocks of numerical plaintext. (If necessary, extra letters can be added to the literal plaintext so that the total number of letters is a multiple of m.) Choose a prime p so that $\frac{25}{99}(10^{2m} - 1) < p < 10^{2m}$ and $\frac{1}{2}(p-1)$ is prime. Choose an exponent j such that $2 \leq j \leq p-2$ and $(j, p-1) = 1$. Now transform each $2m$=digit block P of numerical plaintext into a block C of ciphertext via the congruence

$$C \equiv P^j \pmod{p}$$

where $0 < C < p$. The bounds on the prime p ensure that distinct blocks of numerical plaintext are transformed to distinct blocks of numerical plaintext.

For example, suppose our plaintext message is BEWARE THE IDES OF MARCH, and we encipher it with an exponential cipher with $m = 2$, $p = 9987$, $j = 3$. Then our numerical plaintext is

$$0104\,2200\,1704\,1907\,0408\,0304\,1814\,0512\,0017\,0207$$

We encipher via the congruence $C \equiv P^3 \pmod{9987}$. This procedure yields the numerical ciphertext

$$7633\,7497\,0367\,1798\,3509\,5497\,3312\,1703\,4913\,1104$$

In order to decipher a block C of numerical ciphertext, we need to know k, the multiplicative inverse of $j \pmod{p-1}$. That is, k is the unique integer such that $2 \leq k \leq p-2$ and $jk \equiv 1 \pmod{p-1}$. Then $C^k \equiv (P^j)^k \equiv P^{jk} \equiv P \pmod{p}$, so we decipher by raising each block of ciphertext to the kth power $\pmod{p}$. In the preceding example, $3k \equiv 1 \pmod{9886}$ implies $k \equiv 6591 \pmod{9886}$, so we decipher via

$$P \equiv C^{6591} \pmod{9887}$$

If even one block P of numerical plaintext is known, then decryption requires solving the congruence $C \equiv P^j \pmod{p}$ for j. We call j the logarithm of C to the base $P \pmod{p}$. The best algorithms for finding logarithms $\pmod{p}$ require approximately $exp(\sqrt{\log p \log \log p})$ bit operations unless all factors of $p-1$ are small. We avoid this exception by insisting that $\frac{1}{2}(p-1)$ be prime. As in the case of factoring a large integer, the number of bit operations required to find j for large p renders the decryption algorithm computationally infeasible.

Section 12.4   Exercises

1. What one-time pad will encipher the plaintext WARTOENDWARS as PLAN-FARAHEAD?

2. What one-time pad will encipher the plaintext KEEPTHEFAITH as PLAN-FARAHEAD?

3. Use an exponential cipher with $m = 2$, $p = 9949$, $j = 7$ to encipher the plaintext DAMN THE TORPEDOES FULL SPEED AHEAD.

4. Use an exponential cipher with $m = 2$, $p = 9973$, $j = 5$ to encipher the plaintext GIVE ME LIBERTY OR GIVE ME DEATH.

5. The ciphertext 28 08 15 20 31 14 03 21 05 19 has been enciphered with an exponential cipher with $m = 1$ and $p = 37$. It is also known that the ciphertext 20 corresponds to the plaintext 19. Decipher it.

6. The following ciphertext has been enciphered with an exponential cipher with $m = 2$, , $p = 9967$, $j = 5$:
9219 2474 7105 2804 6178 7554 2790 4198 1189 9219 3441 2416 9565. Decipher it.

12.5 Public-Key Cryptography

Under the various cryptosystems described heretofore, the problem of key management becomes unwieldy if secure communications are required in a network of users. Each pair of participants needs an enciphering key that is kept secret from all the others. The keys themselves must be transmitted through a secure channel, and they must be changed frequently.

The key-management problem is effectively solved by a new type of cipher system known as the *public-key cryptosystem*, which is a variation of the exponential cipher.

Let $n = pq$, where p and q are distinct, large (over 100 digits) suitably chosen primes. In particular, in order to prevent decryption, is it advisable to choose primes p and q such that (i) $p/q > 1000$, and (ii) both $(p-1)/2$ and $(q-1)/2$ are also prime. Let j be an integer such that $2 < j < \phi(n)$ and $(j, \phi(n)) = 1$. Let the numerical plaintext be split into blocks P of equal length. We obtain a corresponding block C of numerical ciphertext via $C \equiv P^j \pmod{n}$, where $0 < C < n$.

Let k be the multiplicative inverse of $j \pmod{\phi(n)}$, that is, $2 < k < \phi(n)$ and $jk \equiv 1 \pmod{\phi(n)}$. We decipher via $P \equiv C^k \pmod{n}$. This works because $C^k \equiv (P^j)^k \equiv P^{jk} \equiv P \pmod{n}$.

The enciphering key, namely, the values of n and j, is made *public*, while k, p and q are kept secret. An unintended listener needs the value of k in order to decipher a ciphertext message. However, this requires knowing $\phi(n) = (p-1)(q-1)$. Since $p + q = n - \phi(n) + 1$ and $p - q = \sqrt{(p+q)^2 - 4n}$, this is equivalent to knowing p and q. Therefore, in order to decipher, the unintended listener must factor n. For n sufficiently large, this is computationally infeasible.

For example, let us use the RSA cryptosystem to encode the message

DON'T STEP ON ME

First, we convert the literal plaintext into numerical plaintext, arranged in 4-digit blocks. This yields

0314 1319 1819 0415 1413 1204

Next, let $n = 83 * 107 = 8881$, so $\phi(n) = 82 * 106 = 8692$. We pick an exponent, j, such that $(j, \phi(n)) = 1$. In particular, let $j = 3$. We generate the ciphertext by 4-digit blocks, using the formula

$$C \equiv P^j \pmod{n}, \quad \text{that is,} \quad C \equiv P^3 \pmod{8881}$$

This yields the numerical ciphertext

8859 0931 0321 7968 4656 8010

In order to decipher this ciphertext, one needs to know k, the positive integer such that $jk \equiv 1 \pmod{\phi(n)}$, that is, $3k \equiv 1 \pmod{8692}$. We have $k \equiv 5795 \pmod{8692}$, so we decipher using the formula

$$P \equiv C^k \pmod{n} \quad \text{that is} \quad P \equiv C^{5795} \pmod{8881}$$

Remarks

Note that n is only a 4-digit number, and hence easy to factor by a variety of methods. Therefore the example presented above is unrealistic. As mentioned previously, a realistic example would require $n = pq$ where the primes p, q each have over 100 decimal digits.

■

Signatures

How does the intended recipient know that a message received originated from an authorized source? This problem is also disposed of in public-key cryptography by what are known as *signatures*. Let participant Alice have encryption key E_A and decryption key D_A. Similarly, let participant Bob have encryption key E_B and decryption key D_B. Let M be a brief plaintext message that identifies Alice. Alice transmits $E_B(D_A(M))$ to Bob. Bob deciphers this message via $E_A(D_B(E_B(D_A(M)))) = E_A(D_A(M)) = M$. Only Bob can decipher $E_B(D_A(M))$, since only Bob possesses D_B. Only Alice can transmit $E_B(D_A(M))$ to Bob, since only Alice possesses D_A. Therefore Bob can indeed be sure that the author of the message was Alice.

Public-key cryptosystems were first proposed by W. Diffie and R. Hellman in 1975. The public-key cipher described above is known as the *RSA cryptosystem*. The initials are those of its inventors, Ronald Rivest, Adi Shamir, and Leonard Adleman, who published their results in 1977. Since that time, the RSA cryptosystem, whose security is based on the computational infeasibility of factoring large integers, has been under intense scrutiny by numerous investigators. It has survived all attacks and is still in use.

Section 12.5   Exercises

1. If $n = pq = 19939$ and $\phi(n) = 19656$, find p and q.

2. If $n = pq = 63083$ and $\phi(n) = 62568$, find p and q.

3. The ciphertext 7482 6330 4952 4707 0705 has been enciphered using RSA with $n = 8051$ and $j = 5$. Decipher it.

4. Encipher the plaintext message ENDINSIGHT using RSA with $n = 8633$ and $j = 5$.

5. Suppose that Alice's RSA enciphering key is $n = 7663$ with $j = 5$, while Bob's RSA enciphering key is $n = 8881$ with $j = 3$.

   (a) Encipher Alice's signature to Bob, namely, I AM ALICE.

   (b) Encipher Bob's signature to Alice, namely, BOBS BIG BOY.

Review Exercises

1. The message IWGA IU ZWU has been encoded with a Caesar cipher. Decipher it using exhaustive cryptanalysis.

2. Use frequency analysis to decipher the ciphertext DQ QNZ UJR ZPZV WQDZ LVQYZ LO EDNZV-ZRXKBJXKDW XUZ KDXZAAKWZDMZ QS XUZ JBZVKMJD TELAKM.

3. Encode the message MIND YOUR STEP using the RSA cryptosystem with $n = 9853$ and $j = 5$.

Appendix A

Some Open Questions in Elementary Number Theory

Elementary number theory is notorious for containing statements whose meaning is easily understood, but whose truth or falseness is difficult to ascertain. We end this book by presenting a list of 14 such open questions.

1. Does there exist an odd perfect number, that is, an odd natural number n such that $\sigma(n) = 2n$?

Remarks

This question, which dates back to Euclid (300 B.C.), is the oldest unsolved problem in mathematics. Many necessary conditions for the existence of an odd perfect number are known. For example, according to Euler, if n is an odd perfect number, then

$$n = p^\alpha \prod_{i=1}^{r} q_i^{2\beta_i}$$

where $p \equiv \alpha \equiv 1 \pmod 4$ and p and the q_i are primes. Furthermore, n has at least 8 distinct prime factors, that is, $r \geq 7$ in the formula above. In addition, $n > 10^{300}$ and n has a prime factor that exceeds 10^7. ■

2. Are there infinitely many Mersenne primes, that is, primes of the form $2^p - 1$, where p is prime?

Remarks

We saw in Chapter 5 that if n is even, then n is perfect if and only if $n = 2^{p-1}(2^p - 1)$ where $2^p - 1$ is prime. Therefore, a positive answer to Question 2 would imply the existence of infinitely many even perfect numbers. At present, 42 Mersenne primes are known, the largest corresponding to $p = 25964951$. ∎

---

3. Are there infinitely many n such that $(105, \binom{2n}{n}) = 1$?

Remarks

In other words, are there infinitely many n such that $\binom{2n}{n}$ is not divisible by 3, 5, or 7? This question was posed by Ronald Graham of the University of California at San Diego, a distinguished mathematician who is a past president of both the American Mathematical Society and the Mathematics Association of America. ∎

---

4. Are there any primes of the form $2^{2^n} + 1$ where $n \geq 5$?

Remarks

Recall that primes of the form $f_n = 2^{2^n} + 1$ are called *Fermat primes*. Fermat thought (incorrectly) that f_n is prime for all $n \geq 0$. (This is the only incorrect mathematical assertion that Fermat ever made.) It is now known that f_n is prime for $0 \leq n \leq 4$ and composite for $5 \leq n \leq 32$. ∎

---

5. Are there infinitely many primes of the form $n^2 + 1$?

6. Are there infinitely many primes of the form $n^2 + n + 1$?

Remarks

Recall that Dirichlet's Theorem states that $an+b$ is prime for infinitely many values of n, provided that $(a, b) = 1$. That is, an irreducible polynomial of degree 1 in n is prime infinitely often. For irreducible polynomials in n of degree 2 or higher, such as appear in Questions 5 and 6 above, nothing is known. ∎

7. Are there infinitely many primes p such that $2p+1$ is also prime? (Such primes are called *Sophie Germain* primes.)

Remarks

Sophie Germain (1776–1830) was an outstanding mathematician of her time. She proved that if p and $2p+1$ are odd primes and $p \nmid xyz$, then $x^p+y^p \neq z^p$. This was a significant contribution (at the time) in the direction of proving Fermat's Last Theorem. She corresponded with Lagrange and Gauss, who nominated her for an honorary doctorate.

Sophie Germain primes also arise in cryptography. In the RSA public-key cryptosystem, one needs to choose two large primes, p and q. In order to frustrate attempts at decryption, it is helpful to choose p and q such that $\frac{p-1}{2}$ and $\frac{q-1}{2}$ are also prime, that is, such that $\frac{p-1}{2}$ and $\frac{q-1}{2}$ are both Sophie Germain primes. ∎

8. Are there infinitely many prime Fibonacci numbers?

Remarks

Let F_n denote the nth Fibonacci number. It is known that if F_n is prime, then $n = 4$ or $n \geq 5$ and n is prime. ∎

9. Are there infinitely many pairs of twin primes, that is, are there infinitely many n such that $2n - 1$ and $2n + 1$ are both prime?

Remarks

This statement is known as the *twin prime conjecture*. If x is a positive real number, let $\pi_2(x)$ denote the number of twin prime pairs not exceeding x.

G. H. Hardy and J. E. Littlewood made the following conjecture in 1923: Let

$$\pi_2^*(x) = 2C_2 \int_2^x \frac{dt}{(\ln t)^2}$$

and

$$C_2 = \prod_{p \geq 3} \frac{(p-1)^2}{p(p-2)}$$

Then $\pi_2(x)$ is asymptotic to $\pi_2^*(x)$, that is,

$$\lim_{x \to \infty} \frac{\pi_2^*(x)}{\pi_2(x)} = 1$$

In 1919, V. Brun proved that the sum of the reciprocals of the twin primes converges.   ■

10. (Goldbach's Conjecture) Is it true that for each $n \geq 4$, there exist distinct primes p and q such that $p + q = 2n$?

Remarks

Goldbach's Conjecture has been verified for all $n \leq 10^{17}$.   ■

11. Is it true that for every $n \geq 1$, there is a prime, p, such that $n^2 < p < (n+1)^2$?

Remarks

In other words, is there always a prime between two consecutive squares? Numerical evidence seems to indicate an affirmative answer.   ■

12. Let $\phi(n)$ denote Euler's totient function. Does there exist n such that the equation $\phi(x) = n$ has a unique solution?

13. (Artin's Conjecture) Are there infinitely many primes, p, such that 2 is a primitive root $(\bmod\ p)$?

Remarks

Emil Artin (1898–1962) was a distinguished German algebraist. In his early years, he served on the faculty of the University of Hamburg (Germany). In 1933, he emigrated to the United States and joined the faculty of Princeton University.   ∎

14. Is 2^n a sum of distinct powers of 3 for any n other than 2 or 8?

Concluding Remarks

A couple of well-known open questions in number theory have been settled since the first edition of this book was published in 1993, namely,

1. Fermat's Last Theorem (the statement that the equation $x^n + y^n = z^n$ has no solution in positive integers if $n \geq 3$) was finally proved by Andrew Wiles in 1995.

2. In 1994, Alford, Granville, and Pomerance proved the existence of infinitely many Carmichael numbers. (Recall that the positive integer n is called a Carmichael number if n is a composite number such that $b^n \equiv b \pmod{n}$ for all integers b such that $(b, n) = 1$.)

∎

Appendix B

Tables

2	73	179	283	419	547	661	811	947
3	79	181	293	421	557	673	821	953
5	83	191	307	431	563	677	823	967
7	89	193	311	433	569	683	827	971
11	97	197	313	439	571	691	829	977
13	101	199	317	443	577	701	839	983
17	103	211	331	449	587	709	853	991
19	107	223	337	457	593	719	857	997
23	109	227	347	461	599	727	859	
29	113	229	349	463	601	733	863	
31	127	233	353	467	607	739	877	
37	131	239	359	479	613	743	881	
41	137	241	367	487	617	751	883	
43	139	251	373	491	619	757	887	
47	149	257	379	499	631	761	907	
53	151	263	383	503	641	769	911	
59	157	269	389	509	643	773	919	
61	163	271	397	521	647	787	929	
67	167	277	401	523	653	797	937	
71	173	281	409	541	659	809	941	

Table B.2

The least primitive
root $g \pmod{p}$

p	g	p	g	p	g	p	g	p	g	p	g	p	g
2	1	97	5	227	2	367	6	509	2	661	2	829	2
3	2	101	2	229	6	373	2	521	3	673	5	839	11
5	2	103	5	233	3	379	2	523	2	677	2	853	2
7	3	107	2	239	7	383	5	541	2	683	5	857	3
11	2	109	6	241	7	389	2	547	2	691	3	859	2
13	2	113	3	251	6	397	5	557	2	701	2	863	5
17	3	127	3	257	3	401	3	563	2	709	2	877	2
19	2	131	2	263	5	409	21	569	3	719	11	881	3
23	5	137	3	269	2	419	2	571	3	727	5	883	2
29	2	139	2	271	6	421	2	577	5	733	6	887	5
31	3	149	2	277	5	431	7	587	2	739	3	907	2
37	2	151	6	281	3	433	5	593	3	743	5	911	17
41	6	157	5	283	3	439	15	599	7	751	3	919	7
43	3	163	2	293	2	443	2	601	7	757	2	929	3
47	5	167	5	307	5	449	3	607	3	761	6	937	5
53	2	173	2	311	17	457	13	613	2	769	11	941	2
59	2	179	2	313	10	461	2	617	3	773	2	947	2
61	2	181	2	317	2	463	3	619	2	787	2	953	3
67	2	191	19	331	3	467	2	631	3	797	2	967	5
71	7	193	5	337	10	479	13	641	3	809	3	971	6
73	5	197	2	347	2	487	3	643	11	811	3	977	3
79	3	199	3	349	2	491	2	647	5	821	2	983	5
83	2	211	2	353	3	499	7	653	2	823	3	991	6
89	3	223	3	359	7	503	5	659	2	827	2	997	7

Appendix C

Answers to Selected Exercises

1.2/1 $v_n^2 - 2Q_n^2 = (\alpha^n + \beta^n)^2 - 2Q^n = \alpha^{2n} + \beta^{2n} + 2(\alpha\beta)^n - 2Q^n = \alpha^{2n} + \beta^{2n} + 2Q^n - 2Q^n = \alpha^{2n} + \beta^{2n} = v_2.$

1.2/3 n even $\rightarrow n = 2m \rightarrow n^2 = (2m)^2 = 2(2m^2) \rightarrow n^2$ is even.

1.3/19   **Basis Step:** $\alpha F_1 + F_0 = \alpha(1) + 0 = \alpha = \alpha^1.$

         Induction Step: $\alpha^{n+1} = \alpha(\alpha^n) = \alpha(\alpha F_n + F_{n-1}) = \alpha^2 F_n + \alpha F_{n-1} = (\alpha + 1)F_n + \alpha F_{n-1} = \alpha(F_n + F_{n-1}) + F_n = \alpha F_{n+1} + F_n.$

1.3/27 $F_n L_n = (\frac{\alpha^n - \beta^n}{\alpha - \beta})(\alpha^n + \beta^n) = \frac{\alpha^{2n} - \beta^{2n}}{\alpha - \beta} = F_{2n}.$

1.3/43 $\alpha^n / \sqrt{5} \rightarrow F_n$ as $n \rightarrow \infty.$

1.4/3   (a) $\prod_{k=1}^{n}(2k) = 2^n n!$ (b) $\prod_{k=1}^{n}(2k-1) = \frac{(2n)!}{2^n n!}$

1.5/1c $\begin{pmatrix} 10 \\ 4 \end{pmatrix} = \frac{10*9*8*7}{4*3*2*1} = 210.$

1.5/1g $\begin{pmatrix} 100 \\ 98 \end{pmatrix} = \begin{pmatrix} 10 \\ 2 \end{pmatrix} = \frac{100*99}{2*1} = 4950.$

1.5/3 $\frac{n+k}{n} \begin{pmatrix} n \\ k \end{pmatrix} = (1 + \frac{k}{n}) \begin{pmatrix} n \\ k \end{pmatrix} = \begin{pmatrix} n \\ k \end{pmatrix} + \frac{k}{n}(\frac{n!}{k!(n-k)!}) = \begin{pmatrix} n \\ k \end{pmatrix} + \frac{(n-1)!}{(k-1)!(n-k)!} = \begin{pmatrix} n \\ k \end{pmatrix} + \begin{pmatrix} n-1 \\ k-1 \end{pmatrix}.$

1.5/7   The sum is F_{n+1}.

1.5/9   The first two are 1 and 10.

1.6/1   (a) $[\frac{23}{3}] = 7$ (c) $[10\pi] = 31$.

2.2/1 $20 : 1, 2, 4, 5, 10, 20$ $21 : 1, 3, 7, 21$

2.2/5 $n|(a - 1) \rightarrow a = 1 + cn$ $n|(b - 1) \rightarrow b = 1 + kn$. Thus
          $ab = (1 + cn)(1 + kn) = 1 + (c + k + ckn)n \rightarrow n|(ab - 1)$.

2.2/7   (a) 1, 2, 3, 6   (c) 1   (e) 1, 3, 9

2.2/19   Set $r = 0$ in the identity from Exercise 2.2/18 and then simplify.

2.3/5a $(30, 42, 70) = 2$ $[30, 42, 70] = 210$

2.3/5c $(296, 444, 555) = 37$ $[296, 444, 555] = 4440$

2.3/5e $(20, 30, 60) = 10$ $[20, 30, 60] = 60$

2.3/7   No (See Exercise 5e above, for example.)

2.4/3   (a) $13_{10} = 1101_2$ (c) $127_{10} = 1111111_2$ (e) $1001_{10} = 1111101001_2$

2.4/5   (a) $12345_{10} = 2099_{16}$ (c) $30000_{10} = 7530_{16}$ (e) $777_{10} = 309_{16}$

3.2/1   Hint: Show that if $a > 2$ or n is composite, then $a^n - 1$ is composite.

3.2/5   101

3.2/7   3, 7, 13, 31, 43

3.2/11 $10^{10}/\ln 10^{10} = 4.34(10^8)$ $\pi(10^{10}) = 4.55(10^8)$. Relative error = 6%.

3.3/1   223, 227, 229, 233, 239

3.3/3   (a) $299 = 13 * 23$ (b) $480 = 2^4 * 3 * 5$ (c) $399 = 3 * 7 * 13$
          (d) $777 = 3 * 7 * 37$ (e) $145 = 5 * 29$ (f) $221 = 13 * 17$
          (g) $72 = 2^3 3^2$ (h) $2450 = 2 * 5^2 * 7$ (i) $5005 = 5 * 7 * 11 * 13$
          (j) $191 = 191$ (k) $896 = 2^7 * 7$ (l) $529 = 23^2$
          (m) $104 = 2^3 * 13$ (n) $171 = 3^2 * 19$

3.3/7   2, 3, 4, 6, 8, 12, 18, 24, 30

3.3/19 $[\frac{1000}{5}] + [\frac{1000}{25}] + [\frac{1000}{125}] + [\frac{1000}{625}] = 200 + 40 + 8 + 1 = 249$

4.2/3 (a) R is not symmetric (b) Yes (c) R is not reflexive (d) Yes (e) R is not transitive

4.2/5 26

4.3/1a $145 \equiv 2 \pmod{13} \rightarrow 145^6 + 1 \equiv 2^6 + 1 \equiv 65 \equiv 0 \pmod{13}$

4.3/1b $1855 \equiv 2 \pmod{17} \rightarrow 1855^4 + 1 \equiv 2^4 + 1 \equiv 17 \equiv 0 \pmod{17}$

4.3/3 (a) $x \equiv 8 \pmod{23}$ (b) $x \equiv 27 \pmod{47}$ (c) no solution (d) $x \equiv 22 \pmod{25}$ (e) $x \equiv 20 \pmod{31}$ (f) no solution (g) $x \equiv 4, 11, 18 \pmod{21}$ (h) $x \equiv 1, 9 \pmod{16}$ (i) $x \equiv 4, 10, 16, 22 \pmod{24}$

4.4/1 (b) $(x, y) = (1, 13), (5, 6)$ (d) $(x, y) = (8, 2)$ (f) $(x, y) = (15, 7)$

4.4/2 (d) $x \equiv 17 \pmod{280}$ (f) $x \equiv 51 \pmod{210}$ (h) $x \equiv 347 \pmod{1001}$

4.4/3 497

4.5/1 $x \equiv 35, 83 \pmod{121}$

4.5/3 no solution

4.5/5 $x \equiv 3, 6, 7 \pmod{8}$

4.5/7 $x \equiv 61 \pmod{81}$

4.5/9 $x \equiv 1, 8, 15, 22, 24, 29, 36, 43 \pmod{49}$

4.6/1 (a)

*	0	1	2	3	4
0	0	0	0	0	0
1	0	1	2	3	4
2	0	2	4	1	3
3	0	3	1	4	2
4	0	4	3	2	1

4.6/1   (b) $1^{-1} \equiv 1 \pmod 5$, $2^{-1} \equiv 3 \pmod 5$, $3^{-1} \equiv 2 \pmod 5$, $4^{-1} \equiv 4 \pmod 5$

4.6/3   1, 3, 7, 9 have multiplicative inverses (mod 10)

4.7/3   Hint: Theorem 4.12 is helpful.

4.7/5   Hint: Start by using congruence (mod 3) to show that j is even, then congruence (mod 4) to show that k is odd.

4.8/3   (8, 15, 17) and (17, 144, 145)

4.8/5   (a)   Hint: If $3 \nmid xy$, then $x^2 \equiv y^2 \equiv 1 \pmod 3$.

4.8/5   (b)   Hint: If $5 \nmid xy$, then $x^2 \equiv \pm 1 \pmod 5$, $y^2 \equiv \pm 1 \pmod 5$.

4.8/11   Hint: Start by using congruences to show that a and c must be even.

5.2/1

n	40	41	42	43	44	45	46	47	48	49	50
$\tau(n)$	8	2	8	2	6	6	4	2	10	3	6
$\sigma(n)$	90	42	96	44	84	78	72	48	124	57	93

5.2/7 $n = p^3$ or $n = pq$, where p, q are primes.

5.2/13   28

5.3/1

n	42	79	1000	945	91	144
$\mu(n)$	-1	-1	0	0	1	0
$\phi(n)$	12	78	400	432	72	48

5.3/5   Let $f(n) = \sum_{d|n} \mu(d)\sigma(d)$. If $n = \prod_{i=1}^{r} p_i^{e_i}$, then $f(n) = (-1)^r \prod_{i=1}^{r} p_i$.

5.3/13   (a) 1, 2   (b) 3, 4, 6   (c) 5, 8, 10, 12   (d) 7, 9, 14, 18
(e) 15, 16, 20, 24, 30

6.2/1

t	1	2	3	4	5	6	7	8	9	10	11	12
$o(t)$	1	12	3	6	4	12	12	4	3	6	12	2

Primitive roots (mod 13) are: 2, 6, 7, 11.

6.2/3 $2^1 \equiv 2 \not\equiv 1 \pmod{19}$ $2^2 \equiv 4 \not\equiv 1 \pmod{19}$
$2^3 \equiv 8 \not\equiv 1 \pmod{19}$ $2^6 \equiv 7 \not\equiv 1 \pmod{19}$ $2^9 \equiv 12 \not\equiv 1$
$\pmod{19}$

6.2/5   (a) 2, 3   (b) 5, 7   (c) 11, 13

6.2/7 $(t^2)^{\frac{p-1}{2}} \equiv t^{p-1} \equiv 1 \pmod{p}$ by Fermat's Little Theorem. Therefore t^2 has order at most $(p-1)/2 \pmod{p}$, so t^2 is not a primitive root $\pmod{p}$.

6.2/9 $p = 1 + 2^j m$ where $m = 1$ or m is a product of one or more distinct Fermat primes.

6.2/13 $p \equiv 2 \pmod 3$

6.3/1   (a) $x \equiv 1, 3, 9 \pmod{13}$ (b) $x \equiv \pm 1, \pm 4 \pmod{17}$

6.3/3   3, 5, 12, 18, 19, 20, 26, 28, 29, 30, 33, 34

6.4/1

k	1	2	3	4	5	6	7	8	9	10
$ind_2(k)$	0	1	8	2	4	9	7	3	6	5

6.4/3 $x \equiv \pm 4 \pmod{11}$

6.4/5

k	1	2	3	4	5	6	7	8	9	10	11	12	13	14	15	16
$ind_3(k)$	0	14	1	12	5	15	11	10	2	3	7	13	4	9	6	8

6.4/7 $x \equiv \pm 3, \pm 5 \pmod{17}$

6.4/9   no solution

7.2/1 $x = y - 2$ $x^2 \equiv 2 \pmod{11}$

7.2/3 $x = y + 7$ $x^2 \equiv 3 \pmod{17}$

7.2/5 $x = y - 1$ $x^2 \equiv 6 \pmod{23}$

7.3/1   (a) $\{5, 10, 15, 20, 25\} \to \{5, 10, 4, 9, 3\} \to (\frac{5}{11}) = (-1)^2 = 1$

7.3/1   (b) $\{5, 10, 15, 20, 25, 30\} \to \{5, 10, 2, 7, 12, 4\} \to (\frac{5}{13}) = (-1)^3 = -1$

7.3/5   Hint: Rewrite the equation as a congruence (mod 3) and show that n cannot be odd.

7.3/9   (a) $p \equiv 1, 3 \pmod{8}$ (b) $p \equiv 1 \pmod{3}$
         (c) $p \equiv \pm 1 \pmod{5}$ (d) $p \equiv \pm 1, \pm 11 \pmod{60}$

7.4/1 $x \equiv \pm 10 \pmod{37}$ 7.4/3   no solution

7.4/5 $x \equiv \pm 32 \pmod{73}$ 7.4/7 $x \equiv \pm 14 \pmod{37}$

7.5/1 $x \equiv \pm 27 \pmod{11^2}$ 7.5/3 $x \equiv \pm 1075 \pmod{13^3}$

7.5/5   no solution   7.5/7 no solution

7.6/1   (a) $(\frac{30}{1001}) = -1$ (b) $(\frac{91}{561}) = -1$ (c) $(\frac{130}{231}) = 1$

7.6/3 $(\frac{3}{133}) = 1$. No, because $133 = 7 * 19$, and $(\frac{3}{7}) = (\frac{3}{19}) = -1$.

8.2/1   (a) $233 = 13^2 + 8^2$ (b) $317 = 14^2 + 11^2$ (c) $613 = 18^2 + 17^2$
         (d) $1009 = 28^2 + 15^2$ (e) $1409 = 28^2 + 25^2$

8.2/3   (a) $65 = 8^2 + 1^2 = 7^2 + 4^2$ (b) $85 = 9^2 + 2^2 = 7^2 + 6^2$

(c) $221 = 14^2 + 5^2 = 11^2 + 10^2$ (d) $1073 = 32^2 + 7^2 = 28^2 + 17^2$

(e) $1105 = 33^2 + 4^2 = 32^2 + 9^2 = 31^2 + 12^2$

(f) $1885 = 43^2 + 6^2 = 42^2 + 11^2 = 38^2 + 21^2$

(g) $5525 = 74^2 + 7^2 = 73^2 + 14^2 = 71^2 + 22^2 = 70^2 + 25^2 = 62^2 + 41^2 = 55^2 + 50^2$

8.2/7   72,   73,   74

8.3/5   (a) $107 = 9^2 + 5^2 + 1^2 + 0^2$ (b) $191 = 13^2 + 3^3 + 3^2 + 2^2$

(c) $263 = 15^2 + 6^2 + 1^2 + 1^2$ (d) $431 = 19^2 + 6^2 + 5^2 + 3^2$

(e) $1031 = 31^2 + 6^2 + 5^2 + 3^2$ (f) $8831 = 93^2 + 13^2 + 3^2 + 2^2$

9.2/1   (a) $\frac{100}{37} = [2, 1, 2, 2, 1, 3]$ (b) $\frac{1001}{45} = [22, 4, 11]$

9.2/2   (a) $[1, 2, 3, 4, 5] = \frac{225}{157}$ (b) $[1, 1, 2, 2, 3, 3] = \frac{135}{79}$

9.3/1 $n = 6$

9.3/3 $\sqrt{41} = [6, \overline{2, 2, 12}]$ $\sqrt{41} = 6.40$

9.3/5 $\sqrt{n^2 + 1} < C_1 = [n, 2n] = n + \frac{1}{2n}$

9.4/1 $\sqrt{6} = \frac{49}{20} = 2.450,\ n = 3$

9.4/3 $\frac{e+1}{e-1} = [2, 6, 10, 14] = 2.1640$

9.5/1   (a) $\sqrt{2} = [1, \overline{2}]$ (b) $\sqrt{3} = [1, \overline{1, 2}]$ (c) $\sqrt{11} = [3, \overline{3, 6}]$

(d) $\sqrt{22} = [4, \overline{1, 2, 4, 2, 1, 8}]$ (e) $\sqrt{31} = [5, \overline{1, 1, 3, 5, 3, 1, 1, 10}]$

9.6/ 1   (a) $\frac{1+\sqrt{3}}{2}$ (b) $1 + \sqrt{3}$ (c) $\frac{4+\sqrt{37}}{7}$ (d) $\sqrt{3}$

10.3/1   (a) $(x_1, y_1) = (9, 4),\ (x_2, y_2) = (161, 72)$

10.3/1   (b) $(x_1, y_1) = (19, 6),\ (x_2, y_2) = (721, 228)$

10.3/1   (c) $(x_1, y_1) = (640, 189),\ (x_2, y_2) = (842401, 233640)$

10.3/3   (a) $\sqrt{15} = [3, \overline{1, 6}]$. $t = 2 \rightarrow$ there is no solution of the associated Pell's equation by Theorem 10.3

10.3/3   (b) $x^2 - 15y^2 = -1 \rightarrow x^2 \equiv -1 \pmod{3} \rightarrow (\frac{-1}{3}) = 1$, which is false.

10.3/7 $(x, y) = (1, 1), (9, 7), (71, 55)$

11.2/3 $2^{220} \equiv 16 \not\equiv 1 \pmod{221} \rightarrow 221$ is composite.

11.2/5 $(p, q) = (11, 17)$

11.2/7 $(p, q) = (13, 17), (29, 73)$

11.3/1   (a) $2^{140} \equiv 67 \pmod{561}, 2^{280} \equiv 1 \pmod{561}$

11.3/3   (a) $997 - 1 = 996 = 2^2 * 3 * 83$. Use $s = 83$, $c = 2$.

11.3/3   (b) $5003 - 1 = 5002 = 2 * 41 * 61$. Use $s = 122$, $c = 2$.

11.4/1   (a) $299 + 5^2 = 324 = 18^2 \rightarrow 299 = (18 + 5)(18 - 5) = 23 * 13$.

11.4/1   (c) $5141 + 22^2 = 5675 = 75^2 \rightarrow 5141 = (75 + 22)(75 - 22) = 97 * 53$.

11.4/1   (e) $17063 + 19^2 = 17464 = 132^2 \rightarrow$
$17063 = (132 + 19)(132 - 19) = 151 * 113$.

11.4/2   (a) $r_{14} = 25 = 5^2$, $p_{13} = 5479$. $(10511, 5474) = 23$.
$10511 = 23 * 457$.

11.4/2   (c) $r_{10} = 25 = 5^2$, $p_9 = 22611$. $(32639, 22611 - 5) = (32639, 22606) = 127$. $32639 = 127 * 257$.

11.4/2   (e)   c.f. method doesn't work on 141191, so try $2 * 141191 = 282382$. Now $r_2 = 441 = 21^2$, $p_1 = 1063$. $(141191, 1063 - 42) = (141191, 1042) = 571$. $141191 = 571 * 251$.

11.5/1   Only 30941 is prime.

11.5/1   (b) $(192 * 203 * 211)^2 \equiv (2 * 3^4 * 5 * 19 * 31)^2 \pmod{33919} \rightarrow 15538^2 \equiv 2224^2 \pmod{33919}$. Now $(33919, 15538 - 2224) = (33919, 13314) = 317 \rightarrow 33919 = 317 * 107$.

11.5/1   (c) $(262 * 263 * 264 * 265)^2 \equiv (2^3 * 3^2 * 5^2 * 7 * 23)^2 \pmod{68569} \rightarrow 57353^2 \equiv 15524^2 \pmod{68569}$. Now $(68569, 57353 - 15524) = (68569, 41829) = 191 \rightarrow 68569 = 191 * 359$.

11.5/1 (d) $(917 * 923)^2 \equiv (2^2 * 3^2 * 11 * 13)^2 \pmod{838861} \rightarrow 7530^2 \equiv$
5148$^2 \pmod{838861}$. Now $(838861, 7530-5148) = (838861, 2382) =$
$397 \rightarrow 838861 = 397 * 2113$.

11.6/2a

n	p_n	$b_n \pmod{30607}$	$b_n - 1 \pmod{30607}$	d_n
1	2	2	1	1
2	3	4	3	1
3	2	64	63	1
4	5	4096	4095	1
5	7	6747	6746	1
6	2	18797	18796	127

RESULT: $30607 = 127 * 257$

12.2/1 (a) YOUR NUMBER IS UP

12.2/3 TAKE IT EASY

12.2/5 (a) See quotation on p. 185

12.2/7 0, 13

12.3/1 MJKI DLMW EKGU DKRC NFJE

12.3/3 $(a, b, c, d) = (7, 14, 15, 11)$

12.4/1 19, 11, 09, 20, 17, 22, 04, 23, 11, 04, 09, 11

12.4/3 7048 1518 8150 5578 4880 7201 4031 8375 4241 8039 8991 5538
7048 2874 2187

12.4/5 DONT GIVE UP

12.5/1 $p = 157$, $q = 127$

12.5/3 HALLELUJAH

Bibliography

1. W. R. Alford, A. Granville, & C. Pomerance. There are infinitely many Carmichael numbers. *Annals of Math.* (2)**139**. (1994): 703–722.

2. R. Crandall & C. Pomerance. *Prime Numbers: A Computational Perspective.* (2001): Springer Verlag.

3. E. Galois. Demonstration d'un theoreme sur les fractions continues. *Ann. Math. Pures Appliques* **19**. (1828): 294–301.

4. C. F. Gauss. *Disquisitiones Arithmeticae.* (1801).

5. P. Hagis. Outline of a proof that every odd perfect number has at least eight distinct prime factors. *Math. Comp.* **34**. (1980): 1027–1032.

6. I. N. Herstein. *Abstract Algebra* (3rd ed.). (1996): Prentice Hall.

7. J. L. Lagrange. Addition au memoire sur la resolution des equations numeriques. *Mem. Acad. Sci. Berlin* **24**. (1770): 311–352.

8. J. L. Lagrange. *Additions aux elements d'algebre d'Euler.* (1774): Lyon.

9. I. Niven, H. S. Zuckerman, & R. Montgomery. *An Introduction to the Theory of Numbers* (5th ed.). (1991).

10. O. Perron. *Die Lehre von den Kettenbruchen.* (1913): Teubner.

11. M. Polezzi. A geometrical method for finding an explicit formula for the greatest common divisor. *Amer. Math. Monthly* **104**. (1997): 445–446.

12. N. Robbins. Calculating a primitive root $\pmod{p^e}$. *Math. Gazette* **59**. (1975): 195.

13. N. Robbins. On Fibonacci Numbers and Primes of the form $4k + 1$. *Fibonacci Quart.* **32**. (1994): 15–16.

14. N. Robbins. Wilson's theorem via Eulerian numbers. *Fibonacci Quart.* **36**. (1998): 317–318.

15. K. Rosen. *Elementary Number Theory and its Applications* (5th ed.). (2004): Addison-Wesley.

16. E. Trost. *Primzahlen*. (1953): Birkhaeuser.

17. S. Wagon. The Euclidean Algorithm strikes again. *Amer. Math. Monthly* **97**. (1990): 125–129.

18. A. Wiles. Modular elliptic curves and Fermat's Last Theorem. *Annals of Math.* **141**. (1995): 443–551.

Index

— Joseph Louis Lagrange   1736–1813   (p. 208)

John Wilson   1741–1793

Adrien-Marie Legendre 1752–1833

— Carl Friedrich Gauss   1777–1855   (p. 184)

August Ferdinand Moebius   1790–1868

Gabriel Lamé 1795–1870

Carl Gustav Jacobi   1804–1851

Peter Lejeune Dirichlet   1805–1859

Edouard Lucas 1842–1891

Louis Joel Mordell   1888–1972